CONTROL AND DYNAMIC SYSTEMS

Advances in Theory and Applications

Volume 10

CONTRIBUTORS TO THIS VOLUME

MICHAEL FALCO

R. G. GRAHAM

YACOV Y. HAIMES

HENRY J. KELLEY

C. T. LEONDES

DAVID Q. MAYNE

JACK O. PEARSON

DANIEL TABAK

CONTROL AND DYNAMIC SYSTEMS

ADVANCES IN THEORY AND APPLICATIONS

Edited by

C. T. LEONDES

Department of Engineering
University of California
Los Angeles, California

VOLUME 10 1973

ACADEMIC PRESS New York and London

ACADEMIC PRESS, INC.
111 Fifth Avenue, New York, New York 10003

United Kingdom Edition published by
ACADEMIC PRESS, INC. (LONDON) LTD.
24/28 Oval Road, London NW1

LIBRARY OF CONGRESS CATALOG CARD NUMBER: 64-8027

Volumes 1-8 published under the title:
Advances in Control Systems

PRINTED IN THE UNITED STATES OF AMERICA

CONTENTS

CONTRIBUTORS

Numbers in parentheses indicate the pages on which the authors' contributions begin.

Michael Falco, Research Department, Grumman Aerospace Corporation, Bethpage, New York (89)

R. G. Graham, School of Mathematical and Computing Sciences, New South Wales Institute of Technology, Sydney, Australia (1)

Yacov Y. Haimes, Systems Research Center, Case Institute of Technology, Case Western Reserve University, Cleveland, Ohio (435)

Henry J. Kelley, Analytical Mechanics Associates, Inc., Jericho, New York (89), (131)

C. T. Leondes, School of Engineering and Applied Science, University of California, Los Angeles, California (1)

David Q. Mayne, Department of Computing and Control, Imperial College, London SW 7 2BT, England (179)

Jack O. Pearson, Aerospace Group, Hughes Aircraft Company, Culver City, California (255)

Daniel Tabak, Department of Electrical Engineering, University of the Negev, Beer-Sheva, Israel (345)

PREFACE

The tenth volume of the series, *Control and Dynamic Systems: Advances in Theory and Applications,* continues in the purpose of this serial publication in bringing together diverse information on important progress in the field of control and systems theory and applications as achieved and presented by leading contributors. As pointed out in the previous volume, the retitling of this series reflects the growing emphasis on applications to large scale systems and decision making in addition to the more traditional, but still important areas of endeavor, in this very broad field.

The present volume begins with a contribution by R. G. Graham and C. T. Leondes, which explores a number of important engineering design issues in the competitive environment in which control systems must so often function. Algorithms are developed for a broad class of problems beginning with the development of necessary conditions including the effects of multiple subarcs, state and control variable inequality constraints, and discontinuities in the state itself. These results are then applied to some illustrative examples and computational results are developed.

Henry J. Kelley is unquestionably one of the recognized pioneering leaders on the international scene for his many significant contributions to algorithms for complex control optimization problems of engineering significance. Kelley is joined by Michael Falco to explore these issues further. In their contribution, it is pointed out that the continuous gradient/penalty function version seems to remain attractive for applications work. Thus, a mixture of decade-old material, which Kelley pioneered, is compiled here and offered along with some suggestions for further development that may stimulate efforts in the area. Applications are presented to specific engineering example, and it is clearly understood that these specific applications point out issues relevant to a rather broad array of engineering problems.

The next contribution, also by Kelley, explores an attractive option to engineering optimization problems through reduced order approximations. This potentially attractive approach is not without problems. These matters are thoroughly examined with the various considerations carefully elaborated upon, and then the approach is applied to a rather complex engineering problem with the development of computational results.

The contribution by D. Q. Mayne represents a compilation and thorough examination of his pioneering work on differential dynamic programming. At the heart of this approach is the development of exact expressions for the change in cost due to a change in control. The unifying role these expressions can play are then indicated by example. These expressions are then used for obtaining conditions of optimality, particularly, sufficient conditions, and on the other hand, for obtaining optimization algorithms including the powerful differential dynamic programming algorithms.

A wide range of applications in filtering theory have been discovered, following the pioneering and most significant work of Kalman in the late 1950's. The contribution by J. O. Pearson deals with some issues in this area that are most important from an applied engineering point of view. Specifically, the Kalman-Bucy filter gives the unbiased, minimum variance estimate of the state vector of a linear dynamic system disturbed by additive white noise when measurements of the state vector are linear but disturbed by white noise. Such performance is hardly ever realized in actual practice since the information required to construct the Kalman-Bucy filter is known only approximately. Some broad and general techniques for dealing with these issues in the analysis and synthesis of engineering systems are presented by Pearson.

Among the problems which have grown in interest in the past few years in the area of societal problems is that of transportation. Heretofore this broad and important area received little attention by systems engineers, and as a result the substantive contributions their expertise could make were lacking. The contribution by D. Tabak represents a rather comprehensive treatment of the issues in this area with the goal of permitting the systems engineer to gain essential knowledge and insight. Another goal is the development in a concise manner and under unified notation of some mathematical models of transportation systems which have been used in conjunction with application of modern control and optimiation techniques to these systems. This information should also be useful to the control specialist starting to work in the transportation areas. Additionally, the notions developed in this area could perhaps serve as an aid to structuring thinking in systems problems in other societal areas.

An essential precursor to the analysis and optimization of any system problem, whether it be of small, medium, or large scale, is the issue of system identification. The contribution by Y. Y. Haimes represents an in-depth and comprehensive treatment of these issues, including many of his pioneering contributions to this essential area of systems analysis and synthesis.

CONTENTS OF PREVIOUS VOLUMES

Volume 3

Volume 4

Volume 5

Volume 6

Volume 7

Volume 8

Volume 9

CONTROL AND DYNAMIC SYSTEMS

Advances in Theory and Applications

Volume 10

The Evaluation of Suboptimal Strategies[1] Using Quasilinearization

R. G. GRAHAM
School of Mathematical and Computing Sciences,
New South Wales Institute of Technology,
Sydney, Australia

AND

C. T. LEONDES
School of Engineering and Applied Science,
University of California,
Los Angeles, California

[1]This research was supported in part by the Air Force Office of Scientific Research under AFOSR Grant 72-2166.

I. INTRODUCTION

In engineering practice the need for actual mechanization of a strategy nearly always results in one which is not mathematically optimal. The engineer/designer is usually forced to accept a suboptimal strategy because of the difficulty in determining the form of a truly optimal strategy in a realistic model of the environment and also because of the complexity of implementation even if he could determine the optimal strategy. The problem then becomes one of evaluating just how good the suboptimal strategy is.

A possible evaluation procedure is to first evaluate the worst case performance associated with the suboptimal strategy and to then develop a quantitative measure of how much the worst case performance could be improved by changes in the strategy.

The problem of determining worst case performance is conceptually straight-forward. One first assumes that the opponent knows the suboptimal strategy and behaves optimally against it. The worst case performance is then determined by solving the opponent's problem of finding the optimal open-loop controls to minimize your performance. If the worst case performance is known for a number of sets of boundary conditions which are typical of those likely to occur, the efficacy of the suboptimal strategy can be quantitatively assessed. The difficult problem is that of finding the optimal closed-loop strategies. For reasonably simple games, the necessary conditions stated by Isaacs [7] and justified in a heuristic fashion can lead to the optimal strategies and the associated performance; however, for more complex situations the subtle distinction between open-loop and closed-loop strategies becomes important. For example, Starr and Ho [8], [9] have shown that in nonzero-sum games it is not know *a priori* whether or not an optimal closed-loop strategy generates an optimal open-loop strategy, especially if state variable inequality constraints are present. Examples of simple differential games can be constructed where the optimum open-loop strategy is mixed, but the optimum closed-loop strategy can be shown to be pure. When the state itself is discontinuous, there is very little general theory to assist in the determination of the optimal strategies. Berkovitz [10] has developed a set of necessary and sufficient conditions under fairly restrictive assumptions which are not always satisfied in actual situations. Furthermore, the determination of an optimal closed-loop strategy often involves the solution of a set of partial differential equations with difficult boundary conditions. At present, such problem statements are not easily amenable to numerical solution. The unfortunate fact is that the determination of optimal closed-loop strategies in complicated situations is beyond the current state of the art.

Faced with this dilemma, the engineer/designer may recognize that he is not interested so much in the optimal closed-loop strategies as he is in the performance associated with these strategies. An alternate approach then is to determine the optimal open-loop strategies (as opposed to the optimal closed-loop strategies) which yield a saddle-point solution. If the game is a zero-sum differential game and if the open-loop strategies are found to be pure, it is reasonable to speculate that the performance associated with the optimal open-loop strategies is equal to that associated with optimal closed-loop strategies. Such a speculation at least permits the formulation of a set of necessary conditions involving ordinary differential equations. The solution of these equations then provides a quantitative measure of the possible improvement in the worst case performance.

Once the worst case performance and the margin for improvement in that performance have been found, the engineer/designer can judge whether or not he has a problem and, if he does, how his problem must be resolved. No amount of refinement of his strategy or guidance law will enable the vehicle to perform the mission if the performance associated with typical initial conditions and optimal open-loop controls is inadequate. It is the vehicle which must be improved and not the guidance law. If the worst case performance associated with the sub-optimal strategy is unacceptable but that associated with the optimal open-loop strategy is, it is the guidance law and not the vehicle which needs improvement.

The structure of this paper involves three basic parts. In the first part, a variational approach is used to develop the necessary condition. The difficulties arise from the complexity of realistic models of the situation. Specifically, these models can involve multiple subarcs, state and control variable inequality constraints, and even discontinuities in the state itself. While the necessary conditions for optimality in continuous problems have been well developed by Bliss [1] and Hestenes [2] and those for discontinuous problems have been formulated by Mason, Dickerson, and Smith [3] and Vincent and Mason [4], the resulting problem statements are not always in a form easily amenable to numerical solution. The necessary conditions for optimality, then, need to be restated in a form which anticipates the method for obtaining numerical answers. The work by Bryson, Denham, and Dreyfus [5] is a notable example of this type of approach. Instead of the gradient method used by Bryson, Denham, and Dreyfus, the solution method suggested in this paper is an extended version of the quasilinearization algorithm described by McGill and Kenneth [6].

The difficulties in determining worst case performance, however, are small when compared to the problem of finding a quantitative measure of how much the worst case performance could be improved by changes in the strategy. The seemingly straightforward approach of determining the performance associated with a saddle-point solution to the differential game involving the two players is thwarted by the problem of solving for the first-order necessary conditions to determine the worst case performance of a suboptimal strategy. These conditions are then extended to include the case of a differential game involving optimal open-loop strategies. The necessary conditions include the effects of multiple subarcs, state and control variable inequalities constrains, and discontinuities in the state itself. The necessary conditions are then applied to a few simple examples involving state variable discontinuities to illustrate their usage. In the second part, the quasilinearization method is extended to handle the general discontinuous, multiple subarc multi-point boundary value problem resulting from the necessary conditions. The extended algorithm is then applied to a discontinuous version of the brachistochrone problem, and its efficacy is discussed.

Finally, in the third part the quasilinearization technique is used to evaluate a specific suboptimal strategy for a ground to air interceptor.

II. FIRST ORDER NECESSARY CONDITIONS

A. Problem Statement

The system under the consideration is governed by the following set of first order, ordinary differential equations:

$$\dot{x}^{(a)} = f^{(a)}(x,\psi,\theta,t) \quad \text{for} \quad t_{a-1} < t < t_a \qquad a = 1,\ldots,A \tag{1}$$

where t is the independent variable, time; $(\dot{\ })$ is $d(\ )/dt$; x is an n-dimensional vector of state variables; ψ is an r_1-dimensional vector of control variables available to player I; θ is an r_2-dimensional vector of control variables available to player II; and $f^{(a)}$ is an n-dimensional vector of known functions of x, ψ, θ, and t.

The superscript (a) notation in equation (1) is used because the derivative expressions are not necessarily continuous throughout the entire solution. Discontinuities can result from changes in the algebraic form of the derivatives, from a change in some parameter of the system, from discontinuities in ψ or θ, or from a discontinuity in the state, x. Because of these discontinuities, a complete solution to equation (1) consists of a number of segments. Each segment is referred to as a subarc, and the junction of two subarcs is called a corner. The length or duration of a subarc is defined by the corner times; specifically, the ath subarc begins at t_{a-1} and ends at t_a. Because discontinuities in x are permitted at the corner times, the derivative expressions are defined over open intervals. It is assumed that $f^{(a)}(x,\psi,\theta,t)$, ψ, and θ are sectionally continuous and that they possess continuous first and second derivatives with respect to all arguments on each subarc.

The initial conditions for x at $t = t_0$ are assumed to be given. This assumption can be easily relaxed, but it is made here in the interests of expediency.

The solution must satisfy the following terminal constraints:

$$\Omega[x(t_A), t_A] = 0 \tag{2}$$

where Ω is a q-dimensional vector of constraint functions.

At each corner, a discontinuity in the state may occur. The magnitudue of the discontinuity is defined as follows:

$$\lim_{\varepsilon\to 0} x(t_a + \varepsilon) = \lim_{\varepsilon\to 0} \{x(t_a - \varepsilon) + \xi_a[x(t_a - \varepsilon), t_a]\} \quad \text{for} \quad a = 1,\ldots,A-1 \tag{3}$$

where ξ_a is an n-dimensional vector which specifies the magnitude of the discontinuity in x at $t = t_a$. It is assumed that ξ_a is a sectionally continuous and twice differentiable function of x and t. The occurrence of a corner may be at the discretion of one of the players, or it may be determined by satisfaction of the following condition.

$$\lim_{\varepsilon\to 0} \{T_\alpha[x(t_\alpha - \varepsilon), \psi(t_\alpha - \varepsilon), t_\alpha - \varepsilon]\} = 0 \tag{4}$$

where α belongs to a known subset of the index set $\{1,2,\ldots,A-1\}$. It is assumed that all T_α are sectionally continuous and twice differentiable functions of x and t.

The solution of equation (1) is subject to inequality constraints throughout the interval $[t_0,t_A]$. These constraints are of one of the following forms:

$$C(x,\psi,\theta,t) \leq 0 \tag{5}$$

or

$$S(x,t) \leq 0 \tag{6}$$

where C and S are vector valued functions. The restriction must be made that no more than $r_1 + r_2$ inequality constraints can be satisfied as equalities at any one time. Inequality constraints given by equation (5) are called control variable inequality constraints, and those given by equation (6) are called state variable inequality constraints.

Player I tries to maximize the criterion function, $\varphi[x(t_A),t_A]$. Two cases are considered. The first case assumes that player II uses a prespecified strategy, $\theta(x,t)$, and that this strategy is known to player I. This case is not truly a game, but it does correspond to the worst possible thing that could happen to player II. The solutions of this first case will provide valuable information for the evaluation of sub-optimal strategies. The second case assumes that both players use optimal open-loop strategies. Player II tries to minimize $\varphi[x(t_A),t_A]$ as player I tries to maximize it. In both cases, it is assumed that pure strategy solutions exist.

B. Play Against a Known Strategy

In this section the problem is one of determining an optimal strategy against an opponent whose strategy is known. Although this problem is not truly a game, its solution provides the answer concerning worst case performance.

Assume that player II uses a known strategy, $\theta(x,t)$. The problem then is to determine the $\psi(t)$ which maximizes φ. As in classical problems in variational calculus, a set of first order necessary conditions which the optimal ψ must satisfy can be found. To derive these necessary conditions, the equality constraints are adjoined to the criterion function, φ, by means of Lagrange multipliers.

$$J = \varphi + \nu^T\Omega + \sum_{\alpha} \mu_\alpha T_\alpha + \lim_{\varepsilon\to 0} \sum_{a=1}^{A} \int_{t_{a-1}+\varepsilon}^{t_a-\varepsilon} \lambda^T(\dot{x}^{(a)} - f^{(a)})dt \tag{7}$$

where ν is a q-dimensional column vector of Lagrange multipliers associated with the terminal constraints, Ω; the μ_α are multipliers associated with the corner defining relationships, T_α; and λ is an n-dimensional column vector of multipliers associated with the differential constraints.

For a maximizing solution, J must be stationary with respect to small permissable variations in x, ψ, and t_a. A permissable variation for this problem is one which does not violate the inequality constraints. The first variation of J is:

$$\begin{aligned}\Delta J = & \,|(\partial\varphi/\partial x)\Delta x + (\partial\varphi/\partial t)\Delta t_A + \nu^T[(\partial\Omega/\partial x)\Delta x \\ & + (\partial\Omega/\partial t)\Delta t_A]|_{t_A} + \sum_\alpha \mu_\alpha[(\partial T_\alpha/\partial x)\Delta x \\ & + (\partial T_\alpha/\partial t)\Delta t_\alpha + (\partial T_\alpha/\partial\psi)\Delta\psi]_{t_\alpha} \\ & + \lim_{\varepsilon\to 0}\sum_{a=1}^{A}\int_{t_{a-1}+\varepsilon+\Delta t_{a-1}}^{t_a-\varepsilon+\Delta t_a} \lambda^T[\delta\dot{x}^{(a)} - (\partial f^{(a)}/\partial x)\delta x \\ & - (\partial f^{(a)}/\partial\theta)(\partial\theta/\partial x)\delta x - (\partial f^{(a)}/\partial\psi)\delta\psi]dt\end{aligned} \tag{8}$$

where $\Delta(\ )$ denotes the total variation of ( ), including the variation of the independent variable, t; $\delta(\ )$ denotes the variation at a fixed time point; and $(\ )^T$ denotes the matrix transpose. All vectors are column vectors unless otherwise specified. The matrices of partial derivatives, for example $(\partial f^{(a)}/\partial x)$, are defined in accordance with the following.

$$(\partial f^{(a)}/\partial x) \equiv \begin{bmatrix} (\partial f_1^{(a)}/\partial x_1) & (\partial f_1^{(a)}/\partial x_2) & \cdots & (\partial f_1^{(a)}/\partial x_n) \\ (\partial f_2^{(a)}/\partial x_1) & (\partial f_2^{(a)}/\partial x_2) & \cdots & (\partial f_2^{(a)}/\partial x_n) \\ \cdot\cdot\cdot & \cdot\cdot\cdot & & \cdot\cdot\cdot \\ \cdot\cdot\cdot & \cdot\cdot\cdot & & \cdot\cdot\cdot \\ (\partial f_n^{(a)}/\partial x_1) & (\partial f_n^{(a)}/\partial x_2) & \cdots & (\partial f_n^{(a)}/\partial x_n) \end{bmatrix} \tag{9}$$

The variations at a fixed time point are related to the total variations as follows:

$$(\delta x)_{t_a \pm \varepsilon} = (\Delta x)_{t_a \pm \varepsilon} - (\dot{x})_{t_a \pm \varepsilon} \Delta t_a \, . \tag{10}$$

Neglecting higher ordered terms, integrating the $\lambda^T \delta\dot{x}$ term by parts, and substituting equation (10) into equation (8) one obtains the following result for the first variation.

$$\Delta J = [(\partial\varphi/\partial x) + \nu^T(\partial\Omega/\partial x) + \lambda^T]_{t_A} (\Delta x)_{t_A}$$

$$+ [(\partial\varphi/\partial t) + \nu^T(\partial\Omega/\partial t) - \lambda^T f^{(A)}]_{t_A} \Delta t_A$$

$$- (\lambda^T \Delta x)_{t_0} + (\lambda^T f^{(1)})_{t_0} \Delta t_0$$

$$- \lim_{\varepsilon \to 0} \sum_{a=1}^{A} \int_{t_{a-1}+\varepsilon}^{t_a-\varepsilon} \lambda^T(\partial f^{(a)}/\partial\psi)\delta\psi \, dt$$

$$+ \lim_{\varepsilon\to 0} \sum_{\alpha} \{[\mu_\alpha(\partial T_\alpha/\partial x) + \lambda^T]_{t_\alpha-\varepsilon} (\Delta x)_{t_\alpha-\varepsilon}$$

$$- \lambda^T(t_\alpha + \varepsilon)(\Delta x)_{t_\alpha+\varepsilon} + \mu_\alpha(\partial T_\alpha/\partial\psi)_{t_\alpha-\varepsilon}(\Delta\psi)_{t_\alpha-\varepsilon}$$

$$+ [\mu_\alpha(\partial T_\alpha/\partial t)_{t_\alpha-\varepsilon} + \lambda^T(t_\alpha + \varepsilon) \, f^{(\alpha+1)} \, (t_\alpha + \varepsilon)$$

$$- \lambda^T(t_\alpha - \varepsilon)\, f^{(\alpha)}\, (t_\alpha - \varepsilon)]\, \Delta t_\alpha\}$$

$$+ \lim_{\varepsilon \to 0} \sum_{\substack{a=1 \\ a \neq \alpha}}^{A-1} \{\lambda^T(t_a - \varepsilon)(\Delta x)_{t_a - \varepsilon} - \lambda^T(t_a + \varepsilon)(\Delta x)_{t_a + \varepsilon}$$

$$+ [\lambda^T(t_a + \varepsilon) f^{(a+1)}(t_a + \varepsilon) - \lambda^T(t_a - \varepsilon) f^{(a)}(t_a - \varepsilon)]\Delta t_a\}$$

$$- \lim_{\varepsilon \to 0} \sum_{a=1}^{A} \int_{t_{a-1}+\varepsilon}^{t_a - \varepsilon} \{\dot{\lambda} + [(\partial f^{(a)}/\partial x)$$

$$+ (\partial f^{(a)}/\partial \theta)(\partial \theta/\partial x)]^T\, \lambda\}^T\, \delta x\, dt \tag{11}$$

The variations, $(\Delta x)_{t_a + \varepsilon}$, are related to the variations, $(\Delta x)_{t_a - \varepsilon}$ because of equation (3)

$$(\Delta x)_{t_a + \varepsilon} = (\partial \xi_a/\partial x)_{t_a - \varepsilon}\, (\Delta x)_{t_a - \varepsilon} \tag{12}$$

$$+ (\partial \xi/\partial t)_{t_a - \varepsilon}\, \Delta t_a + (\Delta x)_{t_a - \varepsilon}$$

When equation (12) is substituted into equation (11) and $(\Delta x)_{t_0}$ and Δt_0 are specified as zero, the following result is obtained

$$\Delta J = [(\partial \varphi/\partial x) + \nu^T(\partial \Omega/\partial x) + \lambda^T]_{t_A}\, (\Delta x)_{t_A}$$

$$+ [(\partial \varphi/\partial t) + \nu^T(\partial \Omega/\partial t) - \lambda^T\, f^{(A)}]_{t_A}\, \Delta t_A$$

$$- \lim_{\varepsilon \to 0} \sum_{a=1}^{A} \int_{t_{a-1}+\varepsilon}^{t_a - \varepsilon} [\![\lambda^T(\partial f^{(a)}/\partial \psi)\delta\psi$$

$$+ \{\dot{\lambda} + [(\partial f^{(a)}/\partial x) + (\partial f^{(a)}/\partial \theta)(\partial \theta/\partial x)]^T \lambda\}^T \delta x]\!] dt$$

$$+ \lim_{\varepsilon \to 0} \sum_{\alpha} [\![\{\lambda^T(t_\alpha - \varepsilon) - \lambda^T(t_\alpha + \varepsilon)\ [I + (\partial \xi_\alpha/\partial x)_{t_\alpha - \varepsilon}$$

$$+ \mu_\alpha (\partial T_\alpha/\partial x)_{t_\alpha - \varepsilon}\} (\Delta x)_{t_\alpha - \varepsilon} + \mu_\alpha (\partial T_\alpha/\partial \psi)_{t_\alpha - \varepsilon} (\Delta \psi)_{t_\alpha - \varepsilon}$$

$$+ \{\lambda^T(t_\alpha + \varepsilon) f^{(\alpha+1)}(t_\alpha + \varepsilon) - \lambda^T(t_\alpha - \varepsilon) f^{(\alpha)}(t_\alpha - \varepsilon)$$

$$- \lambda^T(t_\alpha + \varepsilon)(\partial \xi_\alpha/\partial t)_{t_\alpha - \varepsilon} + \mu_\alpha (\partial T_\alpha/\partial t)_{t_\alpha - \varepsilon}\} \Delta t_\alpha]\!]$$

$$+ \lim_{\varepsilon \to 0} \sum_{\substack{a=1 \\ a \neq \alpha}}^{A-1} [\![\{\lambda^T(t_a - \varepsilon) - \lambda^T(t_a + \varepsilon) [I$$

$$+ (\partial \xi_a/\partial x)_{t_a - \varepsilon}]\} (\Delta x)_{t_a - \varepsilon}$$

$$+ \{\lambda^T(t_a + \varepsilon) f^{(a+1)}(t_a + \varepsilon) - \lambda^T(t_a - \varepsilon) f^{(a)}(t_a - \varepsilon)$$

$$- \lambda^T(t_a + \varepsilon)(\partial \xi_a/\partial t)_{t_a - \varepsilon}\} \Delta t_a]\!] \qquad (13)$$

where I is the $(n \times n)$ identity matrix.

If there were no inequality constraints, equation (13) could be used to deduce the first order necessary conditions for a stationary solution. Since inequality constraints are present, however, the variations must be restricted to be those which do not violate the inequality constraints. The treatment here of these restrictions is very similar to that of Reference 5, however, the treatment here is not restricted to scalar inequalities and control variables.

Consider first the control variable inequality constraint given by equation (5) with θ replaced by $\theta(x,t)$. If p components of the C vector are zero over a subarc, the relationship $C = 0$ can be used to determine p of the

r_1 components of ψ. If not all components of C are simultaneously zero, a new C vector can be constructed of only those components which are zero; and p becomes the dimension of this new C vector. The vector of control variables, ψ, is partitioned into two parts; one part is used to satisfy $C = 0$, and the other part can be freely chosen. The part which is used to satisfy $C = 0$ is denoted as $\hat{\psi}$, and the other part is noted by ψ^*. The dimension of $\hat{\psi}$ is p, and the dimension of ψ^* is $r_1 - p$. Obviously, judgment must be used in making the partition of ψ. It is clearly necessary that the relationship $C = 0$ can be satisfied with the particular choice of $\hat{\psi}$.

The equation $C = 0$ defines a boundary in (x, ψ, t) space. While on this boundary, small variations must satisfy the following relationships:

$$\delta C = (\partial C/\partial x)\ \delta x + (\partial C/\partial \hat{\psi})\ \delta \hat{\psi} + (\partial C/\partial \psi^*)\ \delta \psi^* \tag{14}$$

$$\delta \dot{x}^{(a)} = (\partial f^{(a)}/\partial x)\ \delta x + (\partial f^{(a)}/\partial \theta)(\partial \theta/\partial x)\ \delta x + (\partial f^{(a)}/\partial \hat{\psi})\ \delta \hat{\psi} + (\partial f^{(a)}/\partial \psi^*)\ \delta \psi^* \tag{15}$$

If it is assumed that the components of C are functionally independent in the region of control space of interest, the matrix $(\partial C/\partial \hat{\psi})$ is nonsingular. Using this assumption, equation (14) can be solved for $\delta \hat{\psi}$, and the result can be substituted into equation (15) to yield the following:

$$\begin{aligned} \delta \dot{x}^{(a)} = {} & [(\partial f^{(a)}/\partial x) + (\partial f^{(a)}/\partial \theta)(\partial \theta/\partial x) \\ & - (\partial f^{(a)}/\partial \hat{\psi})(\partial C/\partial \hat{\psi})^{-1}\ (\partial C/\partial x)]\ \delta x \\ & + [(\partial f^{(a)}/\partial \psi^*) - (\partial f^{(a)}/\partial \hat{\psi})(\partial C/\partial \hat{\psi})^{-1}(\partial C/\partial \psi^*)]\ \delta \psi^* . \end{aligned} \tag{16}$$

When equation (16) is substituted into equation (13),

the integral term is modified during that subarc for which the inequality constraints restrict the permissible variations. The modification is as follows.

$$\int_{t_{a-1}+\varepsilon}^{t_a-\varepsilon} [\![\lambda^T(\partial f^{(a)}/\partial\psi)\delta\psi + \{\dot{\lambda} + [(\partial f^{(a)}/\partial x) + (\partial f^{(a)}/\partial\theta)(\partial\theta/\partial x)]^T \lambda\}^T \delta x]\!] dt \qquad \text{becomes}$$

$$\int_{t_{a-1}+\varepsilon}^{t_a-\varepsilon} [\![\lambda^T[(\partial f^{(a)}/\partial\psi^*) - (\partial f^{(a)}/\partial\hat{\psi})(\partial C/\partial\hat{\psi})^{-1}(\partial C/\partial\psi^*)] \delta\psi^* + \{\dot{\lambda} + [(\partial f^{(a)}/\partial x) + (\partial f^{(a)}/\partial\theta)(\partial\theta/\partial x) - (\partial f^{(a)}/\partial\hat{\psi})(\partial C/\partial\hat{\psi})^{-1}(\partial C/\partial x)]^T \lambda\}^T \delta x]\!] dt . \tag{17}$$

When inequality constraints are present, the encountering of the constraint boundary is considered as a corner; therefore, a T_α function must be identified with each of the components of the C vector. The time to leave the constraint boundary is at the player's discretion, and there may or may not be a state variable discontinuity as the constraint boundary is left. A criterion for determining the time to leave the constrain boundary must be found.

When the inequality constraints are of the state variable type; i.e., of the type expressed by equation (6), the control variables cannot be determined directly from the constraint relationships. Also, the inverse matrix, $(\partial C/\partial\hat{\psi})^{-1}$, does not exist, no matter what partition of ψ is used. Since the constraints must be identically zero along the boundary, however, all the time derivatives of the constraints must also be zero along the boundary.

$$\begin{aligned} dS/dt &= (\partial S/\partial x)\, \dot{x}^{(a)} + (\partial S/\partial t) \\ &= (\partial S/\partial x)\, f^{(a)} + (\partial S/\partial t)\ . \end{aligned} \tag{18}$$

Since $f^{(a)}$ is an explicit function of the control variables, equation (18) may contain the control variables explicitly. Even if the first derivative of S is not an explicit function of the control variables, it is clear that an explicit function can be obtained eventually by differentiating S a sufficient number of times. If S must be differentiated Q times to obtain an explicit function of ψ, the following function can be kept zero along the boundary.

$$(d^Q S/dt^Q) = \overline{C}(x, \psi, t) = 0\ . \tag{19}$$

Equation (19) can be used precisely in the same manner equation (5) is used along the boundary.

With state variable inequality constraints, equations (14), (15), (16), and (17) are unchanged except the C is replaced by $\overline{C}$. Henceforth, the notation C is used to denote an inequality constraint with the understanding that it may in fact represent a time derivative of a state variable inequality constraint.

When equation (19) is used, however, S will not be zero along the boundary unless the following conditions are satisfied at the entering corner.

$$S\big|_{t_\alpha} = (dS/dt)\big|_{t_\alpha} = \cdots = (d^{Q-1}S/dt^{Q-1})\big|_{t_\alpha} = 0\ . \tag{20}$$

Equations (20) specify a set of constraints which must be satisfied at a corner point. These constraints can be incorporated into the analysis by treating each T_α and μ_α as a Q-dimensional column vector; i.e.,

$$T_\alpha \equiv \begin{Bmatrix} T_{\alpha 1} \\ T_{\alpha 2} \\ \cdot \\ \cdot \\ \cdot \\ T_{\alpha Q} \end{Bmatrix} \equiv \begin{Bmatrix} S\big|_{t_\alpha - \varepsilon} \\ (dS/dt)\big|_{t_\alpha - \varepsilon} \\ \cdot \\ \cdot \\ \cdot \\ (d^{Q-1}S/dt^{Q-1})\big|_{t_\alpha - \varepsilon} \end{Bmatrix} \tag{21}$$

$$\mu_\alpha = [\mu_{\alpha 1}, \mu_{\alpha 2}, \ldots, \mu_{\alpha Q}]^T . \tag{22}$$

The only modification in the analysis is that wherever μ_α appears it must be replaced by μ_α^T. It should be noted that T_α is a vector function only when S must be differentiated two or more times to yield an explicit function of the control variables. For many problems, one differentiation is sufficient, and T_α remains a scalar quantity.

By modifying equation (13) as indicated by equation (17) and by using the fact that all the first order terms in ΔJ must vanish for an extremal solution, one can deduce the following first-order necessary conditions.

$$\lambda^T(t_A) = -[(\partial\varphi/\partial x) + \nu^T(\partial\Omega/\partial x)]_{t_A} \tag{23}$$

$$(\lambda^T f^{(A)})_{t_A} = [(\partial\varphi/\partial t) + \nu^T(\partial\Omega/\partial t)]_{t_A} \tag{24}$$

$$\begin{aligned} \lambda^T(t_\alpha - \varepsilon) = \lambda^T(t_\alpha + \varepsilon)\ [I + (\partial\xi_\alpha/\partial x)_{t_\alpha - \varepsilon}] \\ - \mu_\alpha^T(\partial T_\alpha/\partial x)_{t_\alpha - \varepsilon} \end{aligned} \tag{25}$$

$$\begin{aligned} (\lambda^T f^{(\alpha)})_{t_\alpha - \varepsilon} = \lambda^T(t_\alpha + \varepsilon)[f^{(\alpha+1)}(t_\alpha + \varepsilon) - (\partial\xi_\alpha/\partial t)_{t_\alpha - \varepsilon} \\ + \mu_\alpha^T(\partial T_\alpha/\partial t)_{t_\alpha - \varepsilon} \end{aligned} \tag{26}$$

$$\mu_{\alpha}^{T}(\partial T_{\alpha}/\partial\psi)_{t_{\alpha}-\varepsilon} = 0 \tag{27}$$

$$\lambda^{T}(t_{a}-\varepsilon) = \lambda^{T}(t_{a}+\varepsilon)[I + (\partial\xi_{a}/\partial x)]_{t_{a}-\varepsilon} \quad a \neq \alpha \tag{28}$$

$$(\lambda^{T}f^{(a)})_{t_{a}-\varepsilon} = \lambda^{T}(t_{a}+\varepsilon)[f^{(a+1)}(t_{a}+\varepsilon) - (\partial\xi_{a}/\partial t)_{t_{a}-\varepsilon}] a \neq \alpha. \tag{29}$$

If the solution is not on any of the inequality constraint boundaries, the following necessary conditions must be satisfied.

$$\lambda^{T}(\partial f^{(a)}/\partial\psi) = 0 \tag{30}$$

$$\dot{\lambda} + [(\partial f^{(a)}/\partial x) + (\partial f^{(a)}/\partial\theta)(\partial\theta/\partial x)]^{T} \lambda = 0\,. \tag{31}$$

If the solution is on one of the inequality constraint boundaries, equations (30) and (31) must be replaced by the following necessary conditions.

$$\lambda^{T}[(\partial f^{(a)}/\partial\psi^{*}) - (\partial f^{(a)}/\partial\hat{\psi})(\partial C/\partial\hat{\psi})^{-1} (\partial C/\partial\psi^{*})] = 0 \tag{32}$$

$$\begin{aligned}\dot{\lambda} + [(\partial f^{(a)}/\partial x) &+ (\partial f^{(a)}/\partial\theta)(\partial\theta/\partial x) \\ &- (\partial f^{(a)}/\partial\hat{\psi})(\partial C/\partial\hat{\psi})^{-1}(\partial C/\partial x)]^{T} \lambda = 0\,.\end{aligned} \tag{33}$$

If the final point, t_A, lies on one of the state variable inequality constraint boundaries, equations (23) and (24) are modified to restrict the variations $(\Delta x)_{t_A}$ and Δt_A. Specifically, the Ω and ν vectors must be augmented to include $T_{\alpha 1}$ and $\mu_{\alpha 1}$, respectively.

Equations (23) and (24) are the familiar transversality conditions. Equations (25) through (29) are analagous to the Weierstrass-Erdmann corner conditions, and they are the necessary conditions arising from the state variable discontinuities and from the corner defining relationships. Equations (30) through (33) correspond to the Euler-Lagrange equations modified for the effects of player II's strategy and of inequality constraints.

Although equations (23) through (33) provide necessary conditions which must be satisfied by an optimal strategy, satisfaction of these equations certainly does not insure that an optimal strategy has indeed been found. In many conventional problems in variational calculus, physical intuition leads one to expect that only one stationary solution exists; and that is the solution which is desired. In game problems, however, physical intuition causes one to expect multiple stationary solutions. For example, in the game of tag one strategy for the evader could minimize the time until he is tagged. Another strategy, which is the desired one, maximizes the time until he is tagged. In this case, both a best and a worst strategy exist, and both strategies satisfy the conditions for stationarity. Some additional conditions must be imposed to insure that the solution is the desired one.

The specific conditions which must be imposed are the Weierstrass - E Condition and the Clebsch Condition. These conditions are higher order conditions which can be used to eliminated the spurious solutions which satisfy the first order necessary conditions.

The Weierstrass - E Condition for this problem states that the Hamiltonian function must be maximized by the choice of ψ. This condition is expressed as follows.

$$\lambda^T f^{(a)}(x,\Psi,\theta,t) - \lambda^T f^{(a)}(x,\psi,\theta,t) \leq 0 . \tag{34}$$

The Ψ function is any permissible choice for player I's control. A permissible choice is one which does not cause violation of the inequality constraints. This statement of the Weierstrass - E Condition is derived from the conditions as stated in Reference 10.

The Clebsch Condition states that the Weierstrass - E Condition must be true for a second order function Taylor series expansion about the optimal solution. Since the first order terms in the expansion vanish if the solution satisfies the first order necessary conditions, the Clebsch Condition for this problem is as follows.

$$\delta\psi^T(\partial^2 H/\partial\psi^2)\delta\psi \leq 0 \ . \qquad (35)$$

The $\delta\psi$ function is any permissible small variation in player I's strategy and $H \equiv \lambda^T f^{(a)}$. If there are no inequality constraints, the Clebsch Condition states that $(\partial^2 H/\partial\psi^2)$ must be negative semi-definite. For the general problem under consideration, a second order expansion also includes terms involving the inequality constraints and the corner times. Since the complete second order expansion is very messy algebraically, it is not presented here. If multiple solutions arise, one must make the expansion to reject the spurious solutions.

C. Play Against an Optimal Strategy

In order to determine a quantitative measure of how much the worst case performance could be improved by changes to the suboptimal strategy, it is assumed that both players use optimal open-loop strategies.

To derive the first order necessary conditions for play against an optimal strategy, it is necessary to make the assumption that the inequality constraints are separable. Separable means that each component of the inequality constraints, or the appropriate time derivative of the constraints, are of the following form.

$$C(x, \psi, t) \leq 0 \tag{36}$$

or

$$C(x, \theta, t) \leq 0\ . \tag{37}$$

Equations (36) and (37) simply state that the inequality constraints restricting each player's choice of control are independent of the opposing player's choice of control. This assumption of separable constraints is often valid in actual practice.

The first-order necessary conditions for play against an optimal strategy follow directly from the necessary conditions for play against a known strategy. Player I must assume that player II chooses a strategy which minimizes φ. For such a strategy, the first variation of J with respect to θ must be zero. To derive the first variation with respect to θ, the integral term in equation (13) must be rewritten in the following manner.

$$\int_{t_{a-1}+\varepsilon}^{t_a-\varepsilon} \Big[\!\Big[\lambda^T(\partial f^{(a)}/\partial\psi)\delta\psi + \{\dot{\lambda} + [(\partial f^{(a)}/\partial x) + (\partial f^{(a)}/\partial\theta)(\partial\theta/\partial x)]^T\}\delta x \Big]\!\Big]\, dt$$

$$= \int_{t_{a-1}+\varepsilon}^{t_a-\varepsilon} \Big[\!\Big[\lambda^T(\partial f^{(a)}/\partial\psi)\delta\psi + \{\dot{\lambda} + (\partial f^{(a)}/\partial x)^T \lambda\}^T \delta x + \lambda^T(\partial f^{(a)}/\partial\theta)\delta\theta \Big]\!\Big]\, dt\ . \tag{38}$$

Because the first variation of J, ΔJ, must be zero for small permissible variations in θ, the coefficient of permissible variation in θ must be zero. By the same procedure as before, the arbitrary variations, δx, $\delta\psi$, and now $\delta\theta$, can be restricted to be permissible variations. The first order necessary conditions for play against an optimal strategy are found by following this procedure and by specifying that the first variation of J is zero. The transversality and corner conditions given by equations (23) through (29) are unchanged. The Euler-Lagrange equations given by equations (30) through (33) must be modified to account for the optimality of player II's strategy. If the solution is not on an inequality constraint boundary, equations (30) and (31) must be replaced by the following necessary conditions.

$$\lambda^T(\partial f^{(a)}/\partial\psi) = 0 \tag{39}$$

$$\lambda^T(\partial f^{(a)}/\partial\theta) = 0 \tag{40}$$

$$\dot{\lambda} + (\partial f^{(a)}/\partial x)^T \lambda = 0 . \tag{41}$$

If the solution is on an inequality constraint boundary, equations (39), (40) and (41) must be modified as follows.

$$\lambda^T\{\partial f^{(a)}/\partial\psi^*) - (\partial f^{(a)}/\partial\hat{\psi})[\partial C(\psi)/\partial\hat{\psi}]^{-1}[\partial C(\psi)\partial/\partial\psi^*]\} = 0 \tag{42}$$

$$\lambda^T\{(\partial f^{(a)}/\partial\theta^*) - (\partial f^{(a)}/\partial\hat{\theta})[\partial C(\theta)/\partial\hat{\theta}]^{-1}[\partial C(\theta)/\partial\theta^*] = 0 \tag{43}$$

$$\dot{\lambda} + \{(\partial f^{(a)}/\partial x) - (\partial f^{(a)}/\partial\hat{\psi})[\partial C(\psi)/\partial\hat{\psi}]^{-1}[\partial C(\psi)/\partial x]$$
$$- (\partial f^{(a)}/\partial\hat{\theta})[\partial C(\theta)/\partial\hat{\theta}]^{-1}[\partial C(\theta)/\partial x]\}^T \lambda = 0 \tag{44}$$

The functions $C(\psi)$ and $C(\theta)$ are the inequality constraints on the choices of ψ and θ, respectively. The vector $\hat{\theta}$ is partitioned into $\hat{\theta}$ and θ^* in a manner analogous to the partitioning of ψ. If $C(\psi)$ and $C(\theta)$ are not both satisfied as equalities at a specific time, the terms corresponding to the non-applicable constraint are deleted from equations (42), (43) and (44).

It is perhaps worth noting that the necessary conditions for an optimal strategy for player I have been derived as a limiting situation in which player II's strategy becomes as effective as it possibly can be. In this limiting case, players I's knowledge that player II is using an optimal strategy does not cause player I to alter his strategy to take advantage of that knowledge.

As before, the Weirstrass - E Condition and the Clebsch Condition must be imposed to eliminate any undesired solutions to the first-order necessary conditions. The necessary conditions for ψ as given by equations (34) and (35) are unchanged; however, there are additional necessary conditions if θ is to be a minimizing strategy. The Weierstrass - E Condition and the Clebsch Condition as applied to the θ strategy are as follows.

$$\lambda^T f^{(a)}(x,\psi,\Theta,t) - \lambda^T f^{(a)}(x,\psi,\theta,t) \geq 0 \tag{45}$$

$$\delta\theta^T(\partial^2 H/\partial\theta^2)\delta\theta \geq 0 . \tag{46}$$

The Θ function is any permissible choice for player II's strategy, and the $\delta\theta$ function is any permissible small variation in player II's strategy. As before, permissible implies satisfaction of the inequality constraints.

When the first-order necessary conditions are combined with Weierstrass - E Condition and the Clebsch Condition, it is seen that the basic necessary conditions for a solution to the differential game specify that the Hamiltonian function, H, is simultaneously maximized by the choice of ψ and minimized by the choice of θ. It can be said, therefore, that the optimal strategies satisfy a two-sided Maximum Principle.

D. Examples of Applying the Necessary Conditions

1. Purpose

The purpose of this section is to demonstrate how the first order necessary conditions can be used to determine solutions to variational and differential game problems. The selected examples are simple, in that the equations of motion and Euler-Lagrange equations can be integrated in closed-form.

The first two examples are not differential games because only one player is involved. Nonetheless, these examples illustrate the effects of state variable discontinuities in variational problems. The first example concerns a minimum time path between two points. A state variable discontinuity occurs when a specified relationship involving the state variables is satisfied; i.e., the corner is defined by $T_\alpha = 0$. The second example is similar to the first except that the corner time can be freely chosen.

The third example is a simple pursuit-evasion game where one of the players experiences a state variable discontinuity when the game is half over. This example is a differential game and is selected to illustrate the effects of state variable discontinuities in differential games.

2. Minimum Time Problem with Specified Corner

As an analytical example of the effects of state variable discontinuities, consider the problem of determining the minimum time path between two points for a particle moving at constant velocity. The solution is not a straight line because one of the position coordinates is discontinuous when a specified condition is satisfied. This problem is not a differential game because there is only one player, but it does illustrate the effects of state variable discontinuities.

The equations of motion are:

$$\dot{y} = V \cos \gamma \tag{47}$$

$$\dot{z} = V \sin \gamma \tag{48}$$

where V is the velocity magnitude (a constant), y and z are the position coordinates, and γ defines the direction of motion. The control variable is γ. The initial and terminal boundary conditions are:

$$y(0) = z(0) = 0 \tag{49}$$

$$y(T) = Y \tag{50}$$

$$z(T) = Z \tag{51}$$

where T is the final time. The performance index is the final time; i.e.,

$$\varphi = T . \tag{52}$$

A state variable discontinuity occurs when the following relationship is satisfied.

$$T_1 = z - aZ = 0 . \tag{53}$$

The magnitude of the discontinuity is:

$$\xi = \begin{Bmatrix} 0 \\ bz \end{Bmatrix}_{t=t_1-\varepsilon} \tag{54}$$

For this problem, the Euler-Lagrange equations are given by equations (30) and (31) with $\theta \equiv 0$.

$$V(-\lambda_y \sin \gamma + \lambda_z \cos \gamma) = 0 \tag{55}$$

$$\dot{\lambda}_y = 0 \tag{56}$$

$$\dot{\lambda}_z = 0 \tag{57}$$

Equations (56) and (57) state that λ_y and λ_z are constant over each subarc. Equation (55) states that γ is also constant over each subarc and is equal to

$$\gamma = \tan^{-1}(\lambda_z/\lambda_y) . \tag{58}$$

The transverality conditions are given by equations (23) and (24); however, only equation (24) yields any real information.

$$V(\lambda_y \cos\gamma + \lambda_z \sin\gamma)_{t=T} = 1\,. \tag{59}$$

The corner conditions are given by equations (25) and (26). These conditions yield the result that the discontinuity in λ_z is:

$$\lambda_z^{(-)} \equiv \lambda_z(t_1 - \varepsilon) = (1+b)\,\lambda_z(t_1+\varepsilon) \equiv (1+b)\,\lambda_z^{(+)}\,. \tag{60}$$

Equation (26) in conjunction with equations (55), (56) and (57) yields the interesting result that equation (59) holds throughout the entire solution. This result can be combined with equation (55) to eliminate λ_y and to express γ as:

$$\gamma = \sin^{-1}(V\lambda_z)\,. \tag{61}$$

Equation (61) holds for either subarc.

Equations (47) through (60) can be combined to obtain the following equation, which must be solved for $\lambda_z^{(+)}$ by trial and error.

$$(Z/Y) = \left[(1+b)^2 a V \lambda_z^{(+)} \Big/ \sqrt{1 - V^2(1+b)^2(\lambda^{(+)})^2}\right] + \left[(1-a)V\lambda_z^{(+)} \Big/ \sqrt{1 - V^2(\lambda_z^{(+)})^2}\right]. \tag{62}$$

Once $\lambda_z^{(+)}$ is determined, $\lambda_z^{(-)}$ can be computed from equation (60) and γ can be computed for both subarcs from equation (61). The minimum final time is:

$$T = (Z/V)\left\{\left[a \Big/ \sqrt{1 - V^2(1+b)^2(\lambda_z^{(+)})^2}\right] + \left[(1-a) \Big/ \sqrt{1 - V^2(\lambda_z^{(+)})^2}\right]\right\} \tag{63}$$

Figure 1 shows the minimum time solutions for various values of the parameter b. The following values are used for the remaining parameters.

$$Y = 100 \tag{64}$$

$$Z = 100 \tag{65}$$

$$V = 1 \tag{66}$$

$$a = 0.5 \tag{67}$$

3. Minimum Time Problem with Free Corners

This example is similar to the preceding one except that the corner time is free and the magnitude of the discontinuity is define in a different manner. The equations of motion,

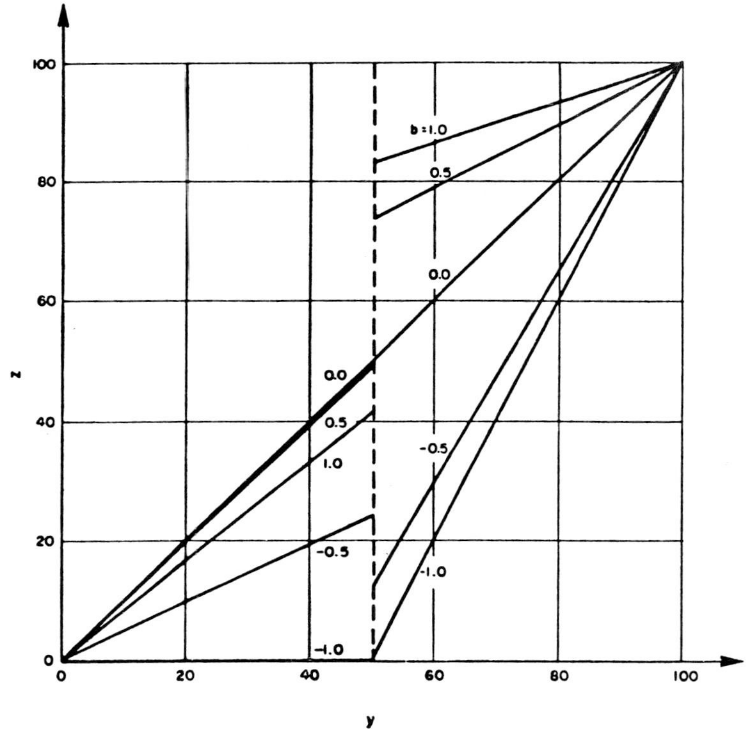

FIG. 1. Minimum time solution with specified corner

boundary conditions, and performance index are identical to those of the preceding example; i.e., they are given by equations (47) through (52). The magnitude of the discontinuity is defined as follows:

$$\xi = \left\{ \begin{matrix} 0 \\ \\ (z - c_1) + c_2(y - c_3)^2 \end{matrix} \right\}. \tag{68}$$

Since the corner time is free, it can be selected to minimize the final time.

The Euler-Lagrange equations and the transverality conditions are identical to those of the preceding example. The solution again consists of straight line segments, the slopes of which are determined by equation (58). It can be seen from equation (29) that equation (59) is again valid throughout the entire solution; therefore, equation (61) is also valid throughout the entire solution. The discontinuity in λ_z at the corner is:

$$\lambda_z^{(-)} = 2\,\lambda_z^{(+)}. \tag{69}$$

In this example, λ_y is discontinuous; but the value of λ_y need not be determined because γ can be computed from equation (59).

The relationship from which the optimum corner time can be determined is derived from equations (28) and (29).

$$\begin{aligned} &\lambda_y^{(+)}[\cos\gamma^{(+)} - \cos\gamma^{(-)}] \\ &\quad + \lambda_z^{(+)}[\sin\gamma^{(+)} - 2\sin\gamma^{(-)} - 2c_2(y_1 - c_3)\cos\gamma^{(-)}] = 0 \end{aligned} \tag{70}$$

where y_1 is the value of y at $t = t_1 - \varepsilon$. By combining equations (59), (55), (69) and (70) y_1 can be expressed in terms of $\lambda_z^{(+)}$.

$$y_1 = c_3 + \left[\sqrt{1 - 4V^2(\lambda_z^{(+)})^2} - \sqrt{1 - V^2(\lambda_z^{(+)})^2} \right] \Big/ \left[2c_2 \lambda_z^{(+)} \right] . \qquad (71)$$

By integrating the equations of motion and performing some algebra, the final boundary conditions can be expressed in terms of $\lambda_z^{(+)}$.

$$\begin{aligned} Z = & \left[4V \lambda_z^{(+)} z_1 \Big/ \sqrt{1 - 4V^2(\lambda_z^{(+)})^2} \right] + c_2(y_1 - c_3)^2 - c_1 \\ & + \left[V \lambda_z^{(+)} (Y - y_1) \Big/ \sqrt{1 - V^2(\lambda_z^{(+)})^2} \right] . \end{aligned} \qquad (72)$$

Equations (71) and (72) can be combined and the result solved for $\lambda_z^{(+)}$ by trial and error. Equations (69) and (61) can be used then to determine γ for both subarcs. The minimum final time is computed as:

$$\begin{aligned} T = & \left[(y_1/V) \Big/ \sqrt{1 - 4V^2(\lambda_z^{(+)})^2} \right] \\ & + \left\{ [Y - y_1)/V] \Big/ \sqrt{1 - V^2(\lambda_z^{(+)})^2} \right\} . \end{aligned} \qquad (73)$$

Figure 2 shows the minimum time solutions for various values of the parameter c_1. The following values are used for the remaining parameters.

$$Y = 100 \qquad (74)$$

$$Z = 100 \qquad (75)$$

$$V = 1 \qquad (76)$$

$$c_2 = 0.02 \qquad (77)$$

$$c_3 = 50 \qquad (78)$$

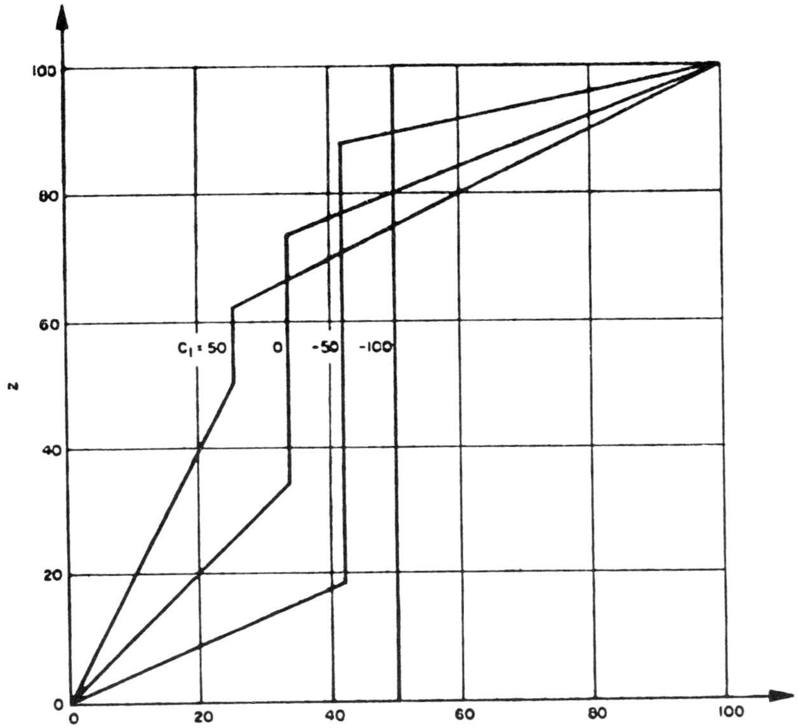

FIG. 2. Minimum time solution with free corner.

4. Differential Game with a State Variable Discontinuity

A very simple pursuit-evasion game involving a state variable discontinuity is considered in this section. This example differs from the preceding two examples in that this one is truly a game. Player I seeks to maximize the performance index while player II seeks to minimize it. The optimal open-loop strategies of both players are to be determined.

Each player moves at constant velocity, but each player can control his direction of motion. The equations of motion are:

$$\dot{x}_1 = V_1 \cos \psi \tag{79}$$

$$\dot{x}_2 = V_1 \sin \psi \tag{80}$$

$$\dot{x}_3 = V_2 \cos \theta \tag{81}$$

$$\dot{x}_4 = V_2 \sin \theta \tag{82}$$

where x_1 and x_2 are the position coordinates of player I; x_3 and x_4 are the position coordinates of player II; V_1 and V_2 are the velocities of players I and II, respectively; and ψ and θ are the respective directions of motion. The initial conditions are specified. The problem is that player I evades player II such that T seconds after the beginning of play, the square of the distance between them will be as large as possible. Player II pursues player I to minimize the square of the final miss distance. The performance criterion is then:

$$\varphi(T) = \tfrac{1}{2}[(x_1 - x_3)^2 + (x_2 - x_4)^2] . \tag{83}$$

Player I experiences a state variable discontinuity when the following condition is satisfied.

$$T_1 = t - \tfrac{1}{2} T = 0 . \tag{84}$$

The magnitude of the discontinuity is:

$$\xi_1 = [0, bx_2, 0, 0]^T_{t=t_1-\varepsilon} . \tag{85}$$

The Euler-Lagrange equations for this problem are derived from equations (39), (40) and (41).

$$-V_1 \lambda_1 \sin \psi + V_1 \lambda_2 \cos \psi = 0 \tag{86}$$

$$-V_2 \lambda_3 \sin \theta + V_2 \lambda_4 \cos \theta = 0 \tag{87}$$

$$\dot{\lambda}_1 = \dot{\lambda}_2 = \dot{\lambda}_3 = \dot{\lambda}_4 = 0 \,. \tag{88}$$

Equations (86) and (87) can be solved for ψ and θ.

$$\psi = \tan^{-1}(\lambda_2/\lambda_1) \tag{89}$$

$$\theta = \tan^{-1}(\lambda_4/\lambda_3) \,. \tag{90}$$

Because of equation (88) ψ and θ are seen to be constants along each subarc.

The transversality conditions are derived from equations (23) and (24). The latter equation provides no real information since it involves the unknown ν_1. Equation (23) provides the following terminal values for λ.

$$\lambda_1(T) = x_3(T) - x_1(T) \tag{91}$$

$$\lambda_2(T) = x_4(T) - x_2(T) \tag{92}$$

$$\lambda_3(T) = x_1(T) - x_3(T) = -\lambda_1(T) \tag{93}$$

$$\lambda_4(T) = x_2(T) - x_4(T) = -\lambda_2(T) \,. \tag{94}$$

These transversality conditions and equations (89) and (90) yield the result that the directions of motion are parallel during the second subarc. The Clebsch Conditions, equations (35) and (46) yield the result that ψ and θ are actually equal during the second subarc. The value of ψ during the second subarc is denoted as ψ_2.

The discontinuity in the λ vector at the corner is derived from equation (25).

$$\lambda_1(t_1 - \varepsilon) = \lambda_1(t_1 + \varepsilon) \tag{95}$$

$$\lambda_2(t_1 - \varepsilon) = (1 + b)\lambda_2(t_1 + \varepsilon) \tag{96}$$

$$\lambda_3(t_1 - \varepsilon) = \lambda_3(t_1 + \varepsilon) \tag{97}$$

$$\lambda_4(t_1 - \varepsilon) = \lambda_4(t_1 + \varepsilon) \tag{98}$$

Equations (90), (97), and (98), show that θ is continuous through the corner. The value of ψ on the first subarc ψ_1, is computed from equations (89), (95), and (96). It is related to ψ_2 in the following manner.

$$\tan \psi_1 = (1 + b) \tan \psi_2 . \tag{99}$$

Equations (85), (89), (90) and (95), through (99) can be substituted into the equations of motion, which can be integrated in closed form. The results for $x_1(T)$, $x_2(T)$, $x_3(T)$, and $x_4(T)$ can be substituted into the transversality conditions, equations (91) through (94), to yield the following equation involving only $\tan \psi_2$.

$$-(x_3 - x_1)_{t_0} \tan \psi_2 + (x_4 - x_2)_{t_0}$$
$$= V_1(\tfrac{1}{2} T)(b^2 + 2b) \tan \psi_2 \Big/ \sqrt{1 + (1+b)^2 \tan^2 \psi_2} . \tag{100}$$

Equation (100) can be solved for $\tan \psi_2$ by a simple iterative procedure. Figure 3 shows the optimizing solutions for several values of b. The other parameters have the following values.

$$T = 100 \tag{101}$$

$$V_1 = V_2 = 1 \tag{102}$$

$$x_1(t = 0) = x_2(t = 0) = 0 \tag{103}$$

$$x_3(t = 0) = x_4(t = 0) = 100 \tag{104}$$

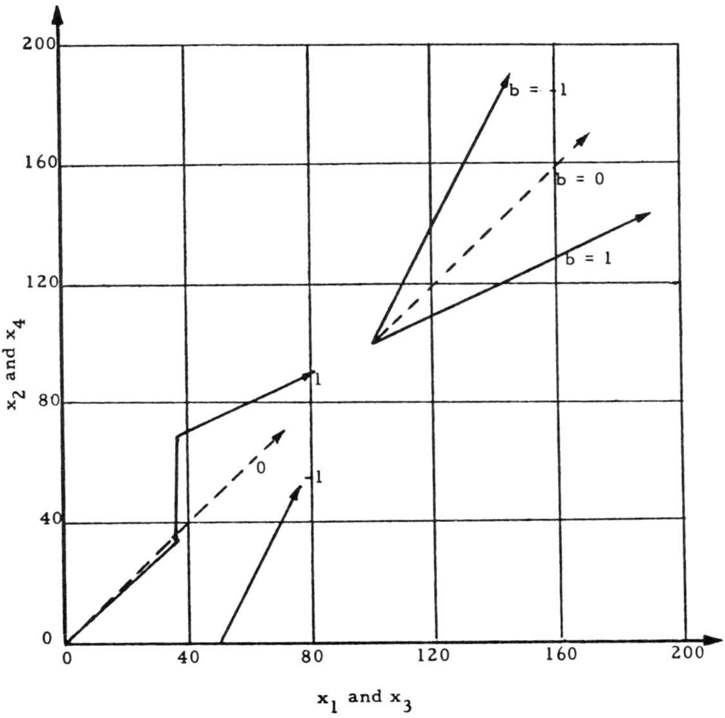

FIG. 3. Solution of differential game with state-variable discontinuity

If player II chooses the suboptimal strategy of $\theta = \pi/4$ it can be easily shown that player I's optimal strategy for $b = 1$ is $\theta = \pi/8$ on the first subarc and $\theta = \pi/4$ on the second subarc. With the suboptimal strategy the value of φ is 137.3. With an optimal open-loop strategy the value of φ is 122.5; therefore, player II could improve his worst case performance by as much as 14.8 by changes in his strategy.

III. AN EXTENDED QUASILINEARIZATION ALGORITHM

A. General Multi-Point Boundary Value Problem

The necessary conditions developed in Sections II, B and II, C comprise a discontinuous, multi-point boundary value problem. Previous work has shown that quasilinearization is a promixing technique for solving two point boundary value problems. In one of the earliest computer applications of quasilinearization, McGill and Kenneth [6] obtained excellent convergence properties in their orbit transfer problems. Moyer and Pinkham [11] compare the convergence properties of quasilinearization to those of gradient and second-order methods. For the orbit transfer problem considered, the convergence properties of quasilinearization are significantly better than those of the other two methods. Paine [12] extends the algorithms to permit bounded control variables, but he does not treat general forms of control variable inequality constraints. McGill [13] uses a penalty function approach to handle state variable inequality constraints. This approach yields a sequence of optimal solutions which should converge to the true optimum. Unfortunately, the approach is inconvenient as it is not likely to yield an optimal solution with a single submittal to the computer. In none of the references are state and adjoint variable discontinuities handled directly.

In this section the quasilinearization algorithm is extended to handle a general class of multi-point boundary value problems involving multiple subarcs, state and/or control variable constraints, and discontinuities in the state and/or adjoint variables. This extended algorithm is then applied to a discontinuous version of the brachistochrone problem to test its efficacy.

In order to minimize notational confusion, the general multi-point boundary value problem is restated here using notation tailored to the quasilinearization algorithm.

The system under consideration is governed by

$$\dot{Y}^{(a)} = F^{(a)}(Y,u,t) \quad \text{for} \quad t_{a-1} < t < t_a \quad a = 1,\ldots,A \tag{105}$$

where t is the independent variable, time; $(\dot{\ })$ is $d(\ )/dt$; Y is an N-dimensional vector of the system variables; u is an r-dimensional vector of control variables; and $F^{(a)}$ is an N-dimensional vector of known functions of Y, u, and t. The Y vector includes the state variables, the adjoint variables, and any unknown multipliers associated with terminal and/or corner constraints. The constant multipliers, denoted as μ, obey the following differential equations.

$$\dot{\mu} = 0 . \tag{106}$$

The superscript (a) notation in equation (105) is used because of the discontinuous nature of the problem as described in Section II, A.

The control variables, u, are determined by numerical solution of the following relationship.

$$G^{(a)}(Y,u,t) = 0 \tag{107}$$

where $G^{(a)}$ is an r-dimensional vector which is explicitly dependent on u. It is assumed that equation (107) implicitly defines unique values for all the components of u if values are specified for Y and t. The components of G can arise from several sources. If the subarc is not affected by inequality constraints, they are the usual algebraic Euler-Lagrange equations. If the subarc is affected by inequality constraints, they consist of a modified subset of the algebraic Euler-Lagrange equations plus the equation of the inequality constraint or a time derivative thereof. If equation (107) can be solved in closed form for any of the components of u, it is desirable to do so; however, equations (105) and (107) represent the more general case.

The system variables, Y, are discontinuous at the corner times. The magnitude of the discontinuity is defined as:

$$\lim_{\varepsilon\to 0} Y(t_a + \varepsilon) = \lim_{\varepsilon\to 0} \{Y(t_a - \varepsilon) + \tilde{\xi}[Y(t_a - \varepsilon),\ t_a - \varepsilon]\} \tag{108}$$

where ξ is an N-dimensional vector function of Y and t which possesses first derivatives with respect to both Y and t. The individual components of $\tilde{\xi}$ correspond to discontinuities in the physical system and in the adjoint variables. These latter discontinuities often arise if state variable inequality constraints are present.

The occurrence of the corner is defined by the satisfaction of a subarc stopping condition.

$$\psi_a[Y(t_a - \varepsilon), Y(t_a + \varepsilon),\ t_a] = 0 \qquad a = 1,\dots,A. \tag{109}$$

The ψ_a function may correspond to the encountering of an inequality constraint, a criterion for a discontinuity in the physical system, to a Weierstrass-Erdmann corner condition, or to constraint at the final time.

In addition to the subarc stopping conditions, the solution must also satisfy additional constraint vectors at each corner.

$$\Omega_a[Y(t_a - \varepsilon),\ Y(t_a + \varepsilon),\ t_a] = 0 \qquad a = 1,\dots,A. \tag{110}$$

These constraints may correspond to time derivatives of the

inequality constraint function, a midpoint constraint imposed on the physical system, additional Weierstrass-Erdmann corner conditions, physical constraints at the final time, or transversality conditions.

It is assumed that n of the initial values of Y are specified. For a well posed problem, therefore, the sum of the dimensions of Ω_a for all a must equal $N - n$. The unknown corner and final times are implicitly determined by the ψ_a functions.

In summary, there are N differential equations with n boundary conditions specified at the initial time and a total of $N - n$ conditions specified at the corner and final times. These corner and final times are implicitly defined by the stopping conditions equation (109). The r components of u are determined by the roots of equation (107), and the discontinuities in the system variables, Y, are defined by equation (108). The problem is to find a solution, $Y(t)$, which satisfies all these conditions.

B. Linear Model

The basic assumption behind the quasilinearization algorithm is that two neighboring solutions, Y_k and Y_{k-1}, are sufficiently close to be related by linear expressions. The theory of linear operators can then be applied to determine a sequence of solutions which satisfy the boundary conditions and hopefully will converge to a solution of the differential constraints. The first step of the analysis, therefore, is the development of linear expressions relating Y_k and Y_{k-1} along a given subarc.

If two neighboring solutions, Y_k and Y_{k-1}, of the differential system, equation (105), are sufficiently close, one can make the following linear approximation.

$$\delta Y^{(a)} \equiv \dot{Y}_k^{(a)} - \dot{Y}_{k-1}^{(a)} = J^{(a)}(Y_{k-1}, u_{k-1}, t)(Y_k - Y_{k-1}) + P^{(a)}(Y_{k-1}, u_{k-1}, t)(u_k - u_{k-1}) \tag{111}$$

where

$$J^{(a)} \equiv \begin{bmatrix} \partial F_1^{(a)}/\partial y_1 & \cdots & \partial F_1^{(a)}/\partial y_N \\ \vdots & & \vdots \\ \partial F_N^{(a)}/\partial y_1 & \cdots & \partial F_N^{(a)}/\partial y_N \end{bmatrix} \tag{112}$$

$$P^{(a)} \equiv \begin{bmatrix} \partial F_1^{(a)}/\partial u_1 & \cdots & \partial F_1^{(a)}/\partial u_r \\ \vdots & & \vdots \\ \partial F_N^{(a)}/\partial u_1 & \cdots & \partial F_N^{(a)}/\partial u_r \end{bmatrix} \tag{113}$$

and $y_1, y_2, \ldots, y_N$ are the components of Y_1 and $u_1, u_2, \ldots, u_r$ are the components of u.

For sufficiently close solutions, equation (107) can also be linearized.

$$\delta G^{(a)} \equiv G_k^{(a)} - G_{k-1}^{(a)} = M_{k-1}^{(a)}(Y_k - Y_{k-1}) + N_{k-1}^{(a)}(u_k - u_{k-1}) \tag{114}$$

where

$$M^{(a)} \equiv (\partial G^{(a)}/\partial Y) \tag{115}$$

$$N^{(a)} \equiv (\partial G^{(a)}/\partial u)\,. \tag{116}$$

The individual elements of the $M^{(a)}$ and $N^{(a)}$ matrices are defined in a manner similar to equations (112) and (113). The subscript notation refers to the solution for which the quantity is evaluated; e.g., $M_{k-1}^{(a)}$ means that $M^{(a)}$ is evaluated using Y_{k-1} and u_{k-1}. Assuming that $N_{k-1}^{(a)}$ is nonsingular, equation (114) can be solved for $u_k - u_{k-1}$.

$$u_k - u_{k-1} = \left(N_{k-1}^{(a)} \right)^{-1} \left[\delta G^{(a)} - M_{k-1}^{(a)} (Y_k - Y_{k-1}) \right] . \tag{117}$$

The $\delta G^{(a)}$ term will normally be zero, but it is retained here in the interests of generality. Equation (117) can be substituted into equation (111) and the result solved for $\dot{Y}_k^{(a)}$.

$$\begin{aligned} \dot{Y}_k^{(a)} = F_{k-1}^{(a)} &+ \left[J_{k-1}^{(a)} - P_{k-1}^{(a)} \left(N_{k-1}^{(a)} \right)^{-1} M_{k-1}^{(a)} \right] Y_k \\ &- \left[J_{k-1}^{(a)} - P_{k-1}^{(a)} \left(N_{k-1}^{(a)} \right)^{-1} M_{k-1}^{(a)} \right] Y_{k-1} \\ &+ P_{k-1}^{(a)} \left(N_{k-1}^{(a)} \right)^{-1} \delta G^{(a)} . \end{aligned} \tag{118}$$

In equation (111) the assumption is made that $\dot{Y}_{k-1}^{(a)} = F_{k-1}^{(a)}$. Unless $Y_{k-1}^{(a)}$ satisfies the differential constraints, this is not strictly true. However, the approximation is consistent with equation (111) and is made in order to obtain a sequence of solutions which converges to one satisfying the differential constraints.

Equation (118) is a linear differential equation with time varying coefficients. It is the first of the linear relationships required by the quasilinearization algorithm.

A complete solution of the differential system contains a number of corners with state variable discontinuities. For convenience of notation, it is assumed that there is only one such corner, but the extension to multiple corners will be obvious. The stopping condition at the single corner is denoted as ψ_c. The following definition is made to more easily define the conditions at the corner.

$$\tau \equiv t_a - \varepsilon \tag{119}$$

where it is understood that ε is an infinitesimally small positive number. With this notation, τ represents the instant immediately prior to the discontinuity, and $\tau + 2\varepsilon$ represents the instant immediately after it. The value of τ is implicitly defined by equation (109). The use of τ permits the suppression of the superscript (a) notation at the corner. The situation at the corner is then depicted in Figure 4. If Y_k and Y_{k-1} are sufficiently close, one can write the following linear approximation.

$$\begin{aligned}\Delta\psi_c &\equiv \psi_{c_k}(\tau_k) - \psi_{c_{k-1}}(\tau_{k-1})\\ &= K_{cm}[Y_k(\tau_{k-1}) - Y_{k-1}(\tau_{k-1})] + K_{cp}[Y_k(\tau_{k-1} + 2\varepsilon)\\ &\quad - Y_{k-1}(\tau_{k-1} + 2\varepsilon)] + \dot{\psi}_c(\tau_k - \tau_{k-1})\end{aligned} \tag{120}$$

where

$$K_{cm} = [\partial\psi_c/\partial Y(\tau)]\big|_{k-1} \tag{121}$$

$$K_{cp} = [\partial\psi_c/\partial Y(\tau + 2\varepsilon)]\big|_{k-1} \tag{122}$$

$$\dot{\psi}_c = [K_{cm}F^{(a)}(\tau) + K_{cp}F^{(a+1)}(\tau + 2\varepsilon) + \partial\psi_c/\partial t]\big|_{k-1}\,. \tag{123}$$

The terms $Y_k(\tau_{k-1} + 2\varepsilon)$ and $Y_{k-1}(\tau_{k-1} + 2\varepsilon)$ can be approximated as follows.

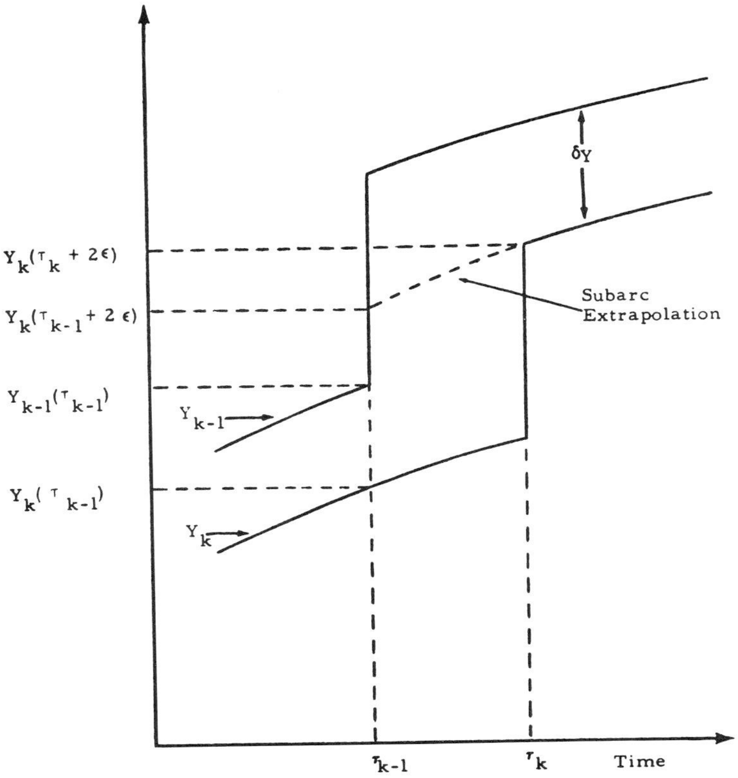

Fig. 4. Geometry of Discontinuity at Corner.

$$
\begin{aligned}
Y_k(\tau_{k-1} + 2\varepsilon) &= Y_k(\tau_{k-1}) + \xi[Y_k(\tau_{k-1})] \\
&= Y_k(\tau_{k-1}) + \xi[Y_{k-1}(\tau_{k-1})] \\
&\quad + R[Y_{k-1}(\tau_{k-1})] \cdot [Y_k(\tau_{k-1}) - Y_{k-1}(\tau_{k-1})] \\
&= \{I + R[Y_{k-1}(\tau_{k-1})]\} Y_k(\tau_{k-1}) + \xi[Y_{k-1}(\tau_{k-1})] \\
&\quad - R[Y_{k-1}(\tau_{k-1})]Y_{k-1}(\tau_{k-1})
\end{aligned}
\tag{124}
$$

$$Y_{k-1}(\tau_{k-1} + 2\varepsilon) = Y_{k-1}(\tau_{k-1}) + \xi[Y_{k-1}(\tau_{k-1})] \tag{125}$$

where I is the identity matrix and

$$R \equiv \partial\xi/\partial Y \, . \tag{126}$$

Substituting equations (124) and (125) into equation (120), assuming $\dot{\psi}_c \neq 0$, and solving the resulting expression for $\tau_k - \tau_{k-1}$, the following result is obtained.

$$\begin{aligned}\tau_k - \tau_{k-1} = &- [\dot{\psi}_c(\tau_{k-1})]^{-1} \Big\{ K_{cm}(\tau_{k-1}) \\ &+ K_{cp}(\tau_{k-1})\{I + R[Y_{k-1}(\tau_{k-1})]\}\Big\} \, Y_k(\tau_{k-1}) \\ &+ [\dot{\psi}_c(\tau_{k-1})]^{-1} \Big\{ K_{cm}(\tau_{k-1}) \\ &+ K_{cp}(\tau_{k-1})\{I + R[Y_{k-1}(\tau_{k-1})]\}\Big\} \, Y_{k-1}(\tau_{k-1}) \\ &+ [\dot{\psi}_c(\tau_{k-1})]^{-1} \, \Delta\psi_c \, .\end{aligned}$$

The $\Delta\psi_c$ term is retained in equation (127) to permit the correction of any errors in ψ_c which occur as a result of the linear approximations used in deriving equation (127).

The boundary conditions at the corner time must also be linearized. For sufficiently close solutions, one can make the following approximation.

$$\begin{aligned}\Delta\Omega_c \equiv\ &\Omega_c[Y_k(\tau_k), Y_k(\tau_k + 2\varepsilon), \tau_k] - \Omega_c[Y_{k-1}(\tau_{k-1}), \\ &Y_{k-1}(\tau_{k-1} + 2\varepsilon), \tau_{k-1}] \\ \cong\ &D_{cm}(\tau_{k-1})[Y_k(\tau_{k-1}) - Y_{k-1}(\tau_{k-1})] +\end{aligned}$$

$$+ D_{cp}(\tau_{k-1})[Y_k(\tau_{k-1} + 2\varepsilon) - Y_{k-1}(\tau_{k-1} + 2\varepsilon)]$$

$$+ \dot{\Omega}_c(\tau_{k-1})(\tau_k - \tau_{k-1}) \tag{128}$$

where

$$D_{cm} = [\partial\Omega_c/\partial Y(\tau)]\Big|_{k-1} \tag{129}$$

$$D_{cp} = [\partial\Omega_c/\partial Y(\tau + 2\varepsilon)]\Big|_{k-1} \tag{130}$$

$$\dot{\Omega}_c = [D_{cm}\ F^{(a)}(\tau) + D_{cp}\ F^{(a+1)}(\tau + 2\varepsilon) + \partial\Omega_c/\partial\tau]\Big|_{k-1} . \tag{131}$$

Using equation (124) to eliminate $Y_k(\tau_{k-1} + 2\varepsilon)$ and equation (127) to eliminate $\tau_k - \tau_{k-1}$, one obtains the following expression for $\Delta\Omega_c$.

$$\Delta\Omega_c = A_c Y_k(\tau_{k-1}) - A_c Y_{k-1}(\tau_{k-1})$$
$$+ \dot{\Omega}_c(\tau_{k-1})[\Delta\psi_c/\dot{\psi}_c(\tau_{k-1})] \tag{132}$$

where

$$A_c \equiv D_{cm}(\tau_{k-1}) + D_{cp}(\tau_{k-1})\{I + R[Y_{k-1}(\tau_{k-1})]\}$$
$$- [\dot{\psi}_c(\tau_{k-1})]^{-1}\ \dot{\Omega}_c(\tau_{k-1})\Big\{K_{cm}(\tau_{k-1}) + K_{cp}(\tau_{k-1})$$
$$\cdot \{I + R[Y_{k-1}(\tau_{k-1})]\}\Big\} . \tag{133}$$

The conditions at the final time must also be linearized. This linearization is performed in a manner similar to that used for the corner times. The final time is denoted as T, and the changes in T and Ω_f are as follows.

$$T_k - T_{k-1} = [\dot{\psi}_f(T_{k-1})]^{-1} \cdot \{\Delta\psi_f - K_f(T_{k-1})[Y_k(T_{k-1}) - Y_{k-1}(T_{k-1})]\} \tag{134}$$

where

$$\Delta\psi_f = \psi_f[Y_k(T_k)] - \psi_f[Y_{k-1}(T_{k-1})] \tag{135}$$

$$K_f \equiv (\partial\psi_f/\partial Y)\Big|_{k-1} \tag{136}$$

$$\dot{\psi}_f = [K_f\, F^{(A)} + (\partial\psi_f/\partial t)]\Big|_{k-1} \tag{137}$$

and

$$\Delta\Omega_f = A_f Y_k(T_{k-1}) - A_f Y_{k-1}(T_{k-1}) + \dot{\Omega}_f(T_{k-1})[\Delta\psi_f/\dot{\psi}_f(T_{k-1})] \tag{138}$$

where

$$D_f \equiv (\partial\Omega_f/\partial Y)\Big|_{k-1} \tag{139}$$

$$\Delta\Omega_f = \Omega_f[Y_k(T_k)] - \Omega_f[Y_{k-1}(T_{k-1})] \tag{140}$$

$$A_f \equiv D_f(T_{k-1}) - [\dot{\psi}_f(T_{k-1})]^{-1}\, \dot{\Omega}_f(T_{k-1})K_f(T_{k-1}) \,. \tag{141}$$

All of the linear relationships required by the quasi-linearization algorithm have now been derived.

C. Computational Procedure

Taken altogether, equations (118), (124), and (118) again can be viewed as a linear equation of the form:

$$L_{k-1}Y_k = b_{k-1} \tag{142}$$

where L_{k-1} is a linear operator, and b_{k-1} is the nonhomogeneous forcing function. The operator L_{k-1} is such that nonzero solutions of the homogeneous version of equation (142) exist. Any linear combination of homogeneous solutions can be added to a particular solution of equation (142) and the result will also be a solution. The specific linear combination of homogeneous solutions can be selected to cause satisfaction of the n initial conditions on Y plus the boundary conditions specified by equations (132) and (140).

If a particular solution of equation (142) is denoted as $Y_k^{(P)}$ and a set of $N-n$ linearly independent solutions of the homogeneous form of equation (142) is denoted as $Y_k^{(H)}$, the following linear combination of solutions also satisfies equation (142)

$$Y_k = Y_k^{(P)} + Y_k^{(H)} c_k \tag{143}$$

where c_k is an $N-n$ vector of as yet undetermined coefficients. These coefficients can be used to find a solution satisfying equations (132) and (140). The initial conditions are satisfied by choosing intial values for $Y_k^{(P)}$ and $Y_k^{(H)}$ such that equation (143) satisfies the initial conditions for any c_k. Substituting equation (143) into equations (143) into equations (132) and (140) yields the following results.

$$\Delta\Omega_c - \dot{\Omega}_c(\tau_{k-1})[\Delta\psi_c/\dot{\psi}_c(\tau_{k-1})] + A_c Y_{k-1}(\tau_{k-1})$$
$$= A_c Y_k^{(H)}(\tau_{k-1})c_k + A_c Y_k^{(P)}(\tau_{k-1}) \tag{144}$$

$$\Delta\Omega_f - \dot{\Omega}_f(T_{k-1})[\Delta\psi_f/\dot{\psi}_f(T_{k-1})] + A_f Y_{k-1}(T_{k-1})$$
$$= A_f Y_k^{(H)}(T_{k-1})c_k + A_f Y_k^{(P)}(T_{k-1}) . \tag{145}$$

In equations (144) and (145) $\Delta\Omega_c, \Delta\psi_c, \Delta\Omega_f$, and $\Delta\psi_f$ can be specified to correct any constraint and stopping condition errors which exist for Y_{k-1}. Equations (144) and (145) can then be solved for c_k as follows.

$$c_k = \begin{bmatrix} A_c Y_k^{(H)}(\tau_{k-1}) \\ \text{-------} \\ A_f Y_k^{(H)}(T_{k-1}) \end{bmatrix}^{-1} \left\{ \Delta\Omega + \begin{bmatrix} A_c[Y_{k-1}(\tau_{k-1}) - Y_k^{(P)}(\tau_{k-1})] \\ \text{-------------} \\ A_f[Y_{k-1}(T_{k-1}) - Y_k^{(P)}(T_{k-1})] \end{bmatrix} \right\} \tag{146}$$

where

$$\Delta\Omega \equiv \left\{ \begin{matrix} \Delta\Omega_c - \dot{\Omega}_c(\tau_{k-1})[\Delta\psi_c/\dot{\psi}_c(\tau_{k-1})] \\ \text{-------------} \\ \Delta\Omega_f - \dot{\Omega}_f(T_{k-1})[\Delta\psi_f/\dot{\psi}_f(T_{k-1})] \end{matrix} \right\} \tag{147}$$

The step-by-step computational procedure for the extended quasilinearization algorithm can now be specified.

1. Guess a complete time history for the Y vector. The initial guess is denoted as Y_0. It need not satisfy any of the boundary or stopping conditions at the corner or final times, but it must satisfy the n boundary conditions at the initial time. The discontinuity in Y_0 at the corner need not satisfy equation (108). Although there is considerable latitude in choosing Y_0, it should be as close as possible to the actual solution. The guessed Y_0 becomes the first nominal solution.

2. Using the nominal solution, Y_{k-1}, compute $F_{k-1}^{(a)}$, $J_{k-1}^{(a)}$, $P_{k-1}^{(a)}$, $N_{k-1}^{(a)}$, $M_{k-1}^{(a)}$, and $\delta G^{(a)}$ as functions of time. Usually $\delta G^{(a)}$ will be identically zero. Specify $\Delta\psi_c$, $\Delta\Omega_c$,

$\Delta\psi_f$, and $\Delta\Omega_f$ to correct any errors in these quantitites for the current nominal solution. Compute K_{cm}, K_{cp}, $\dot{\psi}_c$, $\xi[Y_{k-1}(\tau_{k-1})]$, and R.

3. Generate a set of $N - n$ linearly independent solutions of the homogeneous form of equation (118), i.e., of the following differential equation.

$$\dot{Y}_k^{(H)} = [J_{k-1}^{(a)} - P_{k-1}^{(a)} (N_{k-1}^{(a)})^{-1} M_{k-1}^{(a)}]Y_k^{(H)} . \tag{148}$$

The initial conditions on $Y_k^{(H)}$ should be selected such that the addition of any linear combination of them to the particular solution does not change the initial conditions on the first n components of $Y_k^{(P)}$. One such set of initial conditions for $Y_k^{(H)}$ is as follows.

$$Y_k^{(j)}(t_0) = \lfloor 0,\ldots,0,y_{n+j} = 1,0,\ldots,0\rfloor^T \qquad j = 1,\ldots,N - n \tag{149}$$

where $Y_k^{(j)}$ represents the jth homogeneous solution. Equation (148) is integrated from $t = t_0$ to $t = \tau_{k-1}$. At τ_{k-1} the homogeneous solutions are discontinuous, the magnitudes of the discontinuities being computed as follows.

$$Y_k^{(H)}(\tau_{k-1} + 2\varepsilon) = \{I + R[Y_{k-1}(\tau_{k-1})]\} Y_k^{(H)}(\tau_{k-1}) . \tag{150}$$

After computing $Y_k^{(H)}(\tau_{k-1} + 2\varepsilon)$, the integration of equation (148) is continued from $t = \tau_{k-1} + 2\varepsilon$ to $t = T_{k-1}$.

4. Generate a particular solution of equation (118). The initial conditions for this particular solution must

satisfy the initial conditions specified for the first n components of Y. One such set of initial conditions is:

$$Y_k^{(P)}(t_0) = Y_{k-1}(t_0) . \tag{151}$$

There are other possible choices for $Y_k^{(P)}(t_0)$, but this choice has the nice property that the particular solution converges to the actual solution. Because the initial guess satisfies the boundary conditions at t_0, all subsequent nominal solutions satisfy the boundary conditions at t_0. Equation (118), including the nonhomogeneous terms, is integrated from t_0 to τ_{k-1}. At τ_{k-1} the particular solution is discontinuous. The value of $Y_k^{(P)}$ at $(\tau_{k-1} + 2\varepsilon)$ is computed from equation (124). After the discontinuity, equation (118) is integrated from $t = \tau_{k-1} + 2\varepsilon$ to $t = T_{k-1}$.

5. Compute D_{cm}, D_{cp}, $\dot{\Omega}_c$, A_c, K_f, $\dot{\psi}_f$, D_f, and A_f. Compute the coefficients of the homogeneous solutions using equations (146) and (147). Use the coefficients and equation (143) to compute $Y_k(t_0)$, $Y_k(\tau_{k-1})$, and $Y_k(T_{k-1})$. Compute τ_k and T_k from equations (127) and (134), respectively.

6. Using $Y_k(t_0)$, generate a complete time history for Y_k by integrating equation (118) from $t = t_0$ to $t = \tau_{k-1}$. Extrapolate (or truncate) the terminal portion of this subarc to τ_k using the following relationship.

$$Y_k(\tau_k) = Y_k(\tau_{k-1}) + F_{k-1}^{(a)}(\tau_{k-1})(\tau_k - \tau_{k-1}) . \tag{152}$$

Compute the discontinuity in Y_k at τ_{k-1} using equation (124). After computing $Y_k(\tau_{k-1} + 2\varepsilon)$, continue the integration of equation (118) from $t = \tau_{k-1} + 2\varepsilon$ to $t = T_{k-1}$. Extrapolate (or truncate) the initial portion of the second subarc to τ_k as follows.

$$
\begin{aligned}
Y_k(\tau_k + 2\varepsilon) &= Y_k(\tau_{k-1} + 2\varepsilon) \\
&\quad + F_{k-1}^{(a+1)}(\tau_{k-1} + 2\varepsilon)(\tau_k - \tau_{k-1}) .
\end{aligned} \tag{153}
$$

Extrapolate (or truncate) the terminal portion of the second subarc to T_k as follows.

$$
Y_k(T_k) = Y_k(T_{k-1}) + F_{k-1}^{(a+1)}(T_{k-1})(T_k - T_{k-1}) . \tag{154}
$$

7. Compute a measure of the difference between Y_k and Y_{k-1}. One such measure is as follows.

$$
\rho(k, k-1) = \max_{t \in [t_0, T]} |Y_k(t) - Y_{k-1}(t)| \tag{155}
$$

where ρ is an N-dimensional vector. If $\rho(k, k-1) \leq \bar{\varepsilon}$, where $\bar{\varepsilon}$ is an N-dimensional vector of small positive numbers, the solution has converged. If $\rho(k, k-1) > \bar{\varepsilon}$, let Y_k be the new nominal solution and return to step 2.

Steps 1 through 7 comprise the straightforward computational procedure. Experience has shown that the region of convergence can be significantly expanded if the procedure is slightly modified in the following manner. If $\rho(k, k-1) > \bar{\varepsilon}$, do not use Y_k as the new nominal. Instead, use Y_{new} as the new nominal where Y_{new} is computed from Y_k and Y_{k-1} in the following manner.

$$
Y_{new} = Y_{k-1} + m(k)[Y_k - Y_{k-1}] \tag{156}
$$

where $m(k)$ is a multiplier such that $0 \leq m(k) \leq 1$. If

$m(k) = 1$, Y_k becomes the new nominal solution. If $m(k) < 1$, the new nominal solution is only part way between Y_{k-1} and Y_k. It has been found that convergence can be greatly improved if $m(k)$ is a small number, such as .2 or .3, for the initial iterations. As the iteration number increases, $m(k)$ can be gradually increased to one.

D. Example

A discontinuous version of the brachistochrone problem provides an example of the application of the extended quasilinearization algorithm. The geometry of the problem is depicted in Figure 5. The problem is to find the shape of a frictionless wire connecting the points (x_0, y_0) and (x_f, y_f) which minimizes the time required for a bead with initial velocity V_0 to slide under the action of gravity from (x_0, y_0) to (x_f, y_f). The equations of motion are:

$$\dot{V} = g \sin \gamma \tag{157}$$

$$\dot{x} = V \cos \gamma \tag{158}$$

$$\dot{y} = V \sin \gamma . \tag{159}$$

The velocity is discontinuous at the corner defined by the following stopping condition.

$$\psi_c = x - c_\psi x_f . \tag{160}$$

The magnitude of the discontinuity in velocity is proportional to the velocity just prior to the corner.

$$V(\tau + \varepsilon) = V(\tau - \varepsilon) - c_\xi V(\tau - \varepsilon) \tag{161}$$

where τ is the corner time. The initial values of V, x, and y are specified, and the terminal constraints are:

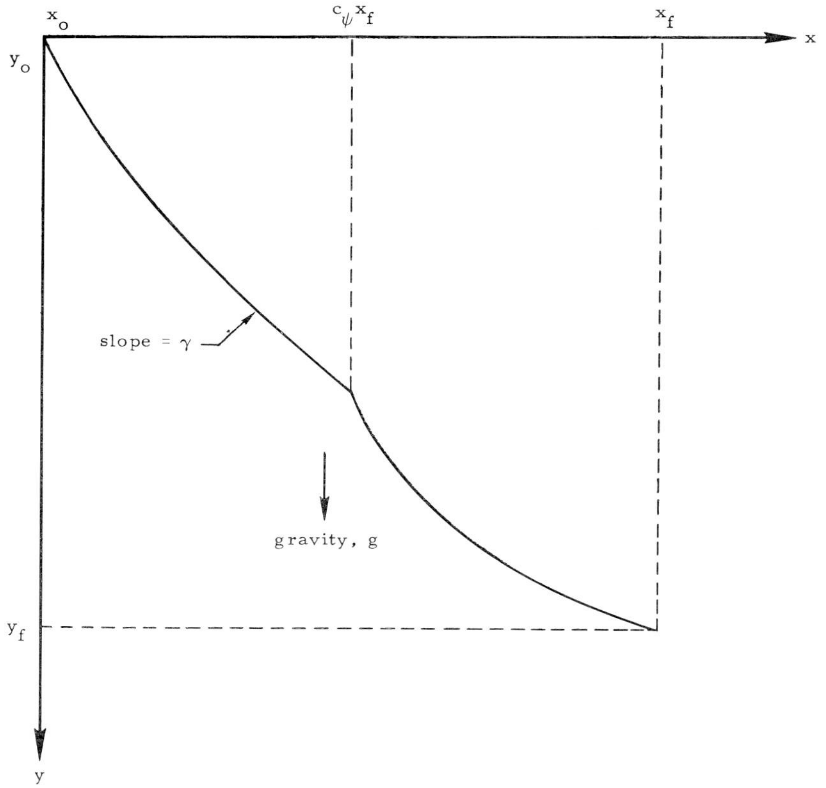

Fig. 5. Geometry of Brachistochrone.

$$\psi_f = x - x_f = 0 \tag{162}$$

$$\Omega_{f_1} = y - y_f = 0 \tag{163}$$

where the constraint on the final x is used as the stopping condition for the second subarc.

The necessary conditions for an optimal solution are derived from the general conditions derived in Section II, B. The differential Euler-Lagrange equations which must be satisfied along each subarc are as follows.

$$\dot{\lambda}_v = -\cos \gamma\, \lambda_x - \sin \gamma\, \lambda_y \qquad (164)$$

$$\dot{\lambda}_x = 0 \qquad (165)$$

$$\dot{\lambda}_y = 0. \qquad (166)$$

The algebraic Euler-Lagrange equation is:

$$G = g \cos \gamma\, \lambda_v - V \sin \gamma\, \lambda_x + V \cos \gamma\, \lambda_y = 0. \qquad (167)$$

The discontinuities in the adjoint variables, λ_v, λ_x, and λ_y, are:

$$\lambda_v(\tau + \varepsilon) - \lambda_v(\tau - \varepsilon) = [c_\xi/(1 - c_\xi)]\, \lambda_v(\tau - \varepsilon) \qquad (168)$$

$$\lambda_x(\tau + \varepsilon) - \lambda_x(\tau - \varepsilon) = \mu \qquad (169)$$

$$\lambda_y(\tau + \varepsilon) - \lambda_y(\tau - \varepsilon) = 0 \qquad (170)$$

where μ is an unknown constant multiplier associated with ψ_c. The Weierstrass-Erdmann corner condition is:

$$\Omega_c = [\lambda_v g \sin \gamma + \lambda_x V \cos \gamma + \lambda_y V \sin \gamma]\big|_{t=\tau+\varepsilon}$$
$$- [\lambda_v g \sin \gamma + \lambda_x V \cos \gamma + \lambda_y V \sin \gamma]_{t=\tau+\varepsilon} = 0 \qquad (171)$$

The transversality conditions at the final time are:

$$\Omega_{f_2} = \lambda_v(T) = 0 \qquad (172)$$

$$\Omega_{f_3} = [\lambda_v g \sin \gamma + \lambda_x V \cos \gamma + \lambda_y V \sin \gamma]\big|_{t=T} - 1 = 0. \qquad (173)$$

In order to apply the extended quasilinearization algorithm, the unknown multiplier, μ, is treated as a component of the Y vector. Its differential equation is:

$$\dot{\mu} = 0 . \tag{174}$$

The complete Y and ξ vectors then are as follows:

$$Y \equiv [V, x, y, \lambda_V, \lambda_x, \lambda_y, \mu]^T \tag{175}$$

$$\xi \equiv \begin{Bmatrix} -c_\xi V \\ 0 \\ 0 \\ [c_\xi/(1-c_\xi)]\ \lambda_V \\ \mu \\ 0 \\ 0 \end{Bmatrix} \tag{176}$$

The data for specific problem are as follows:

$$x(0) = y(0) = 0 \tag{177}$$

$$V(0) = 10 \tag{178}$$

$$c_\psi = c_\xi = .5 \tag{179}$$

$$x_f = y_f = 1{,}000 \tag{180}$$

$$g = 32.2 \tag{181}$$

$$m(k) = 1 \quad \text{for all} \quad k . \tag{182}$$

For the initial guess, x, y, and λ_V were assumed to be linear functions of time.

$$x_0(t) = y_0(t) = 100\ t \tag{183}$$

$$\lambda_V(t) = .03(1 - .1t) . \tag{184}$$

The initial guess for V is displayed in Figure 6. The initial guesses for the corner time, τ, and the final time,

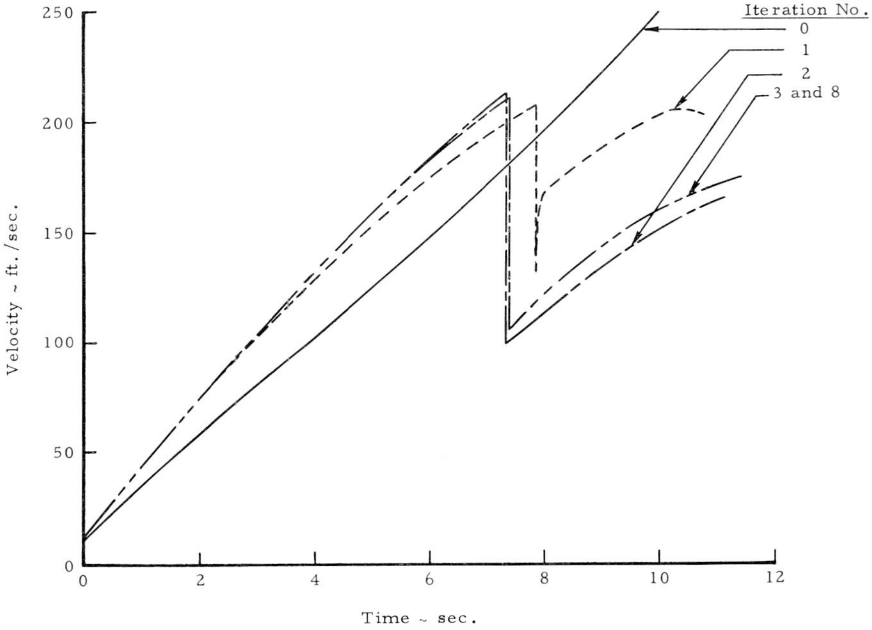

Fig. 6. Convergence Behavior of V.

T, are 5 and 10 seconds, respectively. The intial guesses for λ_y and μ are both zero, and the initial guess for λ_x is a constant .005 for both subarcs.

The convergence behavior in terms of the distances between iterations is displayed in Table 1. The convergence behavior of V and λ_V are depicted in Figures 6 and 7, respectively. The convergence behavior of the space path is depicted in Figure 8. The converged values of the corner and final times are 7.3764 sec. and 11.4247 sec., respectively.

Table 1: Convergence Behavior of Error Indicators

k	$\lvert v_k - v_{k-1}\rvert^*$	$\lvert x_k - x_{k-1}\rvert^*$	$\lvert y_k - y_{k-1}\rvert^*$	$\lvert \lambda_{v_k} - \lambda_{v_{k-1}}\rvert^*$	$\lvert \lambda_{x_k} - \lambda_{x_{k-1}}\rvert^*$	$\lvert \lambda_{y_k} - \lambda_{y_{k-1}}\rvert^*$	$\lvert \mu_k - \mu_{k-1}\rvert^*$
1	4.88×10^{1}	3.71×10^{2}	1.45×10^{2}	9.83×10^{-3}	1.14×10^{-3}	1.39×10^{-3}	1.07×10^{-3}
2	5.46×10^{1}	1.49×10^{2}	1.71×10^{2}	1.54×10^{-2}	1.74×10^{-3}	2.25×10^{-4}	1.03×10^{-3}
3	8.49×10^{0}	6.73×10^{1}	1.48×10^{1}	1.31×10^{-2}	2.16×10^{-3}	5.18×10^{-4}	2.11×10^{-3}
4	6.53×10^{-1}	3.46×10^{0}	2.18×10^{0}	2.36×10^{-3}	4.14×10^{-4}	9.71×10^{-5}	4.28×10^{-4}
5	3.58×10^{-1}	1.65×10^{0}	1.88×10^{0}	2.49×10^{-4}	3.73×10^{-5}	2.44×10^{-5}	5.30×10^{-5}
6	1.74×10^{-1}	8.70×10^{-1}	5.66×10^{-1}	2.01×10^{-5}	6.29×10^{-6}	6.36×10^{-8}	5.59×10^{-6}
7	4.59×10^{-2}	1.59×10^{-1}	2.50×10^{-1}	4.08×10^{-5}	7.05×10^{-6}	2.61×10^{-6}	9.15×10^{-6}
8	1.50×10^{-2}	5.01×10^{-2}	5.39×10^{-2}	1.82×10^{-6}	5.71×10^{-7}	3.02×10^{-7}	8.86×10^{-7}

* Maximum values over $t\in(0,T)$

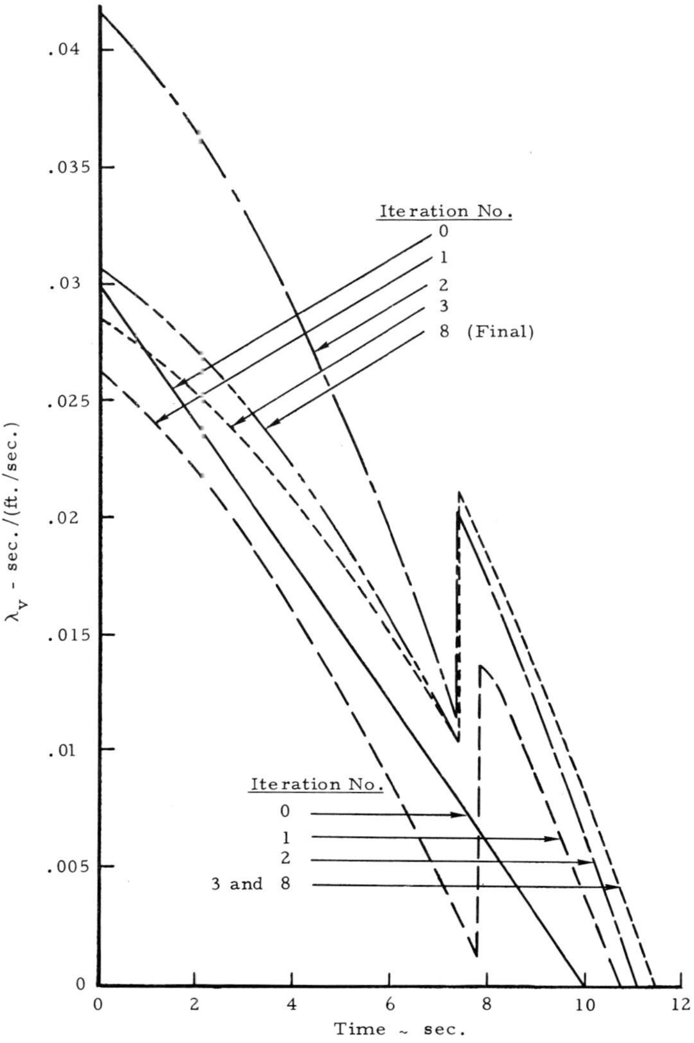

Fig. 7. Convergence Behavior of λ_v.

The converged value of λ_x is .003808 on the first subarc and .005593 on the second subarc. The converged values of λ_y and μ are .001166 and .001785, respectively.

An interesting characteristic of the convergence

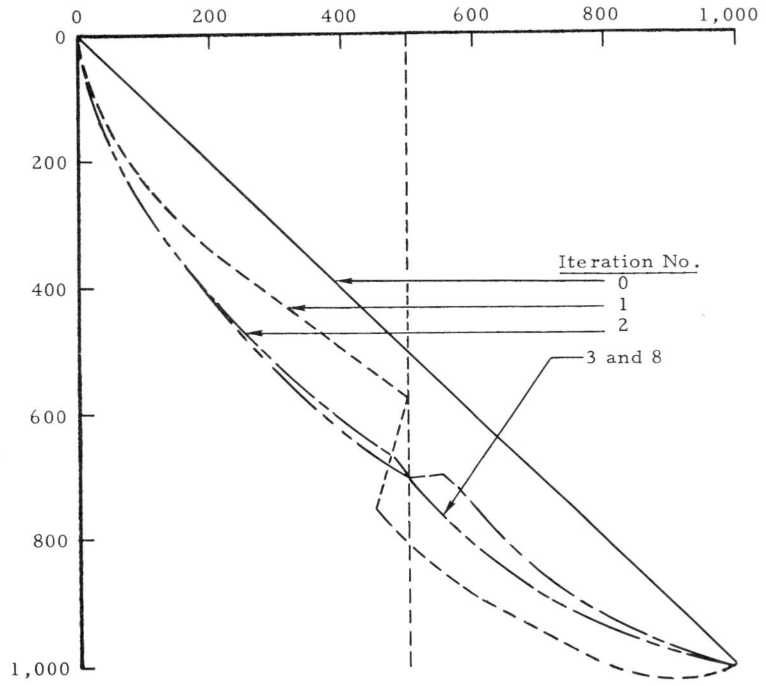

Fig. 8. Convergence Behavior of Physical Coordinates.

behavior is that the implicitly defined corner time introduces discontinuities in state variables which are supposedly continuous; i.e., in x and y. These discontinuities vanish as the procedure converges. The source of this behavior is the manner of linearization and extrapolation at the corner time. Alternate schemes can perhaps be found to avoid this behavior; however, the discontinuities do not adversely effect the convergence of the overall procedure.

IV. ATMOSPHERIC PURSUIT-EVASION PROBLEM

A. Problem Statement

As an example of applying the quasilinearization algorithm to solving a differential game, the problem of determining optimal time histories for an aircraft and an interceptor in a pursuit-evasion game is considered. The geometry of the problem is depicted in Figure 9. It is assumed that the motion

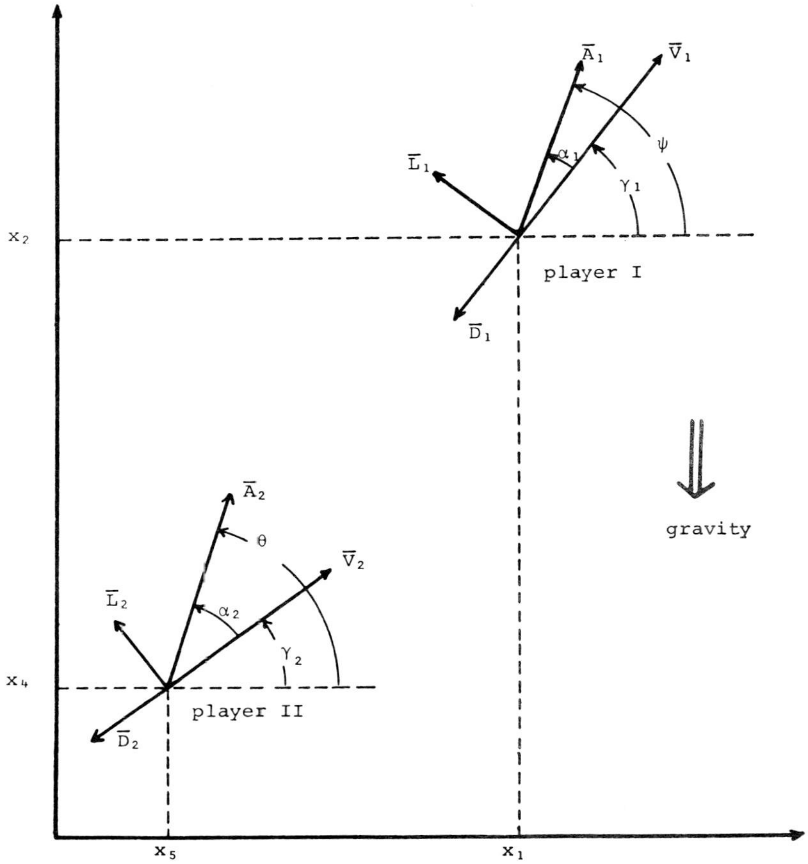

Fig. 9. Geometry of Pursuit-Evasion Game.

occurs in a plane over a flat earth with an exponential atmosphere. The problem is that player I wishes to evade player II such that T seconds after the beginning of play the distance between them is as large as possible. Player II wishes to minimize that distance.

The aerodynamic forces are assumed to be proportional to the velocity magnitude. Such an assumption is not too bad for moderately high supersonic speeds. With this assumption the equations of motion for the pursuer and evader are as follows.

$$\dot{x}_1 = x_3 = F_1 \tag{185}$$

$$\dot{x}_2 = x_4 = F_2 \tag{186}$$

$$\dot{x}_3 = A_1 \cos\psi - [C_{D_1}(\alpha_1)x_3 + C_{L_1}(\alpha_1)x_4](\rho_1 S_1/m_1) = F_3 \tag{187}$$

$$\dot{x}_4 = A_1 \sin\psi - [C_{D_1}(\alpha_1)x_4 - C_{L_1}(\alpha_1)x_3](\rho_1 S_1/m_1) - g = F_4 \tag{188}$$

$$\dot{x}_5 = x_7 = F_5 \tag{189}$$

$$\dot{x}_6 = x_8 = F_6 \tag{190}$$

$$\dot{x}_7 = A_2 \cos\theta - [C_{D_2}(\alpha_2)x_7 + C_{L_2}(\alpha_2)x_8](\rho_2 S_2/m_2) = F_7 \tag{191}$$

$$\dot{x}_8 = A_2 \sin\theta - [C_{D_2}(\alpha_2)x_8 - C_{L_2}(\alpha_2)x_7](\rho_2 S_2/m_2) - g = F_8 \tag{192}$$

where x_1 and x_2 are the position coordinates of vehicle 1, the evader; x_3 and x_4 are the velocity components of vehicle 1; x_5 and x_6 are the position coordinates of vehicle 2, the pursuer; x_7 and x_8 are the velocity components of vehicle 2; A_1 and A_2 are the thrust accelerations of vehicles 1 and 2, respectively; ψ and θ are the attitude angles defined by Figure 9; C_{D_1} and C_{D_2} are the drag coefficients for vehicles 1 and 2; C_{L_1} and C_{L_2} are the lift coefficients; α_1 and α_2 are the angles of attack; ρ_1 and ρ_2 are the atmospheric densities at the positions of vehicles 1 and 2; S_1 and S_2 are the aerodynamic references areas; m_1 and m_2 are the masses; and g is the acceleration due to gravity. For this example, A_1, A_2, S_1, S_2, m_1, m_2, and g are all assumed to be constant. The lift and drag forces are dependent on the angle of attack in the following manner.

$$C_{L_1}(\alpha_1) = (C_{L_0})_1 \sin\alpha_1 \cos\alpha_1 \tag{193}$$

$$C_{D_1}(\alpha) = (C_{D_0})_1 + (C_{D_L})_1 \sin^2\alpha_1 \tag{194}$$

$$C_{L_2}(\alpha_2) = (C_{L_0})_2 \sin\alpha_2 \cos\alpha_2 \tag{195}$$

$$C_{D_2}(\alpha_2) = (C_{D_0})_2 + (C_{D_L})_2 \sin^2\alpha_2 . \tag{196}$$

The angles of attack, α_1 and α_2, are computed in the following manner.

$$\alpha_1 = \psi - \gamma_1 = \psi - \tan^{-1}(x_4/x_3) \tag{197}$$

$$\alpha_2 = \theta - \gamma_2 = \theta - \tan^{-1}(x_8/x_7) . \tag{198}$$

An exponential atmosphere is used to determine the atmospheric density.

$$\rho_1 = K_\rho \rho_0 \, e^{-(x_2/h_0)} \tag{199}$$

$$\rho_2 = K_\rho \rho_0 \, e^{-(x_6/h_0)} \tag{200}$$

where K_ρ is a multiplier between zero and one; ρ_0 is sea level density; and h_0 is the density decay factor.

The initial conditions for $x_1, \ldots, x_8$ are specified. The only terminal boundary condition is that the final time is fixed. The value of the final time can be varied parametrically to determine a value which extremizes the performance. The performance index is one-half the square of the terminal miss distance.

$$\varphi = \tfrac{1}{2}[(x_1 - x_5)^2 + (x_2 - x_6)^2] . \tag{201}$$

The Euler-Lagrange equations are dependent on whether or not the pursuer uses an optimal strategy. In this first case, it is assumed that he does. The differential Euler-Lagrange equations are then derived from equation (41).

$$\dot{\lambda}_1 = 0 = F_9 \tag{202}$$

$$\begin{aligned}\dot{\lambda}_2 = {} & [(\partial\rho_1/\partial x_2)(S_1/m_1)] \cdot [(C_{D_1} x_3 + C_{L_1} x_4)\lambda_3 \\ & + (C_{D_1} x_4 - C_{L_1} x_3)\lambda_4] = F_{10}\end{aligned} \tag{203}$$

$$\begin{aligned}\dot{\lambda}_3 = {} & -\lambda_1 + (\rho_1 S_1/m_1)\{[(\partial C_{D_1}/\partial\alpha_1)(\overline{\partial\alpha_1/\partial x_3})x_3 + C_{D_1} \\ & + (\partial C_{L_1}/\partial\alpha_1)(\overline{\partial\alpha_1/\partial x_3})x_4]\lambda_3 + [(\partial C_{D_1}/\partial\alpha_1)(\overline{\partial\alpha_1/\partial x_3})x_4 \\ & - (\partial C_{L_1}/\partial\alpha_1)(\overline{\partial\alpha_1/\partial x_3})x_3 - C_{L_1}]\lambda_4\} = F_{11}\end{aligned} \tag{204}$$

$$\begin{aligned}\dot{\lambda}_4 = {} & -\lambda_2 + (\rho_1 S_1/m_1)\{[(\partial C_{D_1}/\partial\alpha_1)(\overline{\partial\alpha_1/\partial x_4})x_3 \\ & + (\partial C_{L_1}/\partial\alpha_1)(\overline{\partial\alpha_1/\partial x_4})x_4 + C_{L_1}]\lambda_3 \\ & + [(\partial C_{D_1}/\partial\alpha_1)(\overline{\partial\alpha_1/\partial x_4})x_4 \\ & + C_{D_1} - (\partial C_{L_1}/\partial\alpha_1)(\partial\alpha_1/\partial x_4)x_3]\lambda_4\} = F_{12}\end{aligned} \tag{205}$$

$$\dot{\lambda}_5 = 0 = F_{13} \tag{206}$$

$$\begin{aligned}\dot{\lambda}_6 = {} & [(\partial\rho_2/\partial x_6)(S_2/m_2)] \cdot [(C_{D_2} x_7 + C_{L_2} x_8)\lambda_7 \\ & + (C_{D_2} x_8 - C_{L_2} x_7)\lambda_8] = F_{14}\end{aligned} \tag{207}$$

$$\dot{\lambda}_7 = -\lambda_5 + (\rho_2 S_2/m_2)\ \{[(\partial C_{D_2}/\partial\alpha_2)(\overline{\partial\alpha_2/\partial x_7})x_7 + C_{D_2} +$$

$$+ (\partial C_{L_2}/\partial\alpha_2)(\overline{\partial\alpha_2/\partial x_7})x_8]\lambda_7 + [(\partial C_{D_2}/\partial\alpha_2)(\overline{\partial\alpha_2/\partial x_7})x_8$$

$$- (\partial C_{L_2}/\partial\alpha_2)(\overline{\partial\alpha_2/\partial x_7})x_7 - C_{L_2}]\lambda_8\} = F_{15} \quad (208)$$

$$\dot{\lambda}_8 = -\lambda_6 + (\rho_2 S_2/m_2)\{[(\partial C_{D_2}/\partial\alpha_2)(\overline{\partial\alpha_2/\partial x_8})x_7$$

$$+ (\partial C_{L_2}/\partial\alpha_2)(\overline{\partial\alpha_2/\partial x_8})x_8 + C_{L_2}]\lambda_7$$

$$+ [(\partial C_{D_2}/\partial\alpha_2)(\overline{\partial\alpha_2/\partial x_8})x_8$$

$$+ C_{D_2} - (\partial C_{L_2}/\partial\alpha_2)(\overline{\partial\alpha_2/\partial x_8})x_7]\lambda_8\} = F_{16} \quad (209)$$

where

$$(\overline{\partial\alpha_1/\partial x_3}) = x_4/(x_3^2 + x_4^2) \quad (210)$$

$$(\overline{\partial\alpha_1/\partial x_4}) = -x_3/(x_3^2 + x_4^2) \quad (211)$$

$$(\overline{\partial\alpha_2/\partial x_7}) = x_8/(x_7^2 + x_8^2) \quad (212)$$

$$(\overline{\partial\alpha_2/\partial x_8}) = -x_7/(x_7^2 + x_8^2)\ . \quad (213)$$

The algebraic Euler-Lagrange equations are derived from equations (39) and (40).

$$G_1 = \{-A_1 \sin\psi - (\rho_1 S_1/m_1)[(\partial C_{D_1}/\partial\alpha_1)x_3$$

$$+ (\partial C_{L_1}/\partial\alpha_1)x_4]\}\lambda_3 + \{A_1 \cos\psi$$

$$-(\rho_1 S_1/m_1)[(\partial C_{D_1}/\partial\alpha_1)x_4 - (\partial C_{L_1}/\partial\alpha_1)x_3]\}\lambda_4 = 0 \quad (214)$$

$$G_2 = \{-A_2 \sin\theta - (\rho_2 S_2/m_2)[(\partial C_{D_2}/\partial\alpha_2)x_7$$
$$+ (\partial C_{L_2}/\partial\alpha_2)x_8]\}\lambda_7 + \{A_2 \cos\theta$$
$$- (\rho_2 S_2/m_2)|(\partial C_{D_2}/\partial\alpha_2)x_8 - (\partial C_{L_2}/\partial\alpha_2)x_7]\}\lambda_8 = 0 \ . \quad (215)$$

The transversality conditions for this problem are derived from equation (23).

$$\lambda_1(T) = x_5(T) - x_1(T) \quad (216)$$
$$\lambda_2(T) = x_6(T) - x_2(T) \quad (217)$$
$$\lambda_3(T) = \lambda_4(T) = 0 \quad (218)$$
$$\lambda_5(T) = x_1(T) - x_5(T) = -\lambda_1(T) \quad (219)$$
$$\lambda_6(T) = x_2(T) - x_6(T) = -\lambda_2(T) \quad (220)$$
$$\lambda_7(T) = \lambda_8(T) = 0 \quad (221)$$

The transversality condition given by equation (24) need not be considered because it can always be satisfied by an appropriate choice for ν.

The problem then is to determine a solution to the eight equations of motion plus the Euler-Lagrange equations such that the initial conditions for x_1 through x_8 and the terminal conditions for λ_1 through λ_8 are satisfied. The control variable time histories, ψ and θ, must satisfy the algebraic Euler-Lagrange equations. This problem is a two-point boundary value problem.

B. Application of Quasilinearization Algorithm

The following definitions are made to cast the problem in the form for solution by the quasilinearization algorithm described in Section III.

$$Y \equiv \lfloor x_1, \ldots, x_8, \lambda_1, \ldots, \lambda_8 \rfloor^T \tag{222}$$

$$F \equiv \lfloor F_1, \ldots, F_{16} \rfloor^T \tag{223}$$

$$u \equiv \lfloor \psi, \theta \rfloor^T$$

$$G \equiv \lfloor G_1, G_2 \rfloor^T \tag{225}$$

The matrix G_u is diagonal, and its first term is:

$$\begin{aligned}(\partial G_1 \partial \psi) = &- \{A_1 \cos \psi + (\rho_1 S_1/m_1)[(\partial^2 C_{D_1}/\partial \alpha_1^2)x_3 \\ &+ (\partial^2 C_{L_1}/\partial \alpha_1^2)x_4]\}\lambda_3 - \{A_1 \sin \psi \\ &+ (\rho_1 S_1/m_1)[(\partial^2 C_{D_1}/\partial \alpha_1^2)x_4 - (\partial^2 C_{L_1}/\partial \alpha_1^2)x_3]\}\lambda_4 .\end{aligned} \tag{226}$$

It should be noted at this point that the equation of motion and the Euler-Lagrange equations for the pursuer and evader have precisely the same form. The correspondence between the variables for the evader and pursuer is shown in Table 2. In the interests of brevity, only the equations for the evader are presented, but it must be remembered that a similar set of equations exists for the pursuer.

Examination of equations (226) and (218) reveals a significant problem. At the final time λ_3 and λ_4 must both be zero; therefore, equations (214) is satisfied by any ψ and $(\partial G_1/\partial \psi) = 0$. As a consequence, the G_u matrix is singular at the final time and the linearized equations of motion and the differential Euler-Lagrange equations will have a singularity at the final time. Recognizing that this problem exists, the form of the linearized differential system of equations is as follows.

Table 2. Correspondence of Variables for Evader and Pursuer

Variable for Evader	Corresponding Variable for Pursuer
x_1, x_2, x_3, x_4	x_5, x_6, x_7, x_8
$\lambda_1, \lambda_2, \lambda_3, \lambda_4$	$\lambda_5, \lambda_6, \lambda_7, \lambda_8$
ψ	θ
α_1	α_2
$(C_{L_o})_1, (C_{D_o})_1, (C_{D_L})_1$	$(C_{L_o})_2, (C_{D_o})_2, (C_{D_L})_2$
C_{L_1}, C_{D_1}	C_{L_2}, C_{D_2}
m_1	m_2
S_1	S_2
ρ_1	ρ_2
G_1	G_2

$$\dot{x}_i^{(k)} = \sum_{j=1}^{4} a_{ij} x_j^{(k)} + \sum_{j=1}^{4} b_{ij} \lambda_j^{(k)} + F_i - \sum_{j=1}^{4} a_{ij} x_j^{(k-1)} - \sum_{j=1}^{4} b_{ij} \lambda_j^{(k-1)} \tag{227}$$

$$\dot{\lambda}_i^{(k)} = \sum_{j=1}^{4} c_{ij} x_j^{(k)} + \sum_{j=1}^{4} d_{ij} \lambda_j^{(k)} + F_{(i+8)} - \sum_{j=1}^{4} c_{ij} x_j^{(k-1)} - \sum_{j=1}^{4} d_{ij} \lambda_j^{(k-1)} . \tag{228}$$

The coefficients a_{ij}, b_{ij}, c_{ij}, and d_{ij} are computed using $Y^{(k-1)}$, where the superscript refers to the iteration number. Similar expressions hold for $\dot{x}_5^{(k)}$ through $\dot{x}_8^{(k)}$ and for $\dot{\lambda}_5^{(k)}$ through $\dot{\lambda}_8^{(k)}$.

If the $(k-1)$st iteration satisfies the transversality conditions, it can be shown by application of L'Hospital's Rule that only the b_{ij} and d_{ij} are truly infinite at the final time. The a_{ij} and c_{ij} all possess finite limits as t approaches T. The b_{ij} and d_{ij} go to infinity as $(T-t)^{-1}$.

Assuming for the moment that the $(k-1)$st iteration satisfies the transverality conditions, $\lambda_3^{(k-1)}$ and $\lambda_4^{(k-1)}$ can be approximated near the final time in the following manner.

$$\lambda_3^{(k-1)}(t) \cong -\dot{\lambda}_3^{(k-1)}(T)[T-t] \tag{229}$$

$$\lambda_4^{(k-1)}(t) \cong -\dot{\lambda}_4^{(k-1)}(T)[T-t]. \tag{230}$$

By computing $\dot{\lambda}_3^{(k-1)}$ and $\dot{\lambda}_4^{(k-1)}$ by some numerical differencing method, substituting equations (229) and (230) into equation (214) and dividing the result by $(T-t)$, one can obtain an expression which can be solved for $\psi(T)$. Using this $\psi(T)$ and equations (229) and (230), all the a_{ij} and c_{ij} at $t = T$ can be evaluated. In fact, all the c_{ij} are zero at $t = T$. Although the b_{ij} and d_{ij} cannot be evaluated, the following quantities can be.

$$\tilde{b}_{ij} = -\lim_{t\to T} (T-t)b_{ij} \tag{231}$$

$$\tilde{d}_{ij} = -\lim_{t\to T} (T-t)d_{ij}. \tag{232}$$

If it were certain that the particular solution and all the homogeneous solutions of equations (227) and (228) satisfied the transversality conditions, the terms $b_{ij}\ \lambda_j$ and $d_{ij}\ \lambda_j$ could be computed at the final time in the following manner.

$$\left(b_{ij}\ \lambda_j^{(k)}\right)_{t=T} = \left(\tilde{b}_{ij}\ \dot{\lambda}_j^{(k)}\right)_{t=T} \tag{233}$$

$$\left(d_{ij}\ \lambda_j^{(k)}\right)_{t=T} = \left(\tilde{d}_{ij}\ \dot{\lambda}_j^{(k)}\right)_{t=T} . \tag{234}$$

The singularity at $t = T$ would then be integrable if only solutions satisfying the transversality conditions were considered.

The values of $\tilde{b}_{ij}$ and $\tilde{d}_{ij}$ are computed by replacing $\lambda_3^{(k-1)}(T)$ and $\lambda_4^{(k-1)}(T)$ with $\dot{\lambda}_3^{(k-1)}(T)$ and $\dot{\lambda}_4^{(k-1)}(T)$ in the equations for b_{ij} and d_{ij}. The calculation of $\dot{\lambda}_3^{(k)}(T)$ and $\dot{\lambda}_4^{(k)}(T)$ depends on whether $\lambda_3^{(k)}$ and $\lambda_4^{(k)}$ are homogeneous or particular solutions. If $\lambda_3^{(k)}$ and $\lambda_4^{(k)}$ are for a particular solution, the calculation is as follows.

$$\left\{\begin{matrix} \left(\dot{\lambda}_3^{(P)}\right)^{(k)} \\ \left(\dot{\lambda}_4^{(P)}\right)^{(k)} \end{matrix}\right\}_{t=T} = -\left(M_T\right)^{-1} \left\{\begin{matrix} \left(\lambda_1^{(P)}\right)^{(k)} + \tilde{d}_{33}\dot{\lambda}_3^{(k-1)} + \tilde{d}_{34}\dot{\lambda}_4^{(k-1)} \\ \left(\lambda_2^{(P)}\right)^{(k)} + \tilde{d}_{43}\dot{\lambda}_3^{(k-1)} + \tilde{d}_{43}\dot{\lambda}_4^{(k-1)} \end{matrix}\right\}_{t=T} \tag{235}$$

where $\tilde{d}_{33}$, $\tilde{d}_{34}$, $\tilde{d}_{43}$, and $\tilde{d}_{44}$ are computed using the (k - 1)st iteration, and

$$(M_T) \equiv \begin{bmatrix} (1 - \tilde{d}_{33}) & - \tilde{d}_{34} \\ - \tilde{d}_{43} & (1 - \tilde{d}_{44}) \end{bmatrix} . \tag{236}$$

If $\lambda_3^{(k)}$ and $\lambda_4^{(k)}$ are for a homogeneous solution, the calculation is as follows.

$$\begin{Bmatrix} \dot{\lambda}_3^{(H)} & ^{(k)} \\ \dot{\lambda}_4^{(H)} & ^{(k)} \end{Bmatrix}_{t=T} = - (M_T)^{-1} \begin{Bmatrix} \lambda_1^{(H)} & ^{(k)} \\ \lambda_2^{(H)} & ^{(k)} \end{Bmatrix}_{t=T} \tag{237}$$

Equations (229) through (237) are based on the assumption that the particular and homogeneous solutions satisfy the transversality conditions at the final time. If the quasi-linearization algorithm described in Section II were used without modification, these conditions would not, in general, be satisfied. The algorithm must be modified such that the boundary conditions for the particular and homogeneous solutions are specified at the final time and such that the differential equations are integrated backwards in time from $t = T$ to $t = 0$. The boundary conditions at $t = T$ can then be carefully selected such that they all satisfy the transversality conditions. The constraint vector, Ω, then corresponds to the specified initial conditions on the state variables, x_1 through x_8.

The boundary conditions for the particular solution are specified as follows.

$$x_i^{(P)}(T) = x_i^{(k-1)}(T) \tag{238}$$

$$\lambda_1^{(P)}(T) = x_5^{(k-1)}(T) - x_1^{(k-1)}(T) \tag{239}$$

$$\lambda_2^{(P)}(T) = x_6^{(k-1)}(T) - x_2^{(k-1)}(T) \tag{240}$$

$$\lambda_3^{(P)}(T) = \lambda_4^{(P)}(T) = 0 \tag{241}$$

$$\lambda_5^{(P)}(T) = -\lambda_1^{(P)}(T) \tag{242}$$

$$\lambda_6^{(P)}(T) = -\lambda_2^{(P)}(T) \tag{243}$$

$$\lambda_7^{(P)}(T) = \lambda_8^{(P)}(T) = 0 . \tag{244}$$

The boundary conditions for the eight homogeneous solutions are specified in Table 3. Note that the particular solution and all of the homogeneous solutions satisfy the transversality conditions. The addition of any linear combination of these eight homogeneous solutions to the particular solution will result in a solution which still satisfies the transversality conditions.

The details of generating the homogeneous and particular solutions are as follows.

1. Compute $\dot{\lambda}_3^{(k-1)}(T)$, $\dot{\lambda}_4^{(k-1)}(T)$, $\dot{\lambda}_7^{(k-1)}(T)$, and $\dot{\lambda}_8^{(k-1)}(T)$ by taking numerical differences. Use these values in place of $\lambda_3^{(k-1)}(T)$, $\lambda_4^{(k-1)}(T)$, $\lambda_7^{(k-1)}(T)$, and $\lambda_8^{(k-1)}(T)$ for the computation of $\psi(T)$, $\theta(T)$, a_{ij}, b_{ij}, c_{ij}, and d_{ij}.

2. Initialize the particular solution at $t = T$ according to equations (238) through (244). Initialize the homogeneous solutions at $t = T$ according to Table 3.

3. Compute the time derivatives for the particular and homogeneous solutions at $t = T$ according to equations (235) and (237), respectively. Since the integration is backwards

Table 3. **Boundary Condition for Homogeneous Solutions**

	1	2	3	4	5	6	7	8
$x_1(T)$	1.	0	0	0	-1.	0	0	0
$x_2(T)$	0	1.	0	0	0	-1.	0	0
$x_3(T)$	0	0	1.	0	0	0	0	0
$x_4(T)$	0	0	0	1.	0	0	0	0
$x_5(T)$	1.	0	0	0	0	0	0	0
$x_6(T)$	0	1.	0	0	0	0	0	0
$x_7(T)$	0	0	0	0	0	0	1.	0
$x_8(T)$	0	0	0	0	0	0	0	1.
$\lambda_1(T)$	0	0	0	0	1.	0	0	0
$\lambda_2(T)$	0	0	0	0	0	1.	0	0
$\lambda_3(T)$	0	0	0	0	0	0	0	0
$\lambda_4(T)$	0	0	0	0	0	0	0	0
$\lambda_5(T)$	0	0	0	0	-1.	0	0	0
$\lambda_6(T)$	0	0	0	0	0	-1.	0	0
$\lambda_7(T)$	0	0	0	0	0	0	0	0
$\lambda_8(T)$	0	0	0	0	0	0	0	0

in time, the sign of these derivatives must be reversed.

4. Take an integration step. For times other than $t = T$, there is no problem in computing the time derivatives. Continue integrating from $\tau_i = 0$ to $\tau_i = T$, where $\tau_i = T - t$.

The coefficients of the homogeneous solutions are then computed such that the kth iteration satisfies the initial conditions on the state variables at $t = 0$. These coefficients are computed from equation (146).

The problem of a supersonic airplane attempting to evade a ground-to-air interceptor is used as a specific numerical example. This example is used to examine the convergence characteristics of the quasilinearization algorithm and to determine the best possible guaranteed performance of the interceptor for the selected initial conditions and vehicle parameters. The data for the problem are provided in Table 4.

Table 4. Data for Numerical Example

Parameter	Value	Parameter	Value
$x_1(0)$	50,000.	A_2	320.
$x_2(0)$	30,000.	S_2	16.
$x_3(0)$	1,500.	m_2	125.
$x_4(0)$	0.	$(C_{D_o})_2$	8.
$x_5(0)$	0.	$(C_{D_L})_2$	160.
$x_6(0)$	0.	$(C_{L_o})_2$	75.
$x_7(0)$	10.	ρ_o	.0024
$x_8(0)$	10.	h_o	27,000.
A_1	8.	g	32.2
S_1	6,400.	ρ_p	1.
m_1	5,000.	ρ_v	.1
$(C_{D_o})_1$	1.4583	ρ_λ	.1
$(C_{D_L})_1$	116.666	$\rho_{\dot{\lambda}}$	.1
$(C_{L_o})_1$	170.56	T	See Text

The measure of distance between successive iterations is a modified form of equation (155). The quasilinearization algorithm is assumed to have converged when all of the following relationships are satisfied.

$$e_1 = \left\{ \max_{t\in(0,T)} \left[\left(x_1^{(k)} - x_1^{(k-1)} \right)^2 + \left(x_2^{(k)} - x_2^{(k-1)} \right)^2 \right] \right\}^{1/2} \leq \rho_p \tag{245}$$

$$e_2 = \left\{ \max_{t\in(0,T)} \left[\left(x_3^{(k)} - x_3^{(k-1)} \right)^2 + \left(x_4^{(k)} - x_4^{(k-1)} \right)^2 \right] \right\}^{1/2} \leq \rho_v \tag{246}$$

$$e_3 = \left\{ \max_{t\in(0,T)} \left[\left(\lambda_1^{(k)} - \lambda_1^{(k-1)} \right)^2 + \left(\lambda_2^{(k)} - \lambda_2^{(k-1)} \right)^2 \right] \right\}^{1/2} \leq \rho_\lambda \tag{247}$$

$$e_4 = \left\{ \max_{t\in(0,T)} \left[\left(\lambda_3^{(k)} - \lambda_3^{(k-1)} \right)^2 + \left(\lambda_4^{(k)} - \lambda_4^{(k-1)} \right)^2 \right] \right\}^{1/2} \leq \rho_\lambda \tag{248}$$

$$e_5 = \left\{ \max_{t\in(0,T)} \left[\left(x_5^{(k)} - x_5^{(k-1)} \right)^2 + \left(x_6^{(k)} - x_6^{(k-1)} \right)^2 \right] \right\}^{1/2} \leq \rho_p \tag{249}$$

$$e_6 = \left\{ \max_{t\in(0,T)} \left[\left(x_7^{(k)} - x_7^{(x-1)} \right)^2 + \left(x_8^{(k)} - x_8^{(k-1)} \right)^2 \right] \right\}^{1/2} \leq \rho_v \tag{250}$$

$$e_7 = \left\{ \max_{t\in(0,T)} \left[\left(\lambda_5^{(k)} - \lambda_5^{(k-1)} \right)^2 + \left(\lambda_6^{(k)} - \lambda_6^{(k-1)} \right)^2 \right] \right\}^{1/2} \leq \rho_\lambda \tag{251}$$

$$e_8 = \left\{ \max_{t\in(0,T)} \left[\left(\lambda_7^{(k)} - \lambda_7^{(k-1)} \right)^2 + \left(\lambda_8^{(k)} - \lambda_8^{(k-1)} \right)^2 \right]^{1/2} \right. \leq \rho_\lambda \tag{252}$$

The $m(k)$ function, defined in equation (156), is taken as a ramp function.

$$m(k) = \min[1. , m_0 + m_d k)] \tag{253}$$

where m_0 and m_d are input constants and k is the iteration number.

The efficacy of the quasilinearization algorithm is investigated by setting $T = 24.0$ seconds and using some reasonably crude initial guesses for $x_1(t), \ldots, x_8(t)$ and $\lambda_1(t), \ldots, \lambda_8(t)$. These initial guesses are tabulated in Table 5. The computer program uses quadratic interpolation to determine the values of $x_i(t)$ and $\lambda_i(t)$ at intermediate time points. Using $m_0 = 1.$ and $m_d = 0.$, the algorithm does not converge with these initial guesses. The cause of this divergence seems to be that the distances between successive iterations is so large that $[x_1(T) - x_5(T)]$ and $[x_2(T) - x_6(T)]$ change signs from one iteration to the next. Since successive iterations satisfy the transversality conditions, the signs of $\lambda_1(T)$, $\lambda_2(T)$, $\lambda_5(T)$, and $\lambda_6(T)$ also change, and this behavior in turn affects the complete time histories for λ_3, λ_4, λ_7, and λ_8. The net results are enormous changes between iterations for the ψ and θ time histories.

The divergence problem is easily resolved by using the $m(k)$ factor in equation (153) to reduce the distance between the first few iterations. Specifically, m_0 and m_d are both input as .2. The full correction, then, is not applied until the fifth iteration. Convergence to the optimal solution is achieved in eight iterations. The convergence behavior of x_4, the evader's vertical velocity, and of λ_4 are depicted in Figures 10 and 11, respectively. The time histories of the control angles, ψ and θ, are presented

Table 5. Initial Guesses for State and Multiplier Variables

Variable	Time (sec) 0	14	24
$x_1(t)$	50,000.	65,000.	86,000.
$x_2(t)$	30,000.	30,000.	30,000.
$x_3(t)$	1,500.	1,500.	1,500.
$x_4(t)$	0.	0.	0.
$x_5(t)$	0.	15,000.	80,400.
$x_6(t)$	0.	6,000.	26,240.
$x_7(t)$	10.	3,000.	7,200.8
$x_8(t)$	10.	1,500.	3,600.8
$\lambda_1(t)$	-5,600.	-5,600.	-5,600.
$\lambda_2(t)$	-3,760.	-3,760.	-3,760.
$\lambda_3(t)$	-480,000.	-280,000.	0.
$\lambda_4(t)$	-120,000.	-70,000.	0.
$\lambda_5(t)$	5,600.	5,600.	5,600.
$\lambda_6(t)$	3,760.	3,760.	3,760.
$\lambda_7(t)$	480,000.	280,000.	0.
$\lambda_8(t)$	120,000.	70,000.	0.

in Figure 12. The convergence of the measures of distance between successive iterations, $e_1, \ldots, e_8$, is displayed in Table 6.

The initial guesses for the λ multipliers are in error by approximately one order of magnitude. Despite these poor initial guesses, the algorithm is able to converge rapidly. The use of the $m(k)$ factor is very important in making convergence possible.

The final miss distance associated with $T = 24$ is 4,366.5 feet, and the interceptor is closing at a rate of 5,802.2 feet per second. It is clear that the final miss

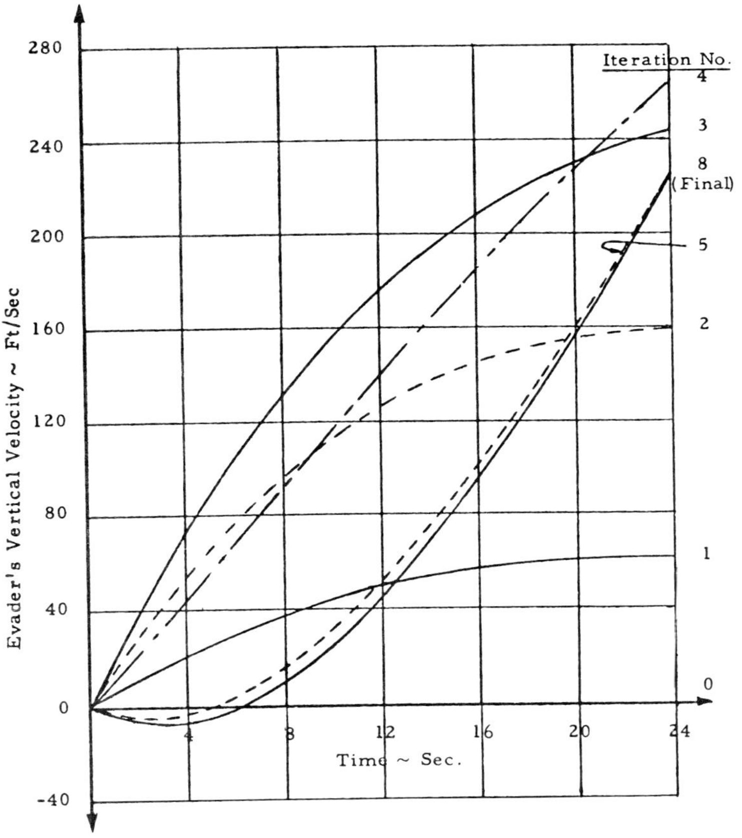

Fig. 10. Convergence Behavior: Evader's Vertical Velocity (Optimal Intercept Strategy).

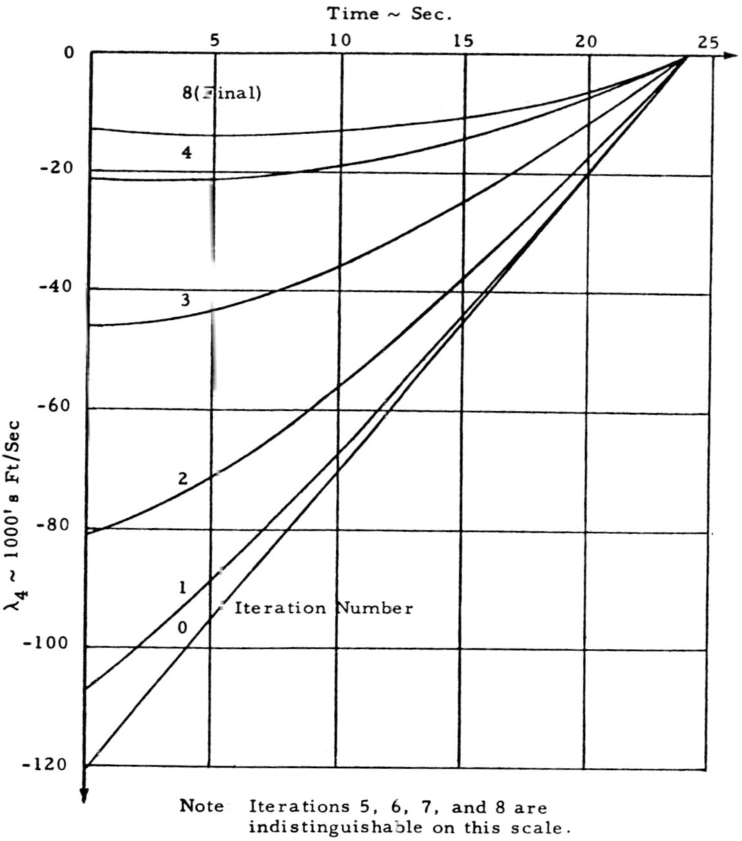

Fig. 11. Convergence Behavior: λ_4 Multiplier (Optimal Intercept Strategy).

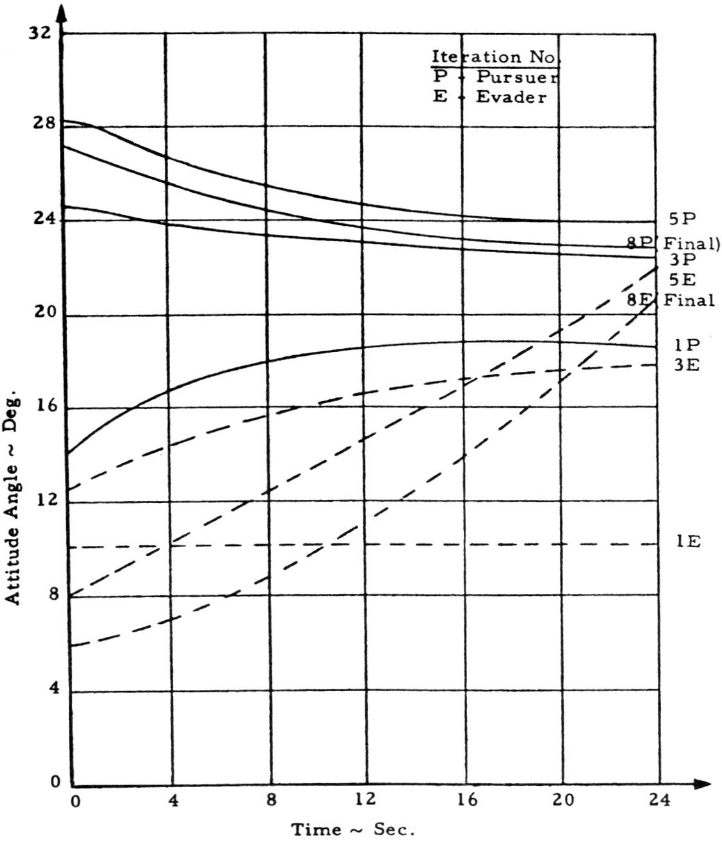

Note: Iterations 6, 7, and 8 are indistinguishable on this scale.

Fig. 12. Convergence Behavior: Control Angle Time Histories (Optimal Intercept Strategy).

Table 6. Convergence Behavior of Error Indicators: 24. Second Flight Time

Iteration No.	e_1	e_2	e_3	e_4	e_5	e_6	e_7	e_8
1	5484.	344.6	6006.	1129.	5201.	446,671.	5200.	482,359.
2	4247.	276.0	5007.	867.5	3191.	342,693.	3142.	357,467.
3	1971.	152.5	3324.	520.7	1425.	183,545.	1420.	177,778.
4	1092.	63.96	265.7	278.8	1119.	63,345.	1096.	55,276.
5	1746.	97.46	1489.	166.2	278.2	13,710.	258.0	12,267.
6	117.5	6.835	119.4	9.192	5.126	222.5	4.666	90.34
7	.2392	.0577	.1828	.0164	.1740	3.643	.1758	4.194
8	.0062	.0051	.0060	.0005	.0005	.0093	.0004	.0097

e_1, e_3, e_5, and e_7 have the units of feet.

e_2, e_4, e_6, and e_8 have the units of feet per second.

distance would be reduced if T were increased slightly. The running of a parameter study to determine the effects of T is greatly facilitated by a special table extrapolation subroutine. By using the converged solution from the preceding case as the input nominal and by using this special subroutine to extrapolate the input nominal to the new final time, the algorithm is able to converge in only four or five iterations for each new value of final time. A total of 33 iterations are required to determine the optimal solutions for 7 different values of T between 24.0 and 24.9. Each solution requires approximately 25 integration steps. The total running time on the central processor of the CDC 6600 computer is 295.505 seconds. This parameter study predicts the result that the miss distance is near zero for $T = 24.738$ seconds. Using this value of T in the program yields an associated miss distance of 3.6 feet. The optimal control angle time histories are depicted in Figure 13, and the spatial coordinates of the optimal solutions are depicted in Figure 14.

C. Evaluation of a Suboptimal Strategy

The results of Section IV B define the best possible guaranteed performance of the interceptor for the specified initial conditions. Now consider the case when the interceptor uses a suboptimal strategy. Specifically, assume that the interceptor uses the strategy which would be optimal in a vacuum. This strategy can be derived from the necessary conditions given in Section II, C, and it is found to be

$$\theta(t) = \tan^{-1}\{[(x_4 - x_8)(T - t) + (x_2 - x_6)] \cdot [(x_3 - x_7)(T - t) + (x_1 - x_5)]^{-1}\}. \tag{254}$$

The guaranteed performance with this suboptimal strategy will,

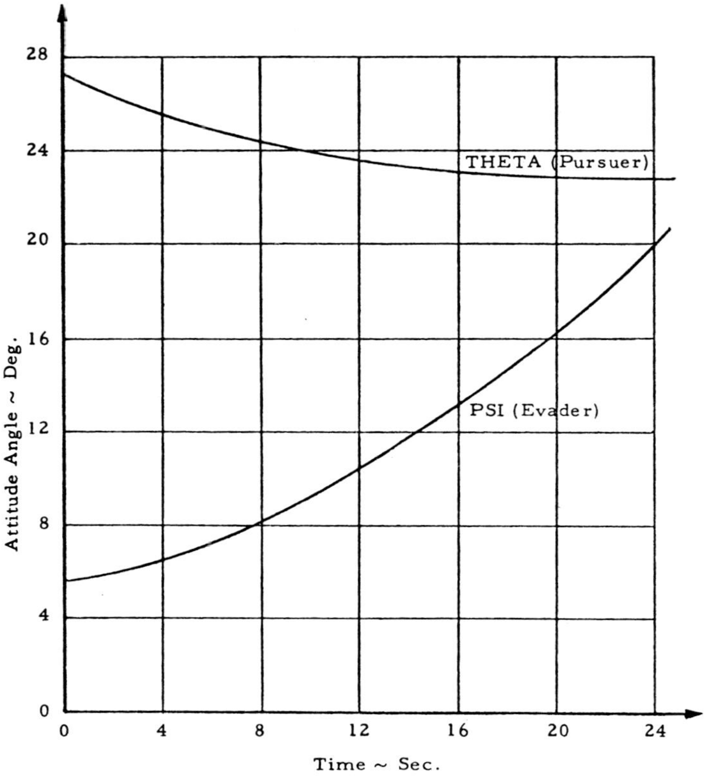

Fig. 13. Control Angle Time Histories for 24.738 Second Flight (Optimal Intercept Strategy).

of course, not be as good as that with the optimal strategy. The question is, just how good is it? The quasilinearization algorithm can be applied to solve the two point boundary value problem posed by equations (23) through (31) and thereby answer the question.

The algebraic forms of the derivative expressions are changed from the previous ones because equation (254) replaces equation (215), and the Euler-Lagrange equations are altered by the addition of the $(f_\theta, \theta_{\underline{x}})^T \lambda$ term in equation (31).

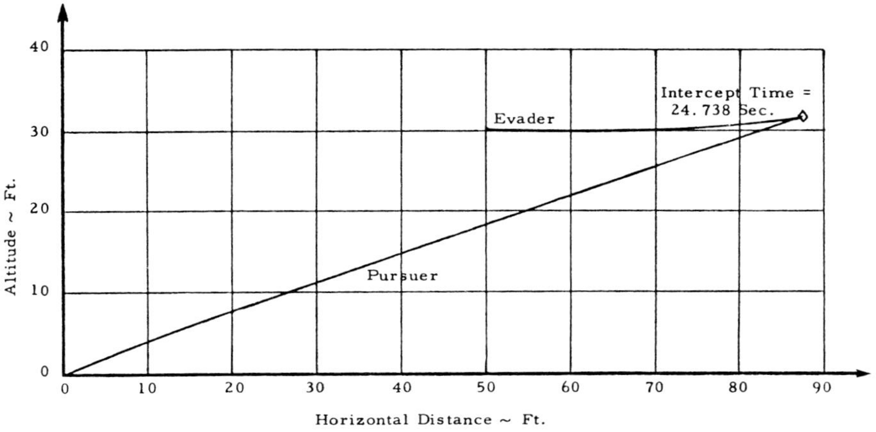

Fig. 14. Optimal Trajectories for 24.738 Second Flight (Optimal Intercept Strategy).

These changes in the derivative expressions do not alter the form of the quasilinearization algorithm described in Section III. The same basic method can be used for determining the optimal time histories.

The problem of a supersonic airplane attempting to evade a ground-to-air interceptor is again used as a specific numerical example. The problem statement is changed from that of Section IV, A only by the specification of a suboptimal intercept strategy. The problem data are again given by Table 4. The final time is 24 seconds.

The converged solution for the case with an optimal intercept strategy is used as the initial guess for this case. The values of m_0 and m_d are 1.0 and 0., respectively. Convergence to the optimal solution is achieved in five iterations. The convergence behavior of x_4, the evader's vertical velocity, is displayed in Figure 15, and that of the control angles, ψ and θ, in Figure 16. The convergence behavior of the error indicators, $e_1,\ldots,e_8$ is displayed in Table 7.

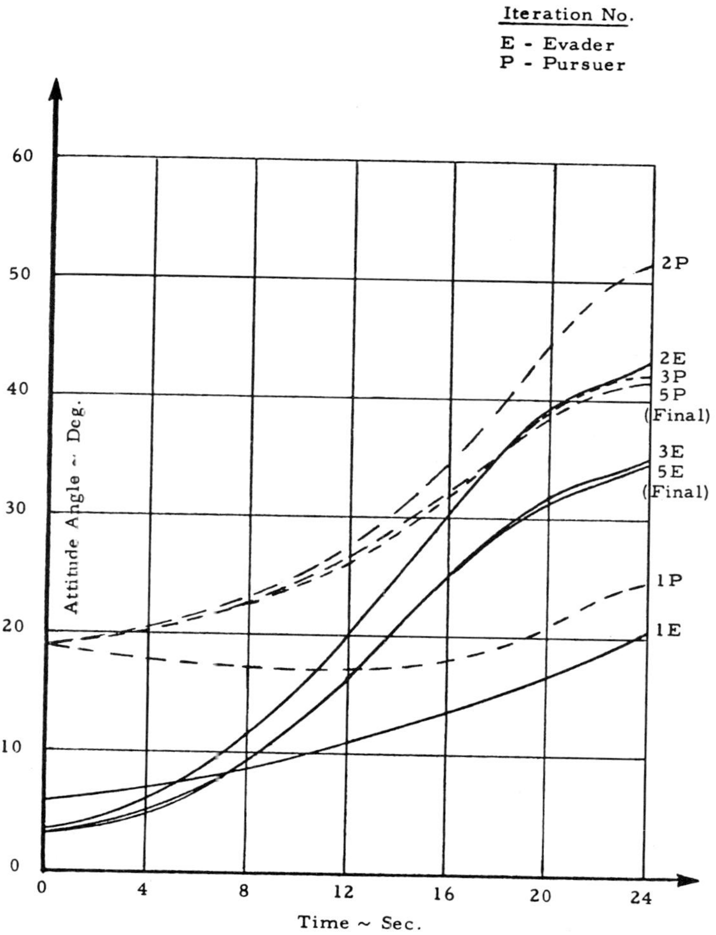

Fig. 15. Convergence Behavior: Control Angle Time Histories (Suboptimal Intercept Strategy).

The final miss distance is 5,204.8 ft., and the interceptor is closing at a rate of 5,522.0.

As in Section IV, B the final miss distance can be reduced by increasing the final time slightly. For T = 24.99 seconds, the final miss distance is only 41.3 ft. For this particular suboptimal strategy, the terminal behavior does not

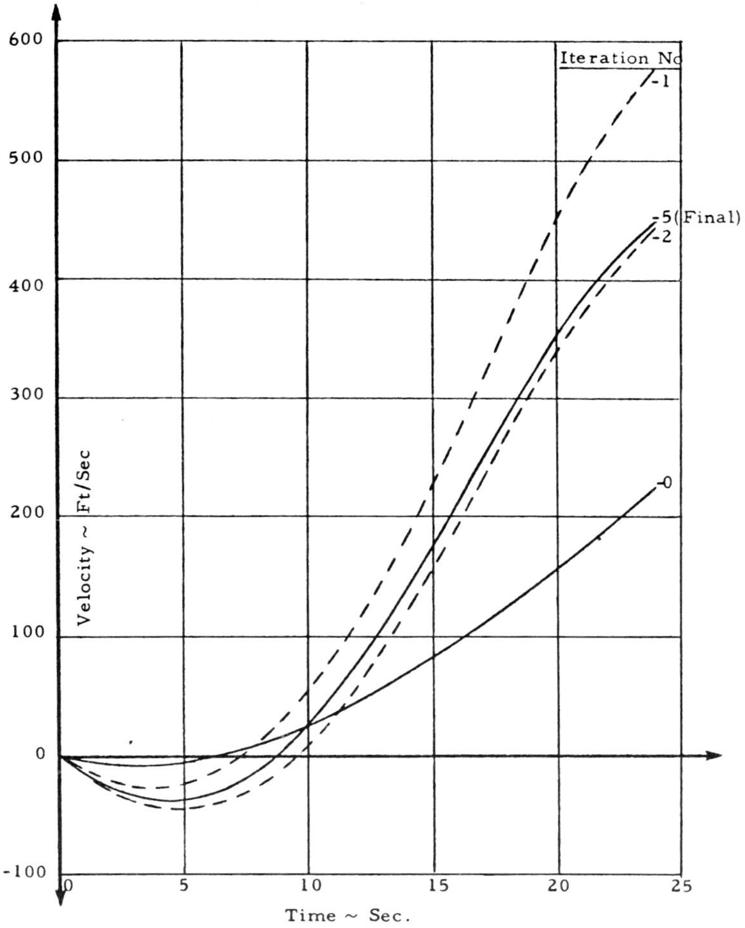

Fig. 16. Convergence Behavior: Evader's Vertical Velocity (Suboptimal Intercept Strategy).

resemble a straight-line chase as it does with an optimal strategy. Instead, it more closely resembles a near miss fly-by. This behavior results in very large pitch rates for the pursuit vehicle; consequently, the integration step sizes must be very small during the terminal portion of the trajectory. The time of closest approach and the associated minimum miss

Table 7. Convergence Behavior of Error Indicators: 24 Second Flight Time With Suboptimal Intercept Strategy

Iteration Number	e_1	e_2	e_3	e_4	e_5	e_6	e_7	e_8
1	3,059.	408.4	2,209.	981.8	2,165.	20,153.	2,169.	21,732.
2	1,529.	146.4	2,320.	364.2	1,692.	33,690.	1,692.	34,218.
3	462.3	43.98	546.8	106.5	127.8	1,069.	129.8	977.0
4	2.852	.2772	2.746	.4477	2.993	25.61	2.996	35.57
5	.0198	.0010	.0342	.0042	.0038	.0692	.0039	.8125

e_1, e_3, e_5, and e_7 have the units of feet

e_2, e_4, e_6, and e_8 have the units of feet per second

distance was not determined to an accuracy of less than .01 seconds. Based on extrapolation of the trajectories, it is estimated that the minimum miss is 32 feet at a final time of 24.9946 seconds.

The ψ and θ control angle time histories for the converged solution to the 24.99 second case are displayed in Figure 17.

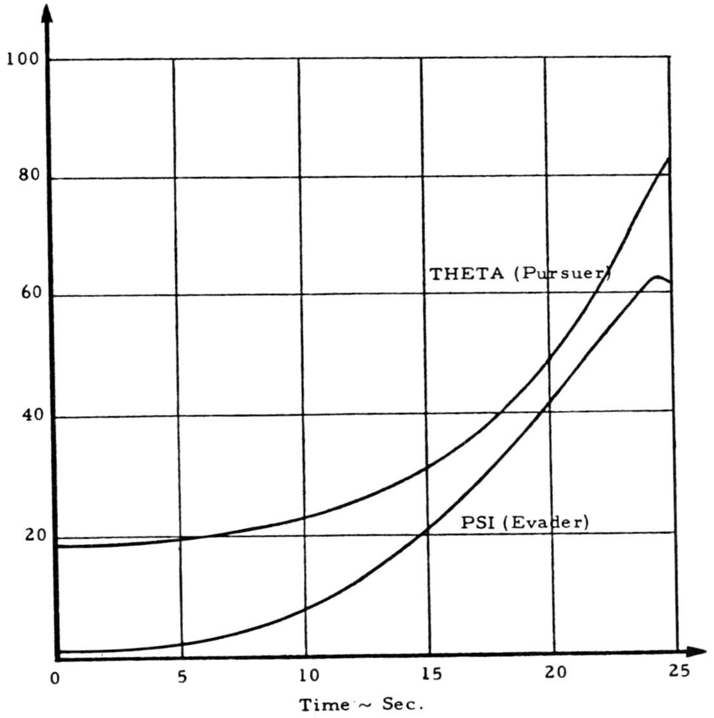

Fig. 17. Control Angle Time Histories for 24.99 Second Flight Time (Suboptimal Intercept Strategy).

D. Discussion

The numerical results described in Sections IV, B and IV, C provide an example of how a specific suboptimal strategy can be quantitatively evaluated. For the vehicle and initial conditions presented in Table 4 and a flight time of 24.0 seconds, the best possible guaranteed performance for the interceptor vehicle is a miss distance Of 4,366.5 feet. This performance assumes that the interceptor uses the best of all possible strategies. If this performance is not acceptable, no amount of guidance sophistication will improve it. The only way to improve this performance is to upgrade the propulsive and aeordynamic characteristics of the interceptor. This performance also provides a bench mark against which suboptimal strategies can be measured. For the specific suboptimal strategy considered, the guaranteed performance for 24.0 second flight is 5,204.8 feet. The maximum amount of performance which could be gained by upgrading of the guidance equations is 838.3 feet. If a miss distance of 5,204.8 feet is unacceptable but 4,8000 feet (for instance) is, it is reasonable to attempt upgrading the intercept guidance law. A comparison of Figures 12 and 16 provides a clue as to how such upgrading might be accomplished. Compared to the truly optimal intercept strategy, the suboptimal strategy tends to lead the target too much. An improved strategy, then, might be one of the following form.

$$\theta(t) = \tan^{-1}\{[K(x_4 - x_8)(T - t) + (x_2 - x_6)] \cdot [K(x_3 - x_7)(T - t) + (x_1 - x_5)]^{-1}\} \tag{255}$$

where K is a gain factor between 0 and 1. A value of K around .8 or .9 might yield significantly better performance than does a K of 1.0.

The convergence behavior of the quasilinearization algorithm is exceptionally good. One reason for this good behavior may be the manner in which the boundary conditions for the homogeneous and particular solutions are specified at the final time. The difficulties with the singularity at the final time are easily handled in this manner, and each iteration is tangent to some optimal solution at the final time. Each individual iteration, therefore, reflects at least the character of the optimal solution.

The use of the $m(k)$ factor is very effective for increasing the range of convergence. It can be used to prohibit overly large changes between one iteration and the next. Since the Newton-Raphson method often displays a tendency to over-correct during the early iterations, the effectiveness of the $m(k)$ factor is not surprising.

An additional aid for obtaining convergence is the use of K_ρ in equations (199) and (200). For $K_\rho = 0$, the differential system for an optimal intercept strategy simplifies to a linear system because of the transversality conditions. For such a linear system, the quasilinearization algorithm always converges By starting with $K_\rho = 0$, using the converged solution from one case as the initial guess for the next, and incrementing K_ρ from 0 to 1 in small steps, convergence to the optimal solution can nearly always be obtained.

REFERENCES

1. G. A. BLISS, "Lectures on the Calculus of Variations," The University of Chicago Press, Chicago, 1946.
2. M. R HESTENES, "Calculus of Variations and Optimal Control Theory," John Wiley and Sons, New York, 1966.
3. J. D MASON, W. D. DICKERSON, and S. B. SMITH, "A Variational Method for Optimal Staging," AIAA Journal, Vol. 3, No. 11, 1965.

4. T. L. VINCENT, and J. D. MASON, "Disconnected Optimal Trajectories," Journal of Optimization Theory and Applications, Vol. 3, No. 4, pp. 263-281, 1969.

5. A. E. BRYSON, W. F. DENHAM, and S. E. DREYFUS, "Optimal Programming Problems with Inequality Constraints I: Necessary Conditions for Extremal Solutions," AIAA Journal, Vol. 1, No. 11, pp. 2544-2550, 1963.

6. R. McGILL, and P. KENNETH, "Solutions of Variational Problems by Means of a Generalized Newton-Raphson Operator," AIAA Journal, Vol. 2, No. 10, pp. 1761-1766, October 1964.

7. R. ISAACS, "Differential Games," John Wiley and Sons, New York, 1965.

8. A. W. STARR, and Y. C. HO, "Nonzero-Sum Differential Games," Journal of Optimization Theory and Applications, Vol. 3, No. 3, pp. 184-206, 1969.

9. A. W. STARR, and Y. C. HO, "Further Properties of Nonzero-Sum Differential Games," Journal of Optimization Theory and Applications, Vol. 3, No. 4, pp. 207-219, 1969.

10. L. D. BERKOVITZ, "A Survey of Differential Games," in Mathematical Theory of Control, Edited by A. V. Balakrishnan and L. W. Neustadt, Academic Press, New York, 1967.

11. H. G. MOYER, and PINKHAM, "Several Trajectory Optimization Techniques, Part II: Application," in Computing Methods in Optimization Problems, Edited by A. V. Balakrishnan and L. W. Neustadt, Academic Press, New York, 1964.

12. G. PAINE, The Application of the Method of Quasi-linearization to the Computation of Optimal Control, Ph.D. Dissertation, Department of Engineering, University of California, Los Angeles, 1966.

13. R. McGILL, "Optimal Control Inequality State Constraints and the Generalized Newton-Raphson Algorithm," SIAM Journal on Control, Series A, Vol. 3, No. 2, pp. 291-298, 1965.

Aircraft Symmetric Flight Optimization*

MICHAEL FALCO
Research Department,
Grumman Aerospace Corporation,
Bethpage, New York

AND

HENRY J. KELLEY
Analytical Mechanics Associates, Inc.,
Jericho, New York

*The research reported was supported by USAF under Contracts AF29(600)-2671 and AF49(638)-1207, and by NASA under Contract NAS9-11532.

I. INTRODUCTION

The development and application of gradient techniques to aircraft optimal performance computations in the vertical plane of flight will be reviewed in the following. Results obtained using the method of gradients will be presented for attitude and throttle control programs that extremize the fuel, range, and time performance indices subject to various trajectory and control constraints including boundedness of engine throttle control. Penalty function treatment is sketched of state inequality constraints that generally appear in aircraft performance problems: a minimum altitude constraint; a Mach number-altitude constraint due to power plant and structural operational limitations; and a ground overpressure "sonic boom" constraint that might be imposed on flight operations of supersonic commercial aircraft. Numerical results for maximum range, minimum fuel, and minimum time climb paths for a hypothetical supersonic turbojet interceptor will be presented and discussed. In addition, minimum fuel climb paths subject to various levels of ground overpressure intensity constraint for a representative supersonic transport will be shown. Discussions of the general character of these optimal flight profiles and the workings of the computational technique will be given.

The computational techniques are those of Refs. 1 and 2, penalty function and projection versions of the gradient algorithm. A variant of the Gel'fand-Tsetlin "Method of Ravines" [3] due to R. E. Kopp will also be reviewed. Numerical results are taken from Refs. 2 and 4. Two possibilities for further development of continuous gradient processes will then be sketched: a projection version of conjugate gradients and a curvilinear search.

The optimization of flight paths for atmospheric vehicles is a subject that has received considerable attention in the aeronautical literature. Yet the means generally available for treatment of this class of problems are rather limited from the viewpoint of practical numerical computations. One reason for this is simply complexity — a typical problem of flight performance in the vertical plane involves two control variables, vehicle attitude angle and throttle, aerodynamic coefficients as tabular functions of Mach number, and engine thrust and fuel consumption as tabular functions of both Mach number and altitude. There are usually overpressure and over-temperature limitations on the airframe and/or engine, which appear as boundaries in the Mach number-altitude chart, and these must be dealt with as subsidiary conditions of inequality type. A second, and perhaps more important, reason is that the system of Euler equations arising from classical variational treatment of optimal atmospheric flight problems is, in most cases, not well suited to numerical solution — the well-known two-point boundary value difficulty is a serious one.

This chapter describes some results obtained on the air vehicle problem employing the method of gradients, the simple extension of the "steepest descent" idea into function space. The considerations of simplicity and stability that originally encouraged the use of first-order direct methods in continuous control format still apply to the lifting atmospheric flight optimization problem, and quite generally to large high-order complex problems. Computer developments of the past decade — increases in speed, internal storage, and precision — have, if anything, tended more to favor competing methods: the classical indirect method, dynamic programming, and the various second-order methods. Yet the continuous gradient scheme in original and in conjugate gradient/penalty function version seems to remain attractive for applications work. And so the following mixture of decade-old material, compiled mainly for the record, and to illustrate symmetric flight phenomena, is offered along with some suggestions for further development for possible stimulation of efforts in the area.

II. TRAJECTORY EQUATIONS FOR A POWERED LIFTING VEHICLE

The following trajectory equations apply to flight in the equatorial plane of a rotating earth. The coordinate system is shown in Figure 1. The equations for the acceleration components are (Figure 2)

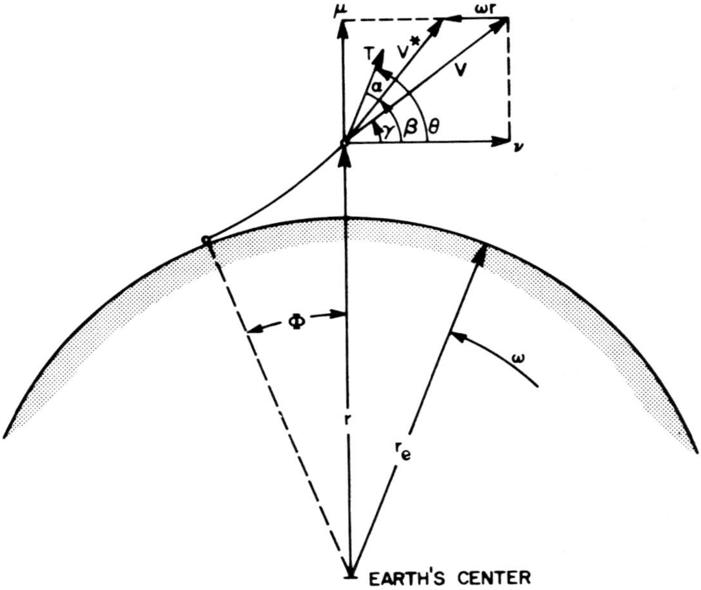

Fig. 1. Coordinate System

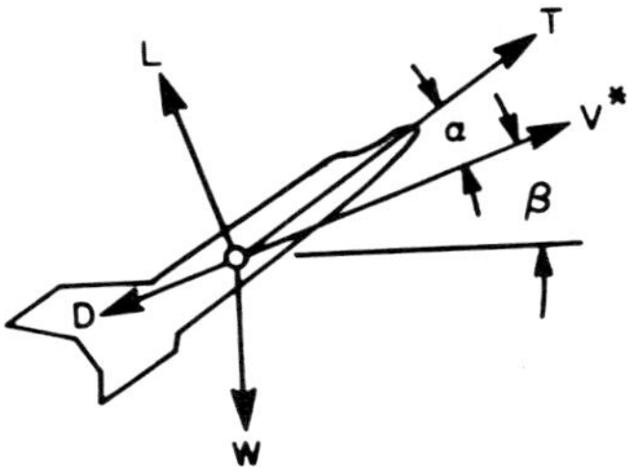

Fig. 2. Resolution of Forces

- Radial Acceleration

$$u = \frac{V^2}{r} - g_e \left(\frac{r_e}{r} \right)^2 + g_e \frac{T}{W} \sin \theta - g_e \frac{D}{W} \sin \beta + g_e \frac{L}{W} \cos \beta \qquad (1)$$

- Circumferential Acceleration

$$\dot{v} = -\frac{uv}{r} + g_e \frac{T}{W} \cos\theta - g_e \frac{D}{W} \cos\beta - g_e \frac{L}{W} \sin\beta . \qquad (2)$$

The relationships between the airspeed components, vehicle path angle, angle of attack, and attitude angle are (Figure 1)

$$V^* = [u^2 + (v - \omega r)^2]^{\frac{1}{2}} \qquad (3)$$

$$\beta = \tan^{-1} \frac{u}{v - \omega r} \qquad (4)$$

$$\alpha = \theta - \beta . \qquad (5)$$

Three additional first-order differential equations for radius rate, range rate, and fuel flow rate are

- Radius Rate

$$\dot{r} = u \qquad (6)$$

- Range Rate

$$\dot{\Phi} = \frac{v}{R} \qquad (7)$$

- Fuel Flow Rate

$$\dot{W} = -Q . \qquad (8)$$

The descriptions of applied forces, L, D, T, and the quantity Q are employed in a form typically employed for a lifting vehicle with an air-breathing propulsion system. The aerodynamic drag, D, is represented as

$$D = qSC_D , \qquad (9)$$

where

$$C_D = C_{D_{min}} + C_{D_i} \qquad (10)$$

An approximation for induced drag valid in the speed range of interest is

$$C_{D_i} = C_{L_\alpha} \alpha^2 . \tag{11}$$

Combining (10) and (11), we have

$$D = qS\left(C_{D_{min}} + C_{L_\alpha} \alpha^2\right). \tag{12}$$

The aerodynamic lift, L, is given by

$$L = qSC_L \tag{13}$$

where

$$C_L = C_{L_0} + C_{L_\alpha} \alpha \tag{14}$$

The quantities T and Q are represented in tabular form as functions of Mach number and altitude for full throttle setting. For computations that include throttle control, T and Q will be taken as linear functions of a throttle variable, η, with $0 \leq \eta \leq 1$. The aerodynamic coefficients C_{L_α} and $C_{D_{min}}$ are represented as tabular functions of Mach number. The atmospheric density and sound speed are represented tabularly as functions of altitude according to the ARDC standard atmosphere.

III. OPTIMIZATION CRITERIA

The optimization criteria of main interest are minimum time, minimum fuel, and maximum range. The function to be minimized may be written as a linear combination of final value of these variables, with coefficients to be chosen in specific cases according to the following selection scheme:

The performance quantity to be minimized is taken, in general, as

$$P = C_{t_f} t_f + C_{W_f} W_f + C_{\Phi_f} \Phi_f , \tag{15}$$

where for minimum time

$$C_{\Phi_f} = C_{W_f} = 0, \quad C_{t_f} = 1, \tag{16a}$$

for minimum fuel

$$C_{\Phi_f} = C_{t_f} = 0, \quad C_{W_f} = -1, \tag{16b}$$

and for maximum range,

$$C_{t_f} = C_{W_f} = 0, \quad C_{\Phi_f} = -1. \tag{16c}$$

A study somewhat similar to that presently reported was carried out by Bryson and Denham [5]; however, this investigation was restricted to full-throttle climb performance in the absence of Mach-altitude boundaries.

IV. PENALTY FUNCTION TREATMENT OF CONSTRAINTS

A. Terminal Constraints

Fixed initial conditions are introduced at the outset into the digital computer program and, during subsequent optimal path approximations, are held at the specified values. One, however, may relax the requirement for meeting fixed terminal conditions on the state variables in favor of an approximation: the addition of terms to the function P of the form

$$\begin{aligned} P' = P &+ \frac{K_u}{2}(u_f - \bar{u}_f)^2 + \frac{K_v}{2}(v_f - \bar{v}_f)^2 + \frac{K_r}{2}(r_f - \bar{r}_f)^2 \\ &+ \frac{K_\Phi}{2}(\Phi_f - \bar{\Phi}_f)^2 + \frac{K_W}{2}(W_f - \bar{W}_f)^2 . \end{aligned} \tag{17}$$

The K_i values are selected in accordance with the following scheme:

$K_i = 0$	if associated variable is to be minimized or if associated variable terminal is unspecified.	(18a)
$K_i = \text{constant} > 0$	if associated variable terminal value is specified	(18b)

The approximation is based on the idea that the solution of the problem min P' will, under appropriate conditions, tend toward the solution of the original problem as the magnitudes of the K_i are increased. This idea was originated by Courant [6]. A means for estimating the error incurred by the minimization of P' instead of P is given in Ref. 1.

B. Inequality Constraints on State Variables

In applications work, inequality constraints enter that involve the altitude and airspeed components, namely, a minimum-altitude limit and an airspeed/altitude-envelope boundary arising from structural and power plant limitations. We introduce additional state variables satisfying the first-order differential equations

$$\dot{x}_6 = \tfrac{1}{2}(r - \tilde{r})^2 \, H(\tilde{r} - r) \,, \quad x_6(t_0) = 0 \,, \tag{19}$$

corresponding to a constraint on minimum altitude, $\tilde{r} < r$, and

$$\dot{x}_7 = \tfrac{1}{2}(V^* - \tilde{V}^*)^2 \, H(V^* - \tilde{V}^*) \,, \quad x_7(t_0) = 0 \tag{20}$$

corresponding to an airspeed constraint, $V^*(h) < \tilde{V}^*(h)$, incorporating dynamic pressure and temperature limits. The sonic boom constraint is given by

$$\dot{x}_8 = \tfrac{1}{2}(\Delta p_g - \Delta\tilde{p}_g)^2 \, H(\Delta p_g - \Delta\tilde{p}_g) \,, \quad x_8(t_0) = 0 \,. \tag{21}$$

Here Δp_g† is the peak ground incremental overpressure caused by supersonic aircraft operation and (21) corresponds to the constraint $\Delta p_g < \Delta\tilde{p}_g$ with $\Delta\tilde{p}_g$ a specified limit overpressure. In these expressions, H is the Heaviside unit step function. The additional variables x_6, x_7, and x_8 represent integral squares of the violations accumulated over those segments of the flight path that violate the inequalities. The terminal values of both variables are to be multiplied by positive constants, K_6, K_7, and K_8, and added to P' as additional penalty terms:

$$P' = P + \sum_{i=1}^{5} \frac{K_i}{2} (x_{i_f} - \bar{x}_{i_f})^2 + \sum_{j=6}^{8} K_j \, x_{j_f} \tag{22}$$

where

$$x_{6_f} = \frac{1}{2} \int_{t_0}^{t_f} (r - \tilde{r})^2 \; H(\tilde{r} - r)dt \tag{23}$$

$$x_{7_f} = \frac{1}{2} \int_{t_0}^{t_f} (V^* - \tilde{V}^*)^2 \; H(V^* - \tilde{V}^*)dt \tag{24}$$

$$x_{8_f} = \frac{1}{2} \int_{t_0}^{t_f} (\Delta p_g - \Delta\tilde{p}_g)^2 \; H(\Delta p_g - \Delta\tilde{p}_g)dt \; . \tag{25}$$

The $\tilde{r}$ and $\tilde{V}^*(h)$ boundaries for the hypothetical turbojet vehicle are illustrated in Figure 3, and for the transport configuration in Figure 14.

V. THROTTLE VARIABLE INEQUALITY CONSTRAINT

For reference in choice of representation of a turbojet throttle control, some basic thrust and propellant-flow characteristics of supersonic turbojet engines over a spectrum

---

† See Appendix.

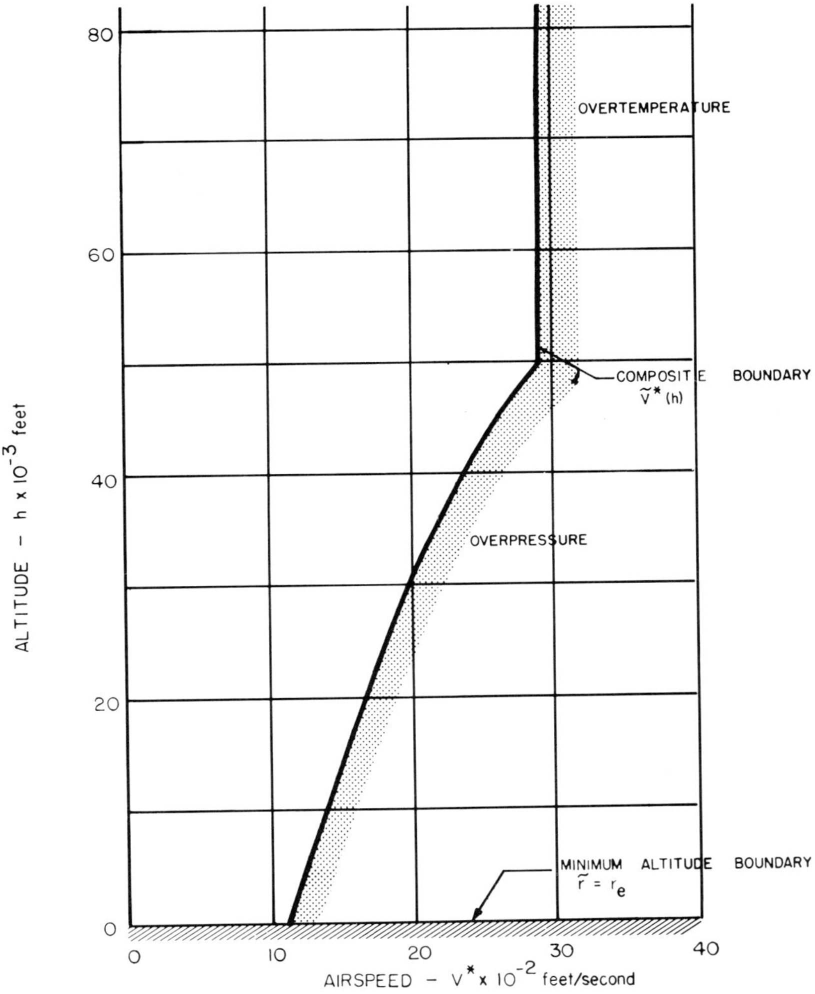

Fig. 3. Limit Boundaries for the Supersonic Interceptor in the Altitude-Airspeed Diagram.

of flight conditions and thrust levels were investigated. These characteristics are depicted in Figure 4 for a representative Mach 3 turbojet equipped with a variable-area tailpipe nozzle at four flight conditions. These conditions represent the extremes of the flight envelope within which

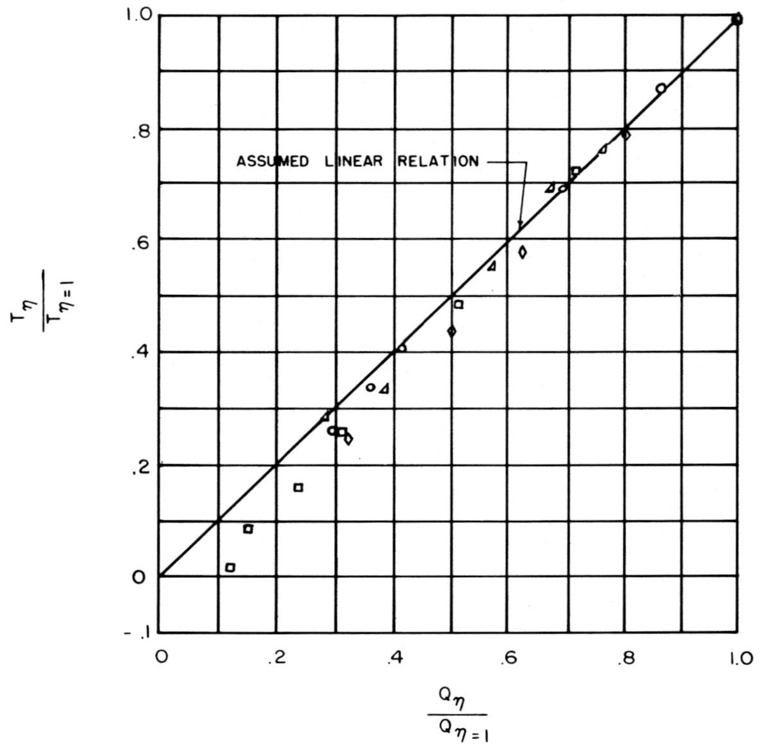

Fig. 4. Supersonic Interceptor Turbojet Characteristics.

the vehicle may operate. From Fig. 4, one may notice the relatively linear relationship between thrust ratio and fuel-flow ratic over the spectrum of flight conditions. A linear relationship between full throttle, $\eta = 1$, and idle throttle, $\eta = 0.1$, values is suggested by the data; however, if power-off operation at $\eta = 0$ is also permitted, there is a convexity difficulty requiring "relaxation" and further analytical complications. So a linear approximation between $T_\eta/T_{\eta=1}$ and $Q_\eta/Q_{\eta=1}$ was assumed, as shown in Fig. 4, in terms of a control variable $\eta = T_\eta/T_{\eta=1}$ bounded above and below according to $0 \leq \eta \leq 1$.

To impose the bounds on η and to enforce the bounds in the gradient process, a parameter, $\zeta(t,\sigma)$, is introduced and a function, $\eta(\zeta)$, defined, as in Reference 1, by

$$\eta = 0 \quad \zeta \leq 0 \tag{26}$$

$$\eta = \zeta \quad 0 \leq \zeta \leq 1 \tag{27}$$

$$\eta = 1 \quad 1 \leq \zeta \, . \tag{28}$$

The function $\eta(\zeta)$ is shown in Figure 5. With the transformation from η to ζ, the integrand in the expression

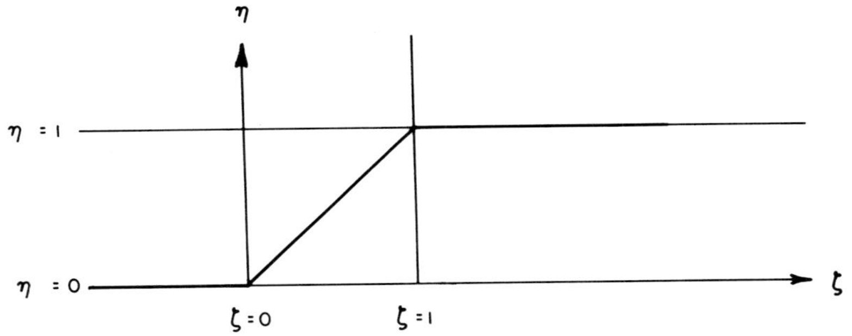

Fig. 5. Variables for Bounded Throttle Control.

for $dP'/d\sigma$ becomes

$$[P']_\eta \frac{\partial \eta}{\partial \sigma} = [P']_\eta \frac{d\eta}{d\zeta} \frac{\partial \zeta}{\partial \sigma} \, . \tag{29}$$

The gradient $[P']_\eta$ is evaluated along the trajectory via adjoint system computations as described in Reference 1. Motion in the negative gradient direction is determined by

$$\frac{\partial \zeta}{\partial \sigma} = - [P']_\eta \frac{d\eta}{d\zeta} \, . \tag{30}$$

The derivative $d\eta/d\zeta$ may be written as

$$\frac{d\eta}{d\zeta} = [H(\zeta) - H(\zeta - 1)] , \tag{31}$$

where H is the Heaviside unit step function. The derivative $d\eta/d\zeta$ is undefined at the points $\zeta = 0$ and $\zeta = 1$. These ambiguities may conveniently be removed by setting

$$\frac{d\eta}{d\zeta} = H(-[P']_\eta) \quad \text{at} \quad \zeta = 0 \tag{32}$$

and

$$\frac{d\eta}{d\zeta} = H([P']_\eta) \quad \text{at} \quad \zeta = 1 , \tag{33}$$

which decides the question on whether the negative gradient direction leads into or out of the interval $0 \leq \zeta \leq 1$.

In a continuous descent process, such a formulation will succeed in holding the control variable η in the desired region, $0 \leq \eta \leq 1$, as the parameter ζ will automatically remain in the region $0 \leq \zeta \leq 1$. Since the right member of Equation (30) is evaluated along the solution $x = \bar{x}(t)$, $\eta = \bar{\eta}(t)$, and with finite step size $\Delta\sigma$, the control parameter ζ will not, in general, remain within limits. Consequently, it will be necessary before each calculation of $\partial\zeta/\partial\sigma(t)$ to alter the function $\zeta(t)$, obtained in the course of the preceding descent computations, to conform to the inequality $0 \leq \zeta \leq 1$; otherwise the right member of Equation (30) will vanish in all subsequent computations at points for which the inequality is violated.

With the assumed linear thrust-fuel flow relationship, the quantitites T and Q become

$$T_\eta = T_{\eta=1}\eta \tag{34}$$

$$Q_\eta = Q_{\eta=1}\eta , \tag{35}$$

where $T_{\eta=1}$ and $Q_{\eta=1}$ are tabular functions of Mach number and altitude.

VI. PENALTY AND PROJECTION VERSIONS OF THE GRADIENT PROCESS

Both penalty and projection versions of the gradient process are described in Reference 1 and were used in essentially the form reported there for the computations of References 2 and 4, some of which are presented in a later section. A combination consisting of alternating cycles of the two was also tried and proved better than either one by itself; it replaced the expensive restoration cycle of the projection version with a penalty cycle, thus functioning to restore gross departures of the constraints without neglecting altogether the main business of optimizing.

The penalty scheme has a certain arbitrariness in choice of penalty coefficients; likewise, the relative weighting of corrections between two control functions must be set arbitrarily in both penalty and projection versions. The question of how large a step to take bedevils projection, as any departure of constraints outside a certain range of linearity requires possibly expensive restoring. These problems were fully appreciated by the writers as well as by A. E. Bryson and his colleagues and several other groups researching gradient techniques; they tended to blur the question of "which version is best," as none of the versions were really precisely stated algorithms without room for artwork.

VII. METHOD OF RAVINES

A conceptually simple method for accelerating the convergence of gradient processes is furnished by an idea due to Gel'fand and Tsetlin [3]. In the context of the unconstrained (penalty version) problem, the procedure is as follows:

(1) Proceed for several (say q) cycles of gradient search as usual. Store the last approximation to the optimal control functions, $\theta_1(t)$ and $\zeta_1(t)$.

(2) Begin the process anew with radically different first approximations to the optimal control functions. Again store the approximations $\theta_2(t)$ and $\zeta_2(t)$ obtained after q cycles.

(3) Connect the two "points" in the function by a straight line:

$$\theta = \theta_1(t) + \xi[\theta_2(t) - \theta_1(t)] \tag{36}$$

$$\zeta = \zeta_1(t) + \xi[\zeta_2(t) - \zeta_1(t)] \tag{37}$$

(4) Perform a one-dimensional search for a minimum versus the interpolation/extrapolation parameter ξ.

This procedure has the effect of smoothing out the "zig-zagging" typically experienced with a gradient process as a "ravine" of the functional is traversed. A variant of this method, suggested by R. E. Kopp of Grumman [2], consists of employing the last approximation $\theta_1(t)$ and $\zeta_1(t)$ produced in Step 1 above as the first approximation for Step 2. With this scheme, it proved convenient to continue the process as far as desired, introducing a "ravine" cycle after every $2q$ gradient cycles.

VIII. SOME COMPUTATIONAL RESULTS FOR OPTIMAL TRAJECTORIES OF SUPERSONIC AIRCRAFT

The first example configuration was that of a hypothetical Mach 3 vehicle powered by a nonafterburning turbojet engine; the second a supersonic transport aircraft. Configuration data are given in Appendix A.

A. Full-Throttle Climbs

Some minimum-time full-throttle climbs to 70,000 ft, initiated at Mach 0.25 and 500 ft., are shown in Figure 6.

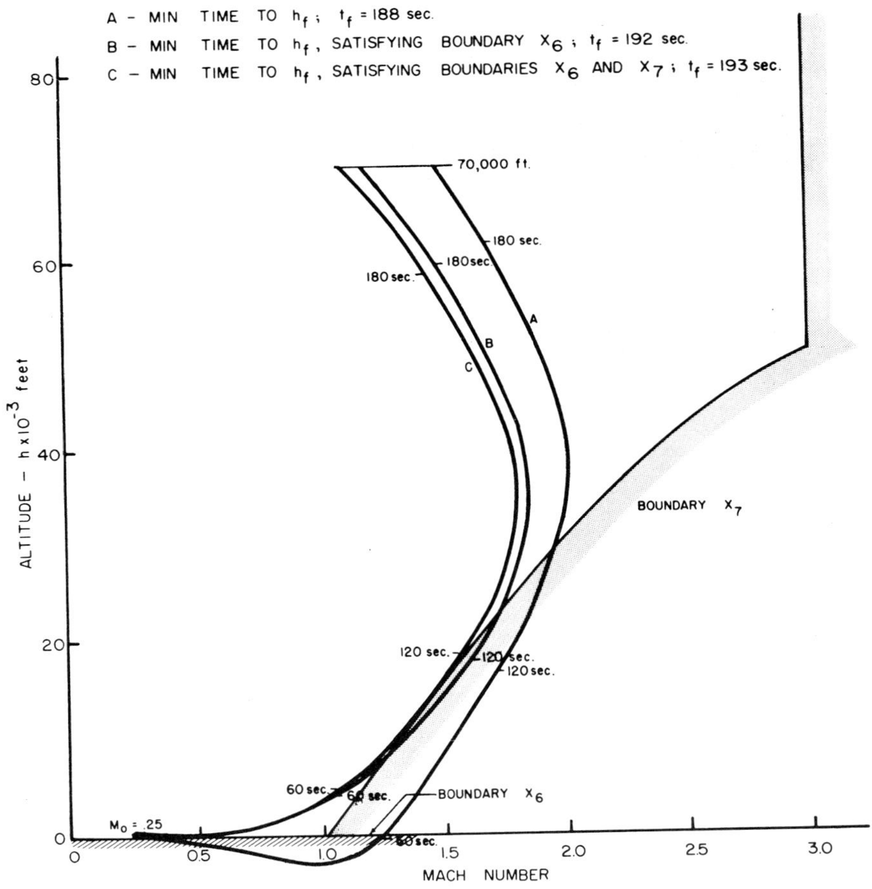

Fig. 6. Supersonic Interceptor Full Throttle Minimum Time Climbs from Low Initial Mach Number.

Terminal values of the weight and velocity components are left unspecified. If altitude and overpressure/overtemperature limits are omitted from consideration, the minimum-time trajectory penetrates both boundaries as a result of an embarrassing initial dive (Trajectory A). Trajectory B is obtained by introduction of a lower limit on altitude (sea level) via the x_6 penalty term. There remains a slight penetration of the overpressure boundary. Trajectory C corresponds to enforcement of the overpressure limit via introduction of the x_7 penalty. While the differences in performance between the three trajectories (188, 192, and 193 sec, respectively) are not large in this example, it is clear that Trajectories A and B are worthless for flight test or operational employment.

Figure 7 shows a similar comparison for the case of a higher initial Mach number $(M = 0.7)$. There is, in this case, no tendency of the unconstrained trajectory to dive initially; however, there is considerable penetration of the overpressure boundary and slightly more appreciable performance loss (6 sec) entailed in enforcing the overpressure restriction.

Figure 8 presents minimum-time climbs to 70,000 ft. altitude with horizontal velocity component specified at Mach 2.6 (Trajectory A) and with zero vertical velocity component specified in addition (Trajectory B). Terminal weight is unspecified in both cases and neither the x_6 (altitude limit) nor x_7 (overpressure/overtemperature) penalities are operative. Shown for comparison is the "excess power schedule" of the simplified climb theory, which corresponds to a maximum of the excess power $V(T - D)$ attained along curves of constant energy $\left(h + \frac{V^2}{2g} = \text{constant} \right)$ with drag evaluated for lift equal to weight. Since the maximum is not sharply defined in the present example, a crosshatched band is employed to convey the idea of a flat maximum. The addition of the x_7 boundary produces a substantial effect on performance in the case of minimum-time climbs with relatively high specified terminal

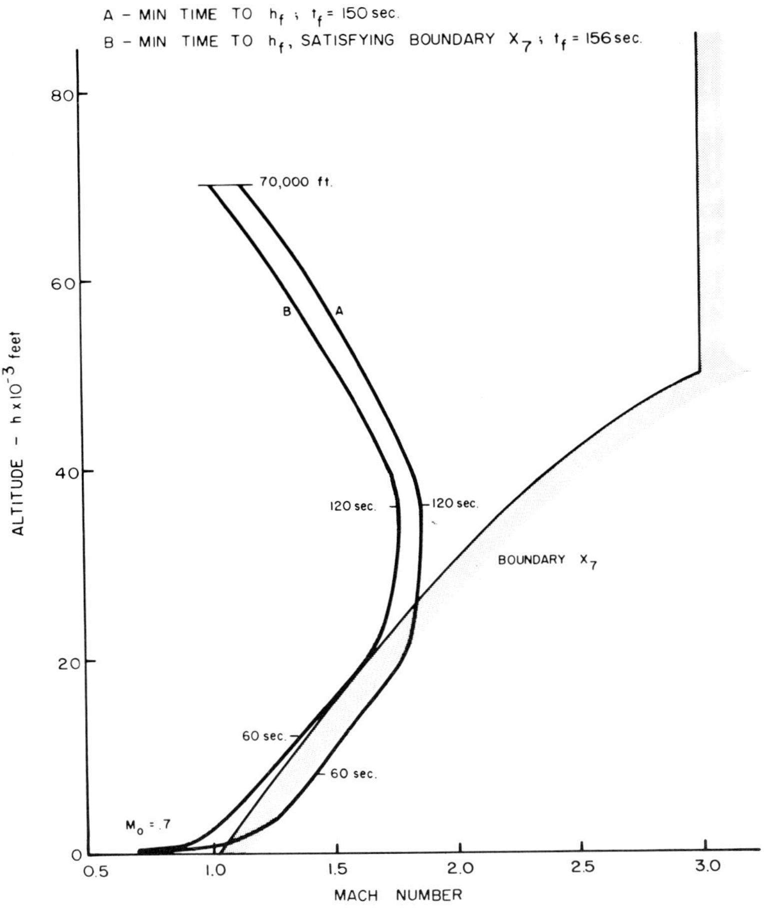

Fig. 7. Supersonic Interceptor Full Throttle Minimum Time Climbs from Higher Initial Mach Numbers.

velocity, as in the case of Figure 8, Trajectory B. Trajectory C represents the corresponding result with the x_7 boundary incorporated. The performance figures for B and C are 236 sec and 262 sec respectively.

Two minimum-fuel (maximum-terminal-weight) trajectories are shown in Figure 9. These correspond to 70,000 ft.

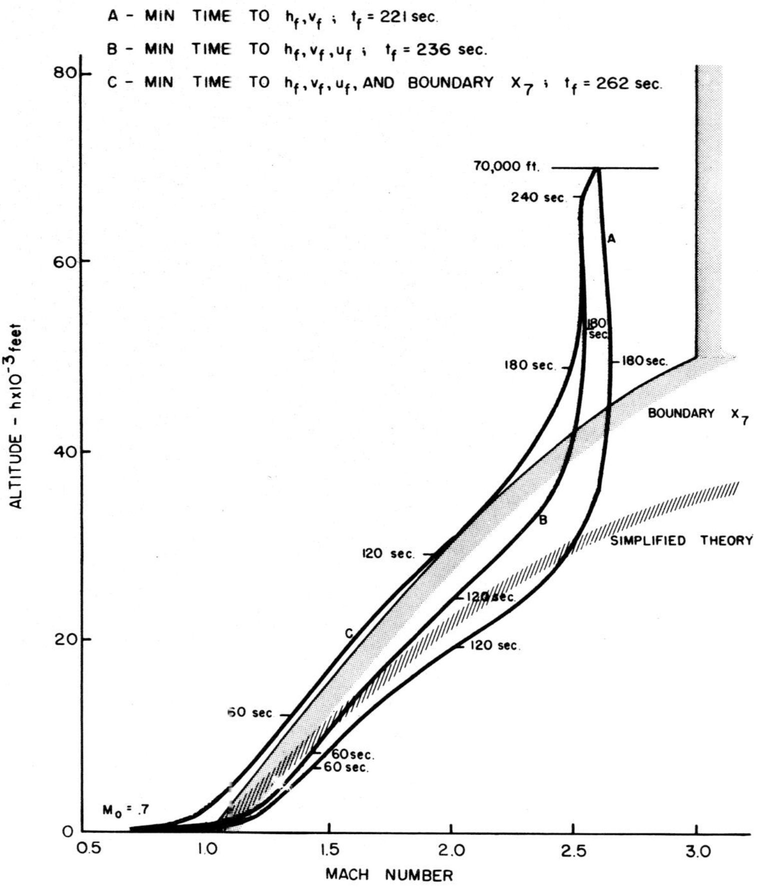

Fig. 8. Supersonic Interceptor Full Throttle Minimum Time Climbs with Final Velocity Specified.

terminal altitude and terminal horizontal velocity Mach 2.6 (Trajectory A) with the additional requirement of zero vertical velocity at the terminal point in the case of Trajectory B. The crosshatched band shown for comparison corresponds again to the simplified theory: it represents a maximum of the excess power/fuel-flow ratio attained along constant energy curves.

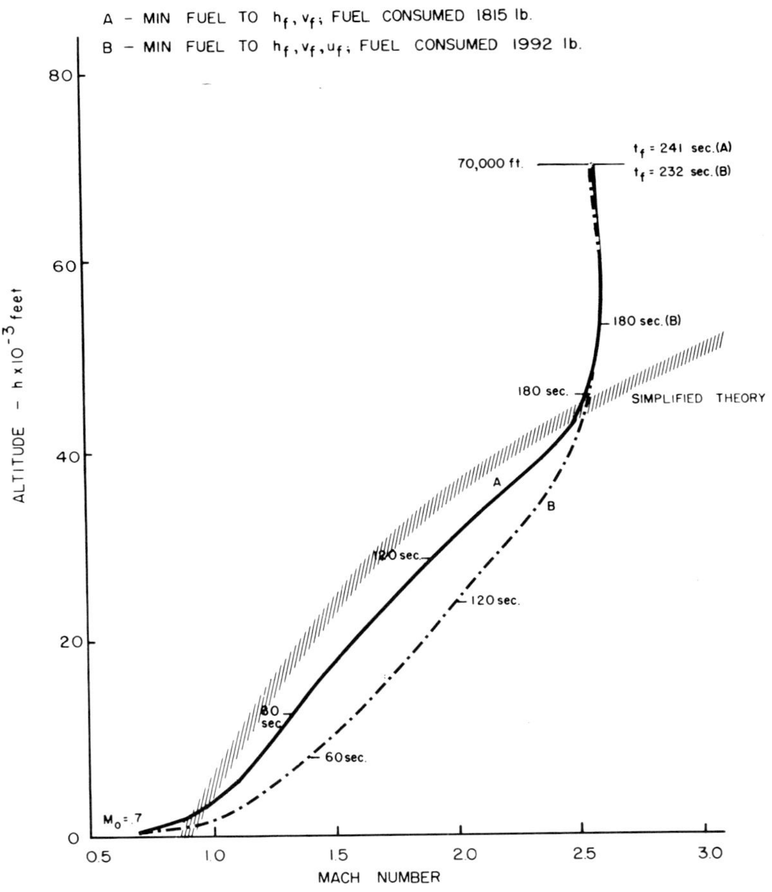

Fig. 9. Supersonic Interceptor Full Throttle Minimum Fuel Climbs.

Results of maximum-range computations for two full-throttle problems are shown in Figure 10. Trajectory A corresponds to maximum range with only terminal weight specified. The full-throttle restriction is unrealistic from an operational viewpoint and results in a porpoising maneuver to a region of low fuel-flow during the latter portion of the flight. Convergence of the descent process was poor in this instance,

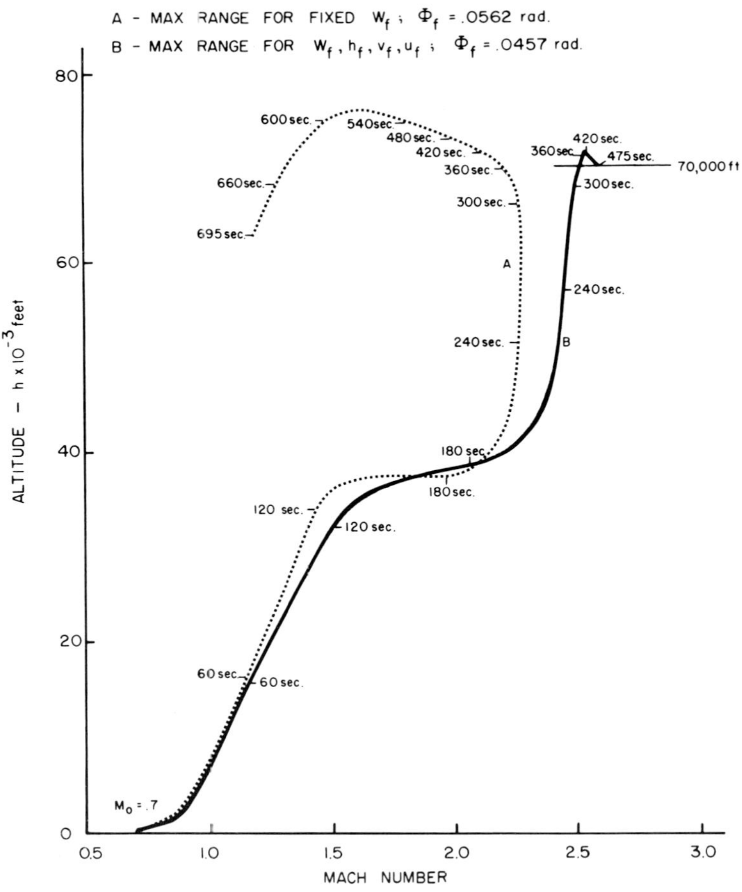

Fig. 10. Supersonic Interceptor Full Throttle Maximum Range Climbs.

as will be discussed subsequently. Trajectory B is a maximum-range path with terminal altitude and velocity components specified in addition to terminal weight.

B. Optimal Glide Paths

Presented in Figure 11 is a family of maximum-range, power-off, glide paths initiated from level flight at various

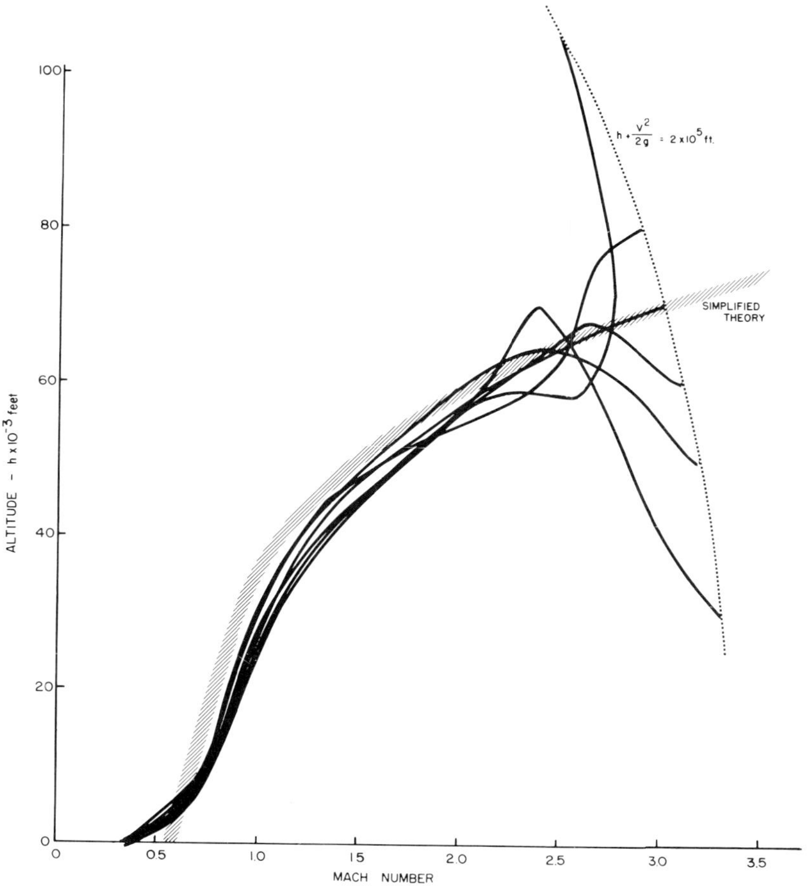

Fig. 11. Supersonic Interceptor Power-Off Maximum Range Glides.

points along a contour of constant energy $\left(h + \frac{V^2}{2g} = \text{constant}\right)$. Also shown for comparison purposes is the result of the simplified "energy" theory, which, in this case, corresponds to a minimum of level-flight drag attained along constant-energy contours. This is very nearly an L/D_{max} glide, differing only slightly from this in the transonic region, owing to the variation of drag coefficient with Mach number.

The trajectories display the general character of a transient, followed by a gliding descent in the vicinity of L/D_{max}, with a terminal flare-out to convert the airspeed remaining at low altitude to range.

C. Variable Throttle Trajectory

A maximum-range trajectory with only terminal altitude and weight specified is shown in Figure 12. The fuel

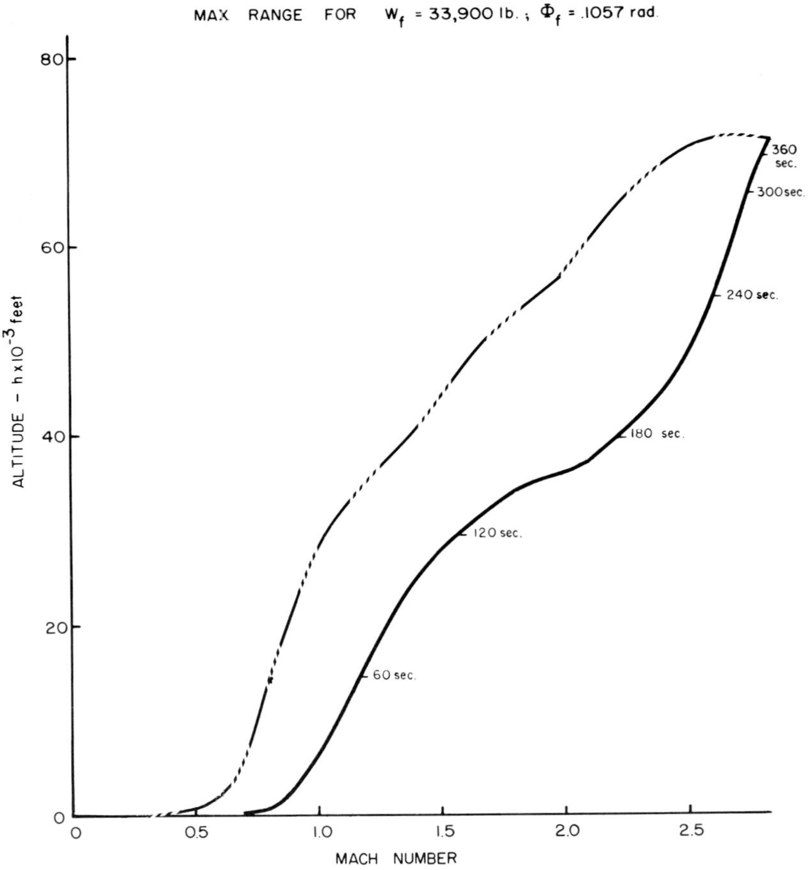

Fig. 12. Supersonic Interceptor Variable Throttle Maximum Range Climb-Glide.

allocation for the trajectory was small enough to avoid violation of the over-temperature boundary. The trajectory is of boost-glide type with bang-bang throttle behavior.

IX. NUMERICAL STUDIES OF SUPERSONIC TRANSPORT MINIMUM FUEL ACCELERATION-CLIMB PATHS

The results presented in this section are typical numerical solutions to the problem of minimum-fuel acceleration climb from post-takeoff initial conditions to begin-cruise terminal conditions for a representative canard/delta transport configuration. The $F(\beta C_L)$ function computed for this configuration is shown in Figure 13. For this particular configuration, F is closely approximated as a linear function of βC_L. The initial conditions (held fixed) are: $u_0 = 0$, $v_0 = 670$ fps, $h_0 = 500$ ft., $\Phi_0 = 0$, and $W_0 = 400{,}000$ lb. The desired terminal conditions are: $\bar{u} = 0$, $\bar{v} = 2900$ fps, $\bar{h} = 65{,}000$ ft., and $\bar{\Phi}$ open. The minimum-fuel trajectory viewed in the Mach

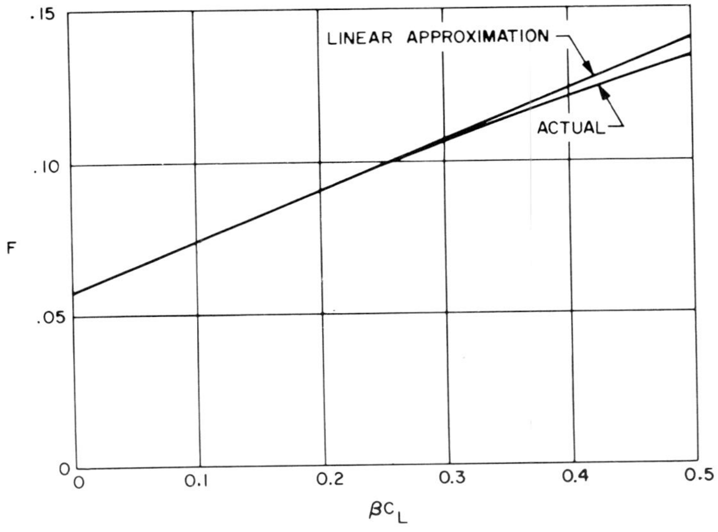

Fig. 13. The Function $F(\beta C_L)$.

number-altitude plane (Figure 14) without any inequality

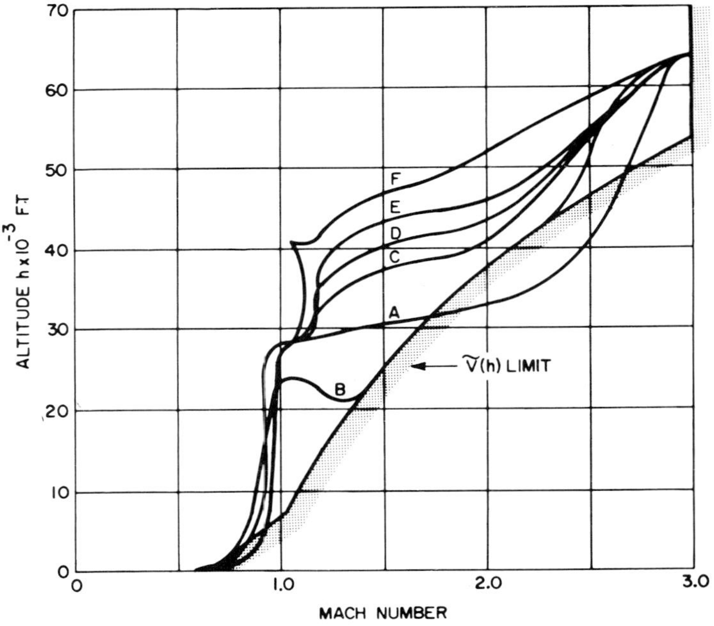

Fig. 14. Supersonic Transport Minimum Fuel Acceleration-Climb Profiles .

constraints is Curve A. The accompanying overpressure history is shown in Figure 15. The trajectory labeled B illustrates satisfaction of a representative engine/airframe limit without regard to overpressure limitations. Trajectories C through F are those paths satisfying particular overpressure limits without engine/airframe limit constraints. The accompanying table serves as a guide in determining the overpressure constraint limits treated in Trajectories C through F and illustrates the strong performance trade-offs computed for lowered sonic-boom intensity limits. The overpressure limit of 2.25 lb/ft^2 (although still above a desired operational limit of 2.0 lb/ft^2) in Trajectory F is near the lowest

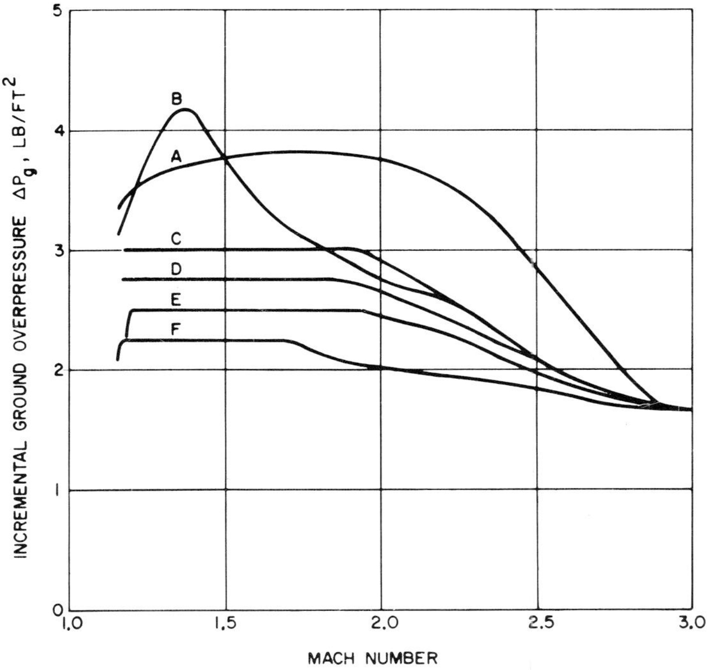

Fig. 15. Overpressure Histories.

attainable with the vehicle considered.

From Trajectories B and F in Table I, one may extract the following important quantitative result; the 44,000 lb. increase in fuel expenditure due to satisfying a reasonable sonic-boom intensity constraint (Trajectory F) over that given by a path that disregards sonic-boom considerations (Trajectory B) becomes the weight equivalent of the vehicle payload. Alternately, the net-range loss experienced in using F compared to B is approximately 500 statute miles. (The range comparison is effected by an extension of Trajectory B to include a cruise leg consuming fuel at approximately 50 lb/statute mile until the total fuel expenditure is equivalent to that of Trajectory F.)

X. CONSIDERATIONS ON COMPUTATIONAL TECHNIQUE

Numerical solution of trajectory optimization problems having the size and complexity of the present example requires careful attention to detail, particularly in regard to matters influencing speed of convergence of the gradient process that may, in some instances, be critical.

A. Accuracy Requirements

The use of a numerical integration scheme incorporating a variable-time interval controlled on the basis of error estimates proved essential with single-precision eight-decimal-digit arithmetic. With such a scheme, the number of intervals for a given accuracy of integration can be held to a reasonable number. Unfortunately, however, complexity then arises in the storage and recall of control variable histories: the numerical values of the independent variable, time t, must be stored and interpolation later performed to obtain control variable values at intermediate times for use in subsequent calculations.

B. Gradient Technique

In many cases, particularly those into which the range variable does not enter, the convergence of the gradient process is inherently rapid, and the penalty function version of the method is quite adequate. It was found that little advantage accrued in employing a "good" first approximation to the optimal control histories in such cases, and that workable values of the penalty coefficients could be settled on with a little experimentation. Generally, several cycles were first run omitting those terminal constraints that have the least effect on the general character of the optimal trajectory; the penalties corresponding to these constraints were added. Usually, the penalty terms corresponding to the boundaries were the last to be added, and, in case the boundary violations were large, the penalty coefficients were increased in two or more stages.

Some examples met with slow convergence, e.g., those shown for the turbojet vehicle with range maximized. This particular type of difficulty for a gradient method may relate to the phenomenon of motions occurring simultaneously on several essentially different time scales, which is the subject of a tailored approximation expounded in another chapter of the present volume. In slowly converging cases, careful selection and adjustment of penalty coefficients was carried out and means of accelerating convergence sought. This was provided by alternating penalty function and projection cycles, as previously discussed, and by the ravine feature. It was found that convergence is accelerated by a factor of approximately two by the alternating-cycle scheme in comparison to that with the penalty function scheme alone. The procedure of introducing constraints sequentially, mentioned in the preceding paragraph, is virtually a necessity in slowly-convergent cases.

Experience with the ravine method was limited to some experiments conducted on one particularly stubborn case. The procedure was to insert one ravine cycle every $2q$ penalty-function cycles, with $q = 6$. Some experimentation indicated that it is advantageous to conduct the one-dimensional ravine search on the function P rather than P', tolerating build-up in terminal errors in favor of performance improvement, later reducing the errors in the course of further penalty-function cycles. The ravine cycle improved the over-all convergence speed by a factor of approximately two. It would appear worthwhile to perform a series of systematic experiments with the ravine method in some less complex problem so as to obtain more conclusive results and improved insight into the mechanism of the technique.

Because further development of essentially first-order successive approximation processes in continuous control format seems attractive, the next two sections are given over to description of two features offering possibilities of acceleration.

XI. CONJUGATE GRADIENT PROJECTION

Conjugate gradient optimization for continuous control problems has been investigated with penalty function treatment of constraints in Reference 7 and with projection treatment in Reference 8; however, the latter is limited to linear constraints. The following sketches an extension of conjugate gradient projection to the case of nonlinear constraints using a device previously employed in a similar extension of projected gradients with the Davidon-Fletcher-Powell algorithm [9]. It is convenient to present the process in the context of an ordinary (finite-dimensional) minimum problem since extension by analogy to the variational case is direct.

The scalar function $f(x)$ is to be minimized subject to the constraint $g(x) = 0$, where g is a row-vector and x is a column-vector. Conventional projection proceeds by first correcting constraint violations as via Rosen's method [10]:

$$\Delta x = - g_x (g_x^T g_x)^{-1} g \, , \tag{38}$$

and/or elaborations thereof.

Gradients f_x and g_x are then calculated and the projection multiplier vector computed

$$\lambda = - (g_x^T g_x)^{-1} \, g_x^T f_x \, . \tag{39}$$

In conventional projection, a series of steps

$$\Delta x = -\alpha(f_x + g_x\lambda) \tag{40}$$

is then taken to a one-dimensional minimum of f versus α, unless constraint violations exceeding preset tolerances are met, in which case the search is terminated short of the minimum. The quantity $f_x + g_x\lambda$ is the projection of the gradient f_x on the subspace of the constraint tangent plane intersection. When large fluctuations cease from cycle to cycle in the magnitude and direction of the λ-vector, the approach to the general vicinity of the constrained minimum may be indicated and conjugate step generation becomes attractive.

With <u>nonlinear</u> constraints, the linear search for f versus α may never terminate at a one-dimensional minimum, even though the reference point be very close to the constrained minimum sought, e.g., if f is linear in x. On the other hand, the conjugacy of successive steps, which is essential to the acceleration property of conjugate gradients, depends crucially on search termination at a minimum. As argued in Reference 9, the function $f + g\lambda$ exhibits a minimum in the subspace of the constraint tangent plane intersection when the reference point is sufficiently close to the constrained minimum, and it is, therefore, appropriate to terminate linear searches on a minimum of $f + g\lambda$. Note that the correction term $g\lambda$ does not contribute during the search if the constraints are linear, hence the procedure proposed in this case reduces to the conventional one.

The algorithm proposed is

$$\Delta x_i = -\alpha p_i \tag{41}$$

$$p_i = \hat{f}_{x_1} \tag{42}$$

$$p_i = \hat{f}_{x_i} + \frac{|\hat{f}_{x_i}|^2}{|\hat{f}_{x_{i-1}}|^2} \left[I - \left(g_x^T g_x \right)^{-1} g_x^T \right] p_{i-1}, \quad i \geq 2 \tag{43}$$

which is the Fletcher-Reeves adaptation of conjugate gradients [11] with

$$\hat{f} \equiv f + g\lambda \tag{44}$$

appearing in place of the unconstrained function whose minimum is sought by that algorithm. The vector λ is calculated each cycle via Equation (39). Ordinary gradient projection is to be employed until two conditions are met: search termination at a one-dimensional minimum, and apparent stabilization of λ as determined by a test such as

$$(1 - \varepsilon)|\lambda_i| \leq |\lambda_{i+1}| \leq (1 + \varepsilon)|\lambda_i| \tag{45}$$

and

$$\frac{\lambda_{i+1}^T \lambda_i}{|\lambda_{i+1}||\lambda_i|} \geq (1 - \varepsilon), \tag{46}$$

Alternate constraint correction cycles as given by Equation (38) are to be performed. If the conditions (45) and (46) are not met subsequently, the process is restarted with an ordinary gradient cycle. In Equation (43), the matrix $\left[I - \left(g_x^T g_x \right)^{-1} g_x^T \right]$ projects p_{i-1} onto the current constraint tangent hyperplane; the operation would be unnecessary with linear constraints, since p_{i-1} would be already in this plane. The value of this correction, which was suggested by Peter Salvato of TRW Systems, is in avoiding early constraint violation build-up during the one-dimensional search versus α.

A similar procedure due to Rosen and Kreuser [12] projects on to linear approximations to the constraints which are held fixed over several cycles, minimizing $f + g\lambda^*$, with λ^* the value computed on the first of the several cycles. The subspace in which the sequence of steps is taken is thus fixed, as is the function whose partial derivatives define conjugacy; both of these features support a legitimate well-defined conjugacy. Disadvantages are the requirement to store initial g_x and the need to correct large constraint violations at the end of each such sequence of several cycles.

In the continuous control case, H_u is the analog of $\hat{f}_x$ in Equations (41), (42), (43), and

$$\int_{t_0}^{t_f} H_u^2 \, dt \tag{47}$$

is the analog of $|\hat{f}_x|^2$. Double the usual control history storage is needed as functions corresponding to the previous cycle's control increments must be available. If final time is a free parameter to be optimized, the gradient is the function/parameter pair (H_u, H_f), a term H_f^2 is added to the expression (47), etc.

XII. PROJECTION WITH CURVILINEAR SEARCH

A projection process is next examined in which the one-dimensional search is carried out along a curve determined so as to follow approximately the nonlinear constraint intersection [13]. Exploratory steps are taken along the direction from $x = \bar{x}$ given by Equation (40). By testing the departures of the constraints from the values $\bar{g}$ at the start of the search, a step-size scalar $\tilde{\alpha}$ is found for which $\tilde{g} - \bar{g}$ is sufficiently large in magnitude to be sensibly unaffected by numerical round-off error, but is within the descriptive range of a Taylor expansion of $g(\alpha)$ truncated at quadratic terms.

A search is then performed along a curve

$$\Delta x = -\alpha(f_x + g_x\lambda) - \frac{\alpha^2}{\tilde{\alpha}^2} g_x(g_x^T g_x)^{-1} (\tilde{g} - \bar{g}) , \qquad (48)$$

which proceeds to a minimum of $f + g\lambda$ versus α or to a build-up of constraint violations to beyond tolerances, whichever arises first. The vector x is then updated and the constraints again restored. The second member of Equation (48) tends to correct constraint drift in the approximate fashion of Equation (41).

It would seem reasonable to operate, using conventional projection with the curved search, until the projection multiplier vector has settled down in magnitude and direction as determined by tests of the form given in (45) and (46), with ε chosen to be somewhere in the range perhaps of $\frac{1}{10} \leq \varepsilon \leq \frac{1}{2}$.

It would appear possible to combine gracefully the curvilinear search feature with acceleration, this because the constraint following correction is orthogonal to the subspace in which conjugacy is desired. The extension of the curved search idea to continuous control cases would appear to be straightforward.

Acknowledgments

The authors are especially indebted to the following people for their various contributions during the development of this work: Donald Ball, Joseph Lindorfer, Alan Schneider, Richard Kopp, of the Grumman Aerospace Corporation; Peter Salvato of TRW Systems; Ivan Johnson of NASA Manned Spacecraft Center; and Joseph Focte of Louisiana State University.

REFERENCES

1. H. J. KELLEY, "Method of Gradients," in Optimization Techniques, (G. Leitmann, ed.), Academic Press, New York, 1962.

2. H. J. KELLEY, M. FALCO, and D. J. BALL, "Air Vehicle Trajectory Optimization," SIAM Symposium on Multivariable System Theory, Cambridge, Massachusetts, November 1-3, 1962.

3. I. M. GEL'FAND and M. L. TSETLIN, "The Organized Search Principle in Systems with Automatic Stabilization," Doklady Akademii Nauk SSR, 137, 2, Moscow (1961), pp. 295-298.

4. M. FALCO, "Supersonic Transport Climb Path Optimization Including a Constraint on Sonic Boom Intensity," AIAA J. Vol. 1, No. 12 (1963).

5. A. E. BRYSON and W. DENHAM, "A Steepest-Ascent Method for Solving Optimum Programming Problems," Raytheon Company, Missile and Space Division Report BR-1303, August 1961. Also J. of Appl. Mech., June 1962.

6. R. COURANT, "Variational Methods for the Solution of Problems of Equilibrium and Vibrations," Bull. Amer. Math. Soc. Vol. 49, No. 1, (1943).

7. L. S. LASDON, S. K. MITTER, and A. D. WAREN, "The Conjugate Gradient Method for Optimal Control Problems," IEEE Trans. Auto Cont. Vol. 12, No. 2, (1967).

8. J. F. SINNOT and D. G. LUENBERGER, "Solution of Optimal Control Problems by the Method of Conjugate Gradients," Preprint Volume, 1967 Joint Automatic Control Conference, pp. 566-573.

9. H. J. KELLEY and J. L. SPEYER, "Accelerated Gradient Projection," Colloquium on Optimization, Nice, France, June 29-July 5, 1969. Proceedings published as Lecture

Notes in Mathematics No. 132, Springer-Verlag, Berlin, 1970.

10. J. B. ROSEN, "The Gradient Projection Method for Nonlinear Programming, Part I: Linear Constraints," J. SIAM, March 1960; "Part II: Nonlinear Constraints," J. SIAM, December 1961.

11. R. FLETCHER and C. M. REEVES, "Function Minimization by Conjugate Gradients," Comp. J., July 1964.

12. J. B. ROSEN and J. KREUSER, "A Gradient Projection Algorithm for Nonlinear Constraints," Conference on Numerical Methods for Nonlinear Optimization, University of Dundee, Scotland, June 28-July 1, 1971.

13. H. J. KELLEY and I. L. JOHNSON, "Curvilinear Projection," Fourth IFIP Colloquium on Optimization Techniques, Santa Monica, California, October 19-22, 1971.

14. H. W. CARLSON, "An Investigation of the Influence of Lift on Sonic Boom Intensity by Means of Wind Tunnel Measurements of the Pressure Fields of Several Wing-Body Combinations at a Mach Number of 2.01," NASA TN D-881, July 1961.

15. F. WALKDEN, "The Shock Pattern of a Wing-Body Combination, Far from the Flight Path," Aeronaut. Quarterly, Vol. 9, pp. 164-194, (1958).

16. D. W. PATTERSON, "Sonic Boom - Limitations on Supersonic Aircraft Flight Operations," Aerosp. Eng., Vol. 19, pp. 20-24, (1960).

APPENDIX A

Parameters Used

The geophysical parameters utilized in these studies were as follows:

$\omega = 0$ (nonrotating earth)

$r_e = \tilde{r} = 20{,}926{,}607$ ft. (equatorial plane)

$g_e = 32.174$ ft/sec^2 (equatorial plane)

a, ρ as a function of h according to the ARDC atmosphere (tabular functions).

The hypothetical turbojet interceptor vehicle parameters employed were as follows:

Fuel flow at full throttle, $Q_{\eta=1}$, as a tabular function of (M,h) shown in Figure A1.

Thrust at full throttle, $T_{\eta=1}$, as a tabular function of (M,h) shown in Figure A2.

Drag and lift coefficients, $C_{D_{min}}$ and C_{L_α} as tabular functions of Mach number shown in Figure A3.

Lift coefficient at zero angle of attack, $C_{L_0} = 0$

$S = 500$ ft^2

V^* as a function of h illustrated in Figure 3.

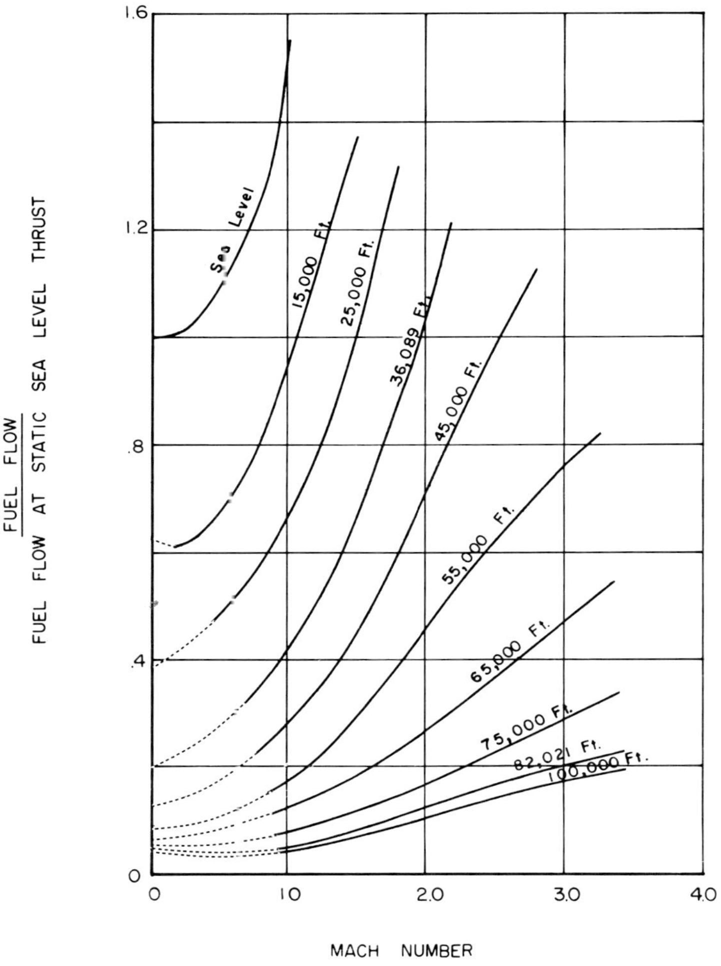

Fig. A1. Supersonic Interceptor Fuel Flow Variation with Mach Number and Altitude at Full Throttle.

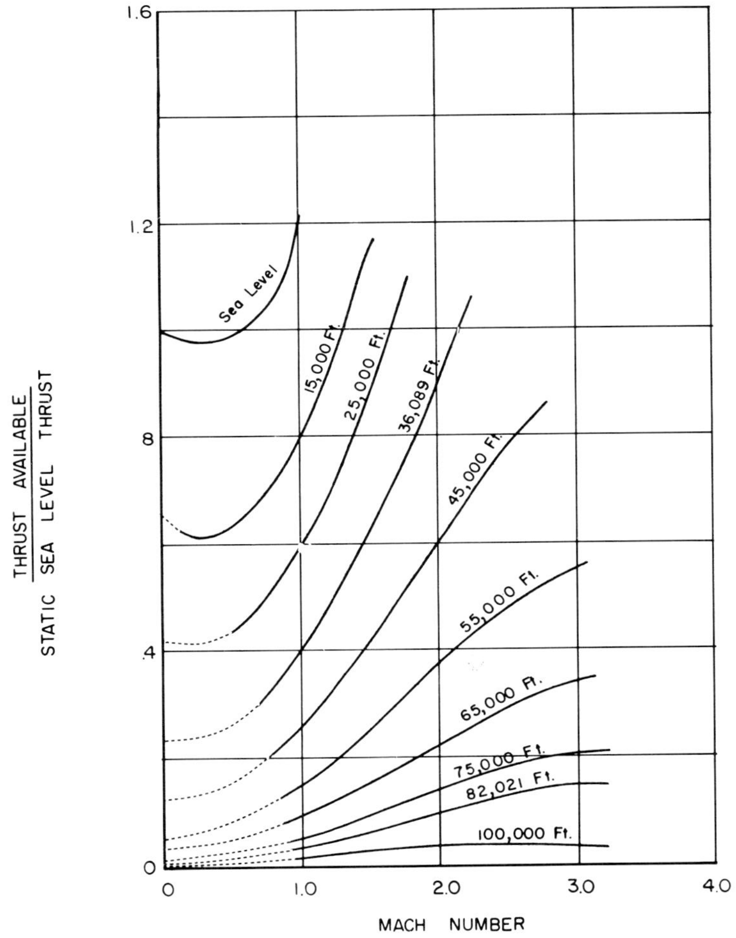

Fig. A2. Supersonic Interceptor Thrust Variation with Mach Number and Altitude at Full Throttle.

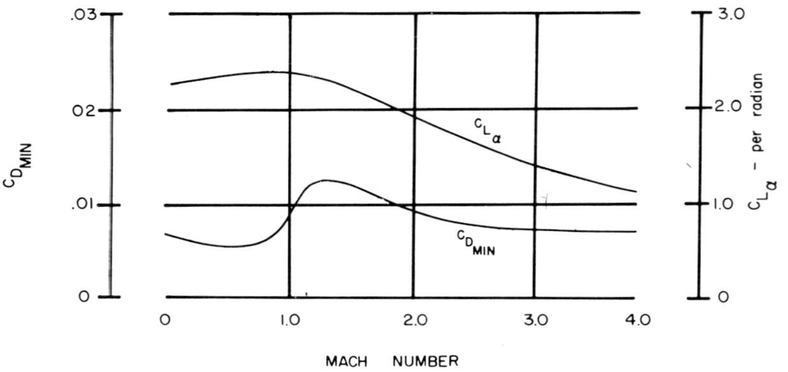

Fig. A3. Supersonic Interceptor Aerodynamic Force Coefficient Variation with Mach Number.

APPENDIX B

The Simplified Overpressure Formula

Carlson [14], following the paper of Walkden [15], has developed the following steady-flight approximation for the peak ground overpressure increment assuming the vehicle configuration to be composed of a slender body of revolution and a thin delta wing operating at small angles of attack:

$$\frac{\Delta p_g}{p_{ref}} = K_1 (M^2 - 1)^{1/8} \ell^{3/4} h^{-3/4} [F(\beta C_L)] \qquad \text{(A-1)}$$

where $p_{ref} = (p_h p_g)^{1/2}$ is taken as the geometric mean of ambient pressures. Here K_1 is a reflection factor, ℓ the body length, $\beta = (M^2 - 1)^{1/2}$, and $F(\beta C_L)$ is determined by lift and cross-sectional area distribution considerations of the wing-body combination. One then seeks to refine Equation (A-1) to account roughly for the effect of complete refraction of a sound ray due to a linear temperature gradient existing in the troposphere and of vehicle flight-path angle. Introducing Snell's law results in the formula

$$\mu + \gamma = (\pi/2) - \cos^{-1}(a_h/a_g) \tag{A-2}$$

where $\mu = \sin^{-1}(1/M)$. One may interpret Equation (A-2) as a definition of "cut-off Mach number", M^*, defined by h and γ as tabularly represented by Patterson [16]. That is, if $M \geq M^*(h,\gamma)$, then Δp_g is computed by Equation (A-1); or if $M < M^*(h,\gamma)$, then $\Delta p_g = 0$. Our formula for Δp_g then becomes

$$\Delta p_g = (p_h p_g)^{1/2} K_1 (M^2 - 1)^{1/8} \ell^{3/4} h^{-3/4} [F(\beta C_L)] H(M - M^*) . \tag{A-3}$$

Table I Trajectory Terminal Values

TRAJECTORY	A	B	C	D	E	F
$\Delta \bar{P}_g$ lb/ft^2	Not Speci-fied	Not Speci-fied	3.00	2.75	2.50	2.25
W_f lb	346,000	343,000	339,000	334,000	326,000	299,000
t_f sec.	690	730	920	1090	1350	2350
Φ_f rad.	0.0525	0.0618	0.0719	.0835	0.1000	0.1610

Remarks: Range Comparison of B and F at W_f = 299,000 lb,
W_0 = 400,000 lb
B $\approx$ 1125 s.m. (and Cruise Leg @ 50 lb fuel/mi)
F $\approx$ 640 s.m.

Aircraft Maneuver Optimization by Reduced-Order Approximation*

HENRY J. KELLEY
Analytical Mechanics Associates, Inc.,
Jericho, New York

The research was performed in part under Contract NAS 12-656 with NASA Electronics Research Center and Contract F 44620-71-C-0123 with USAF Headquarters, Office of the Assistant Chief of Staff for Studies and Analysis.

I. INTRODUCTION

The theory of aircraft "energy climbs" began with the WWII work of Kaiser [1] but remained in the realm of <u>ad hoc</u> engineering approximations for more than two decades. Although a 1962 survey of the consistency of various published approximations [2] held out no particular hope of a solid basis for the theory, engineering interest persisted, spurred by the very considerable difficulties of optimizing atmospheric flight trajectories with an "exact" (i.e., particle-dynamics) system model. The essential feature of the energy type of approximation is reduction in the order of the state differential system. This is a blessing in facilitating solution, which becomes a curse when the problem of transitions from and to specified end conditions is faced: there are embarrassingly many of these after the order reduction, and unthinkably instantaneous jumps in state variables must then be permitted to make things come out right.

Many such problems are suitable for treatment by <u>singular perturbation theory</u> of ordinary differential equations, whose defining feature is precisely reduction in order as a parameter tends toward zero. In this approach, the solution of the reduced-order system is modified near one or both endpoints by a "boundary-layer" correction, also calculated via a system of differential equations of lower order than that of the original system. Higher-order approximations are also offered by existing theory, which is quite well developed, at least for initial-value problems [3]. The present chapter reports recent work in recasting the energy type of approximation to aircraft flight in terms of singular perturbation theory and in extending it to three-dimensional maneuvers. The chapter is mainly a unification of the previously published material of Refs. 4-9 and Ref. 12, and also contains previously unpublished material on selection of variables as well as additional numerical results.

Singular perturbations for differential equations arising in optimal control will first be examined for a system of fairly general form but low order. Attitude dynamics for optimal flight of a rocket in vacuum will then be studied as an introductory, and fairly transparent, example. The question of choice of variables, an often important preliminary, will then be taken up. Finally, optimal aircraft flight in various reduced-order approximations will be investigated, and results of some numerical computations reviewed.

II. SINGULAR PERTURBATIONS IN OPTIMAL CONTROL

A. Introductory Initial-Value Problem of Uncontrolled Motion

Sections 39 and 40 of Wasow's book [3] expound singular perturbation theory of initial-value problems for systems of nonlinear differential equations of the form

$$\dot{x} = f(x,y) \tag{1}$$

$$\varepsilon\dot{y} = g(x,y) \tag{2}$$

due mainly to Tihonov and Vasil'eva. This excellent account, to which the reader is referred, is for the case of scalar x and y, while the development of Tihonov [10] is for vector x and y and for several small parameters, which feature leads to the idea of several time scales, as against two with a single parameter.

For initial values $x(t_0) = x_0$, $y(t_0) = y_0$, it is clear which initial condition is "lost" (i.e., cannot be satisfied) when the approximation $\varepsilon = 0$ is introduced. The solution of the so-called <u>reduced system</u>, which can be generated if $g(x,y) = 0$ can be solved for y, approximates the solution of the original system under suitable stability hypotheses detailed in a theorem of Tihonov [Theorem 39.1 of Ref. 3]. Even under the best conditions, however, the approximation is bound to be poor near the initial time, excepting lucky specification of y_0. It is consistent to study the endpoint transitions in terms of an approximation for y near t_0 which is also based upon an expansion in ε. A time variable

$$\tau = (t - t_0)/\varepsilon \tag{3}$$

is introduced into the system and there is obtained for $\varepsilon = 0$ the <u>boundary-layer system</u>

$$\frac{dx}{d\tau} = 0 \tag{4}$$

$$\frac{dy}{d\tau} = g(x_0, y) \tag{5}$$

whose solution may be designated x^*, y^*.

If the reduced-system solution is designated by superscribed bars, the Vasil'eva composite approximation is given by

$$x = \bar{x}(t) + O(\varepsilon) \tag{6}$$

$$y = \bar{y}(t) + y^*(\tau) - [\bar{y}(t_0) + O(\varepsilon)] \tag{7}$$

from Section 40 of Ref. 3, where higher-order terms in ε are also derived.

B. A General Optimal-Control Problem of Low Order

The present section takes an exploratory look at the simplest variational problem that one might expect to solve approximately in terms of a reduced-order solution plus boundary layers at each end [5]. The primary objective is to adapt the available initial-value machinery to this problem.

The state equations are

$$\dot{x} = f(x,y,u,t) \tag{8}$$

$$\varepsilon\dot{y} = g(x,y,u,t) \tag{9}$$

where x, y, and u are scalars. It is desired to minimize the final value of x, $x(t_f) = x_f$, subject to $x(t_0) = x_0$, $y(t_0) = y_0$, $y(t_f) = y_f$, t_0 and t_f fixed; this is the so-called "classical problem of Mayer" in two state variables. It is assumed that both the original problem, for which $\varepsilon = 1$, and the reduced problem, where $\varepsilon = 0$, have solutions. An approximation to the former is sought in terms of an expansion in powers of ε about the latter.

The Euler-Lagrange equations are given in terms of

$$H \equiv \lambda_x f + \lambda_y g \tag{10}$$

as

$$\dot{\lambda}_x = - \frac{\partial H}{\partial x} \tag{11}$$

$$\varepsilon\dot{\lambda}_y = - \frac{\partial H}{\partial y} \tag{12}$$

$$\frac{\partial H}{\partial u} = 0 \tag{13}$$

and the transversality condition is $\lambda_{x_f} = -1$. For $\varepsilon = 0$, the variable y loses its exalted status; the order of the differential equations drops. It will be assumed, in order to avoid the complications of various singularity phenomena, that the strengthened form of the Legendre-Clebsch condition is satisfied for the reduced problem

$$(H_{yy}g_u^2 - 2H_{uy}g_u\, g_y + H_{uu}g_y^2) > 0 \,. \tag{14}$$

This, together with smoothness assumptions on the functions f and g, insures $\bar{y}(t)$ and $\bar{u}(t)$ continuous — no "corners."

In problems well suited to approximate solution by asymptotic expansion, the reduced system's solution will strongly resemble that of the original except near the endpoints, where jumps in y occur with the reduced-order model to satisfy the end conditions. These abrupt transitions belie the assumption of small $\dot{y}$ which was invoked to obtain the reduction in order.

Again a "stretching" time-transformation is introduced and the state-Euler system becomes, for $\varepsilon = 0$,

$$\frac{d}{d\tau} \lambda_x = 0 \tag{15}$$

$$\frac{dx}{d\tau} = 0 \tag{16}$$

$$\frac{d}{d\tau} \lambda_y = - \frac{\partial H}{\partial y} \tag{17}$$

$$\frac{dy}{d\tau} = g \tag{18}$$

$$\frac{\partial H}{\partial u} = 0 \, . \tag{19}$$

This is the boundary-layer system, whose solution may be denoted by $x^*(\tau)$, $y^*(\tau)$, $u^*(\tau)$. Evidently $x^*(0) = \bar{x}(t_0) = x_0$ and $\lambda_x^*(0) = \bar{\lambda}_x(t_0)$. It is noted that the boundary-layer equations are the state and Euler equations for the minimization of the integral performance index

$$\lim_{\tau_f \to \infty} \int_0^{\tau_f} \lambda_{x_0} f(\bar{x}_0, y^*, u^*, t_0) d\tau \tag{20}$$

subject to

$$\frac{d}{d\tau} y^* = g(\bar{x}_0, y^*, u^*, t_0) \tag{21}$$

and

$$y^*(0) = y_0, \quad y^*(\infty) = \bar{y}(t_0) . \tag{22}$$

The question of initial value for λ_y^* would seem hardly less trivial than for the original nonlinear two-point boundary problem, but, if the value could be determined, however indirectly, the possibility of using the existing singular perturbation theory for initial value problems would then arise. For large τ, both λ_y^* and y^* must approach the initial values $\bar{\lambda}_y$ and $\bar{y}$ of the reduced problem if the right members of Equations (17) and (18) are to approach the zero values required for equilibrium. Thus, attention is directed to the limiting behavior in the boundary layer as $\tau \to \infty$.

To this end, the boundary-layer equations may be linearized

$$y^* = \bar{y}_0 + \delta y, \quad \lambda_y^* = \bar{\lambda}_{y_0} + \delta\lambda_y, \quad u^* = \bar{u}_0 + \delta u$$

$$\frac{d}{d\tau}\,\delta\lambda_y = -\,g_y\,\delta\lambda_y - H_{yy}\,\delta y - H_{yu}\,\delta u \tag{23}$$

$$\frac{d}{d\tau}\,\delta y = g_y\,\delta y + g_u\,\delta u \tag{24}$$

$$H_{uy}\,\delta y + H_{uu}\,\delta u + g_u\,\delta\lambda_y = 0 . \tag{25}$$

Eliminating δu from Equations (23) and (24) gives two first-order equations. The various coefficients are independent of τ; they depend only upon t as a parameter; time "stands still" in the boundary layer. This allows easy examination of stability in terms of the characteristic roots of the quadratic

$$s^2 - (a+d)s + (ad - bc) = 0 \tag{26}$$

where

$$a = -\,g_y + (H_{yu}/H_{uu})g_u \tag{27}$$

$$b = - H_{yy} + (H_{yu}^2 / H_{uu}) \tag{28}$$

$$c = - g_u^2 / H_{uu} \tag{29}$$

$$d = g_y - g_u (H_{yu} / H_{uu}) \, . \tag{30}$$

Noting that $a + d = 0$, one obtains

$$s = \pm (bc - ad)^{\frac{1}{2}} = \pm [(H_{yy} g_u^2 - 2H_{yu} g_y g_u + H_{uu} g_y^2)/H_{uu}]^{\frac{1}{2}} . \tag{31}$$

In view of the assumed definiteness of expression (14) and with the additional assumption $H_{uu} > 0$, the bracketed expression in Equation (31) is positive, and the two roots are nonzero, real, equal in magnitude, and opposite in sign. The bracketed expression becomes negative when the Legendre-Clebsch condition for the reduced problem is violated at the initial point.

Tihonov's theory of the initial-value problem rests heavily upon an assumption of asymptotic stability for the boundary-layer equations. The reader is referred to Sections 39 and 40 of Reference 3 for the analysis and the other assumptions, which are less restrictive from an applications viewpoint. The instability of the Euler boundary-layer equations for large τ, evidenced by the positive member of the pair of real roots, represents a difficulty in adapting the theory. The conclusion implies that the unspecified initial condition $\lambda_y^*(0)$ must be chosen to suppress the unstable component of the solution, if this is possible, and this is closely related to the "matching" problem of matched asymptotic expansions.

If one attempts to proceed with expansion when the roots are imaginary, oscillations are encountered in the boundary layer which cannot be suppressed for large τ by choice of a single initial value, and the procedure fails. The borderline case of repeated zero roots encounters secular growth, corresponding also to failure, and hence it appears that the strengthened Legendre-Clebsch condition at the endpoints is essential for success.

In the terminal boundary layer, stability for reversed time is required by the Tihonov theory but, again, one or the other of the pair of real roots will obviate it, and suppression of instability by the choice of an end condition will again be needed. Zero or imaginary roots again signal failure.

After calculating the solution of the reduced system, and then the initial and terminal boundary-layer solutions, each in turn, and combining them à la Vasil'eva [Theorem 40.1 of Ref. 3], one then has an approximation to the solution. Higher-order approximations are also offered by the Vasil'eva theory. The value of approximations of various order in flight mechanics applications remains to be assessed.

The following simple example is of interest. Take $f = \frac{1}{2}(Ay^2 + Bu^2)$, $g = u$, and end conditions $y(0) = y(t_f) = 1$, $x(0) = 0$. The solution of the reduced problem is $\bar{u} = \bar{y} = \bar{x} = \bar{\lambda}_y = 0$, $\bar{\lambda}_x = -1$. The solution of the initial boundary-layer equations takes the form

$$y^*(\tau) = C\, e^{\gamma\tau} + D\, e^{-\gamma\tau} \tag{32}$$

where $\gamma \equiv (A/B)^{\frac{1}{2}}$ and the constants C and D must be chosen to satisfy $y^*(0) = 1$, and $y^* \to \bar{y} = 0$ for large τ. This illustrates the selection of the undetermined multiplier initial value (buried in C and D) to suppress instability in the boundary layer, i.e., $C = 0$, $D = 1$. The situation in the terminal layer is similar, except that the "stable" component is suppressed. The composite solution, representing an expansion to zero-order in ε, is

$$\hat{y}(t) = e^{-\gamma t} + e^{\gamma(t-t_f)} \tag{33}$$

which may be compared with the exact solution

$$y = \left[\left(1 - e^{-\gamma t_f}\right) e^{\gamma t} + \left(e^{\gamma t_f} - 1\right) e^{-\gamma t}\right] \Big/ \left(e^{\gamma t_f} - e^{-\gamma t_f}\right). \quad (34)$$

The approximation appears to be good for t_f large and/or γ large.

This analysis of a simplified two-state-variable case suggests that, when the Legendre-Clebsch condition for the reduced problem is met in strengthened form at the endpoints, initial values for boundary-layer equations should be chosen to suppress unstable solution components, if possible. The case of vector x and y exhibits more complexities; it is explored for linear differential equations and quadratic index by O'Malley [11].

C. An Attitude Control Example

The equations of motion for planar vacuum flight of a rocket vehicle are

$$\varepsilon\dot{\omega} = \frac{\eta}{I} \quad (35)$$

$$\varepsilon\dot{\theta} = \omega \quad (36)$$

$$\dot{u} = \frac{T}{m}\sin\theta + Y \quad (37)$$

$$\dot{v} = \frac{T}{m}\cos\theta + X \quad (38)$$

$$\dot{y} = u \quad (39)$$

$$\dot{x} = v \quad (40)$$

where u, v are velocity components, y, x position components, T thrust magnitude, m mass, θ thrust attitude angle, Y, X gravitational force components, ω angular rate, I moment of inertia, and η control torque. The interpolation parameter ε has been placed so as to rid the system of attitude dynamics for $\varepsilon = 0$. Such ruthless order-reductions abound in the engineering literature; they correspond to assumptions, often not stated explicitly, that certain motions take place on faster time scales than others. Given an optimized trajectory for the reduced system, the attitude control is then to follow it as closely as possible in some sense. The singular perturbation approach leads to a definite criterion, as will be seen.

The Euler equations corresponding to the complete model are

$$\varepsilon\dot{\lambda}_\omega = -\lambda_\theta \tag{41}$$

$$\varepsilon\dot{\lambda}_\theta = -\frac{T}{m}(\lambda_u \cos\theta - \lambda_v \sin\theta) \tag{42}$$

$$\dot{\lambda}_u = -\lambda_y \tag{43}$$

$$\dot{\lambda}_v = -\lambda_x \tag{44}$$

$$\dot{\lambda}_y = -\lambda_u Y_y - \lambda_v X_y \tag{45}$$

$$\dot{\lambda}_x = -\lambda_u Y_x - \lambda_v X_x \tag{46}$$

$$\eta = \arg\min_{|\eta| \leq \bar{\eta}} \frac{\lambda_\omega \eta}{I} \tag{47}$$

where, for simplicity, corrective torque η has been taken bounded and cost-free as far as propellant consumption is concerned.

If the attitude control motions are essentially fast transients superimposed on a slowly and smoothly-varying motion of the center of mass, the optimization of the trajectory can be done in good approximation with the point-mass model obtained for $\varepsilon = 0$, and the attitude motion treated as a boundary-layer correction [6]. To this end, the state and Euler equations are transformed as usual to an independent variable $\tau = t - t_0/\varepsilon$ for analysis of the transition from specified initial attitude and attitude rate:

$$\frac{d}{d\tau}\omega^* = \eta^*/I \tag{48}$$

$$\frac{d}{d\tau}\theta^* = \omega^* \tag{49}$$

$$\frac{d}{d\tau}\lambda_\omega^* = -\lambda_\theta^* \tag{50}$$

$$\frac{d}{d\tau}\lambda_\theta^* = -\frac{T}{\bar{m}_0}(\bar{\lambda}_{u_0}\cos\theta^* - \bar{\lambda}_{v_0}\sin\theta^*) \tag{51}$$

$$\eta^* = \arg\min_{|\eta| \leq \bar{\eta}} \frac{\lambda_\omega^*\eta}{I}\,. \tag{52}$$

The optimal point-mass trajectory, denoted by superscribed bars, has the familiar bilinear tangent steering solution for a uniform gravity model, requires numerical optimization for the inverse-square gravity law, but, in any case, has been extensively investigated.

The Euler equations for the boundary-layer transient, denoted by superscribed asterisks, are of low order, being decoupled from the point-mass motion, except as the initial values (denoted by subscript zero) enter the right members. These are the Euler equations for the rigid-body motion alone, with the performance index

$$\frac{T}{\bar{m}_0}\int_0^\infty (\bar{\lambda}_{u_0}\sin\theta^* + \bar{\lambda}_{v_0}\cos\theta^*)d\tau \tag{53}$$

so that one may characterize the transient as one maximizing the time integral of the dot product of the attitude direction vector with the desired thrust direction. It is intuitively reasonable to maximize the projection of the thrust vector upon the desired thrust direction in this sense, and the similarity to the usual integral-square error index, which approximates it for small pointing errors, is interesting.

The initial conditions on attitude angle θ^* and angular rate ω^* are those on θ and ω in the original problem, and the unspecified initial values of λ_θ^* and λ_ω^* are to be determined to minimize (53), as discussed in the preceding section. In the version of the problem considered here for simplicity, viz., control torque η cost-free, the transient is bang-bang, of finite duration, and features a chattering junction to a singular arc along which $\lambda_\omega^* \equiv \lambda_\theta^* \equiv \eta^* \equiv 0$, and the right member of Equation (51) vanishes. A similar result is obtained even if the several complicating factors, which were purposely omitted for clarity, are brought in. For example, the point-mass rocket model may be three-dimensional and throttleable, and have any performance index that fits the Mayer mold.

A main point of the example, apart from illustration of technique, is treatment of rigid-body motions in boundary-layer approximation, i.e., the possibility of separate treatment with only one-way coupling with the point-mass trajectory optimization in terms of a performance index inherited from that preceding calculation. It is hoped that the transparent illustration of attitude transients for powered-rocket flight in vacuum may draw interest toward the derivation of performance indices for more complex situations, e.g., nonplanar atmospheric maneuvers.

D. State Variable Selection [12]

Selection of state variables is next considered; this consists of ordering variables into groups ranging from slowly to rapidly-varying, and by performing transformations to obtain simplified forms of the system of state differential equations. An attractive form for the state equations turns out to be one whose equations of variation are block-triangular. This is suggested by a study of the linear 2×2 case which illustrates preservation of eigenvalues by triangularization. Efforts at synthesizing triangularizing transformations will lead, in the nonlinear case, to uninviting partial differential equations, as will be seen.

Between singular perturbation theory and applications to flight mechanics and control, there is a gap, in that the theory requires the system of differential equations to come equipped with one or more parameters whose vanishing reduces the system order. Such parameters occurred quite naturally in the original applications of the theory (e.g., viscosity in flow problems), but they must be introduced artificially in many, if not most, of the situations presently in mind, i.e., they are parameters merely of interpolation between the original and simplified systems.

The following look at the closely-related problem of choosing variables is directed toward imbuing interpolation parameters with some of the attributes of classic, naturally occurring, parameters, this to be done by structuring the form of the system of differential equations. In the case of linear equations, the form sought makes the interpolation parameters scaling factors on eigenvalue ratios, i.e., as close to naturally-occurring parameters as one could wish. In the general nonlinear case there is no similar interpretation, however, and it turns out that the desired form is not easily attained. But a choice of variables must nonetheless be made and a knowledge of desirable features would seem to be of value.

If some motions are clearly fast in some sense and others are clearly slow, one sorts the variables into two groupings, represented by an x-vector and a y-vector

$$\dot{x} = f(x,y) \tag{54}$$

$$\dot{y} = g(x,y) . \tag{55}$$

A parameter ε, $0 \leq \varepsilon \leq 1$, is then introduced as a factor of the left member of Equation (55) and an expansion in ε pursued, as by the Tihonov/Vasil'eva theory [3] for initial-value problems.

If motions on widely separate time scales are present in the solutions of Equations (54) and (55), but both types of motion appear in the solutions for both variables (i.e., neither is "pure"), a transformation of variables suggests itself. Aesthetics aside, there is reason to expect that purifications brought about by such a measure may enhance the faithfulness of approximation by a composite comprised of only low-order terms in ε; this will be taken up further.

Consider a transformation of variables

$$x = X(u,v) \tag{56}$$

$$y = Y(u,v) . \tag{57}$$

The transformation inverse to Equations (56), (57) may be denoted

$$u = U(x,y) \tag{58}$$

$$v = V(x,y) \tag{59}$$

and the equations of state in the variables u, v are

$$\dot{u} = U_x f + U_y g \tag{60}$$

$$\dot{v} = V_x f + V_y g . \tag{61}$$

One may consider choosing the functions U and V so as to make the right member of Equation (61) independent of u:

$$\frac{\partial}{\partial u}(V_x f + V_y g) = V_{xx}X_u f + V_x f_x X_u + V_{yx}X_u g + V_y g_x X_u + V_{xy}Y_u f + V_x f_y Y_u + V_{yy}Y_u g + V_y g_y Y_u = 0 \qquad (62)$$

leaving aside the question of whether the expression is to vanish identically in u or merely along some reference solution.

The resulting form of the state equations may be loosely termed block-triangular in the vector case, although this is properly descriptive only of the linearized counterpart, the system of equations of variation. The attraction of this form is clear in the linear constant-coefficient case where the eigenvalues of the reduced system obtained for $\varepsilon = 0$ plus those of the boundary-layer system obtained, as usual, by time-scale stretching, $\tau = t/\varepsilon$, and setting $\varepsilon = 0$, are the same as the eigenvalues of the original system. The choice of lower-triangular in the condition (62) is arbitrary, as an upper-triangular form also preserves eigenvalues. Any reduction of coupling between fast and slow variables seems desirable, in that preservation of eigenvalues tends to improve approximation by a composite of reduced and boundary-layer solutions comprised of only low-order terms in ε. With more than two groupings of variables — multiple time scales — the desired form is a certain block-sparseness of the matrix on the right of the linearized state system.

The function U of Equation (58) has been held in reserve for possible uses such as making Equation (60) as nearly independent of v as possible, producing a form nearly block-diagonal in the vector case. However desirable in producing purity of fast and slow motions this might be, it seems more important to simplify the horrendous expression (62) and so, retreating in the face of this formidable partial differential equation, one may take $u = x$, i.e., $U_x = 1$, $U_v = 0$, which implies

$$X_v V_y = 0, \quad Y_v V_y = 1 \tag{63}$$

from which it follows that $X_v = 0$ and

$$\begin{aligned} Y_v &= \frac{1}{V_y}, \quad X_u = \frac{1}{U_x} = 1 \\ Y_u &= - Y_v V_x = - V_x/V_y \end{aligned} \tag{64}$$

This amounts to settling for only one "pure" variable, viz., v; even so, the P.D.E. does not become very tractable.

$$\begin{aligned} &V_y(V_{xx} f + V_{yx} g + V_x f_x + V_y g_x) \\ &\qquad - V_x(V_{xy} f + V_{yy} g + V_x f_y + V_y g_y) = 0 . \end{aligned} \tag{65}$$

Equation (65) is, in fact, too complex a form even to fall into any of the usual P.D.E. classifications. Subscripts denote partial differentiation throughout.

Retrenching further to the linear case for scalar u and v where

$$\begin{bmatrix} u \\ v \end{bmatrix} = A \begin{bmatrix} x \\ y \end{bmatrix} \tag{66}$$

and A has the form

$$A = \begin{bmatrix} 1 & 0 \\ a_{21} & a_{22} \end{bmatrix} \tag{67}$$

and the state system is

$$\begin{bmatrix} \dot{x} \\ \dot{y} \end{bmatrix} = B \begin{bmatrix} x \\ y \end{bmatrix} \tag{68}$$

$$\begin{bmatrix} \dot{u} \\ \dot{v} \end{bmatrix} = ABA^{-1} \begin{bmatrix} u \\ v \end{bmatrix} \tag{69}$$

one obtains

$$C=ABA^{-1}=\begin{bmatrix} b_{11}-\dfrac{b_{12}a_{21}}{a_{22}} & \dfrac{b_{12}}{a_{22}} \\ \left[(a_{21}b_{11}+a_{22}b_{21})-\left(\dfrac{a_{21}}{a_{22}}\right)(a_{21}b_{12}+a_{22}b_{22})\right] & \dfrac{a_{21}b_{12}+a_{22}b_{22}}{a_{22}} \end{bmatrix} \tag{70}$$

The vanishing of the element c_{21} corresponds to the P.D.E. (65) and with

$$\alpha \equiv a_{21}/a_{22} \tag{71}$$

one obtains

$$b_{12}\,\alpha^2 + (b_{22}-b_{11})\alpha - b_{21} = 0\ . \tag{72}$$

Solution of the quadratic Equation (72) yields:

$$\alpha = \frac{(b_{11}-b_{22}) \pm \sqrt{(b_{11}-b_{22})^2 + 4b_{12}\,b_{21}}}{b_{12}}\ . \tag{73}$$

The eigenvalues of the original state system matrix B are given by

$$\lambda = \frac{(b_{11}+b_{22}) \pm \sqrt{(b_{11}+b_{22})^2 - 4(b_{11}b_{22}-b_{21}b_{12})}}{2} \tag{74}$$

which should be noted as having the same discriminant as the expression (73):

$$d = (b_{11}-b_{22})^2 + 4b_{12}b_{21}\ . \tag{75}$$

The discriminant (75) shows the importance of two-way coupling (i.e., $b_{12} \neq 0,\ b_{21} \neq 0$) for when the product $b_{12}b_{21}$ is negative and large in magnitude, the eigenvalues (74) approach one another, becoming eventually equal in magnitude as they become complex. At the same time, it becomes impossible to find a real value for α which will produce the desired triangular form. Thus a decoupling transformation can be found in the 2×2 linear case if the eigenvalues are real, but not if they are complex.

The results obtained for this linear 2×2 case are suggestive in spite of limited applicability. For example, a system of variables arising naturally in the original form of a problem may be recognized as a good candidate for singular perturbations by virtue of weak or nonexistent coupling, if such is indeed the case; conversely, oscillatory behavior in the motion of a second-order system is an indication that efforts at decoupling will likely prove unproductive.

With autonomous systems, a change of independent variable may sometimes be carried out advantageously to one of the x components or to a function of them which is monotonic along the solution. The variable t, having become a dependent variable, sits high in the hierarchy since it appears nowhere in the right members of the differential equations, i.e., there is a zero column in the matrix appearing on the right of the linear counterpart.[1] The temptation to shift independent variable may thus be strong, even if monotonicity is confined

[1] One might, with tongue in cheek, christen variables y and x (or v and u) as state and superstate, respectively. A really élite variable such as that just discussed, which does not enter the right members at all, would then surely qualify as hyperstate, for no matter what gerrymandering of groupings might take place, it is not eligible for demotion ($g = 0$ must be solvable for y). True royalty, of course, has a zero row for an insigne, but such virtue is possessed only by a good solid constant of the motion (ultrastate?)! The point is that one may have become accustomed, after a few years of state equations in standard first-order form, to regarding state components as more or less alike; they are not, and vive les différences!

to the neighborhood of an endpoint; however, in such cases it can be done only for purposes of the boundary-layer computations.

The practical synthesis of transformations favorable for particular applications will doubtless be a matter more of artwork than of partial differential equations. Be that as it may, it seems worth knowing that reduction of coupling must be a main consideration, if not the primary objective, in synthesis. It appears that the overall problem of selection of variables and introduction of interpolation parameters, an important preliminary to singular perturbations, has as yet received insufficient research attention.

III. APPLICATIONS TO AIRCRAFT MANEUVER OPTIMIZATION

The idea of time-scale separation in vehicle dynamics is expounded in an excellent paper by Ashley [13] based upon asymptotic expansion techniques from fluid dynamics and celestial mechanics as reported in references cited in the paper. The same intuitive insight regarding multiple time scales underlies the treatment of the present chapter by singular perturbation theory of differential equations; however, in spite of the close conceptual kinship, there appears to be only limited cross-referencing between the two bodies of literature. The present section formulates the problem of three-dimensional aircraft flight for singular perturbation treatment and illustrates possibilities for decoupling into several lower-order problems. One of the models obtained is the analogue of "energy climbs" [1, 2, 14, 15, 16, 17, 18, 19] in three-dimensional flight.

The use of reduced-order approximation facilitates numerical computations in two ways: the obvious one of reducing the number of multiplier initial values that must be determined simultaneously; and further, by improving the conditioning of the differential equations. That is, the wild, undamped, phugoid-like oscillation characteristic of the complete Euler-state system for lifting atmospheric flight is avoided for the most part, being relegated to boundary-layer or sublayer corrections where it may be no more docile but, at least, can be dealt with separately over shorter lengths of arc. The material presented in the following is mainly from References 4-9.

A. Aircraft Motion on Three Time Scales

The equations of motion for three-dimensional aircraft flight are

$$\varepsilon_2 \dot{h} = V \sin \gamma \tag{76}$$

$$\varepsilon_2 \dot{\gamma} = (g/V)[(L + \varepsilon_2 T\eta \sin \alpha)\cos \mu/(W_0 + \varepsilon_2 \Delta W) - \cos \gamma] \tag{77}$$

$$\varepsilon_1 \dot{E} = \{[(T\eta - D)V + \varepsilon_2 T\eta V(\cos \alpha - 1)]/(W_0 + \varepsilon_2 \Delta W)\} \tag{78}$$

$$\varepsilon_1 \dot{\chi} = gL \sin \mu/V(W_0 + \varepsilon_2 \Delta W)\cos \gamma \tag{79}$$

$$\dot{x} = V \cos \gamma \sin \chi \tag{80}$$

$$\dot{y} = V \cos \gamma \cos \chi \tag{81}$$

$$\Delta \dot{W} = -\eta Q(V,h) . \tag{82}$$

These apply for zero side-force flight over a flat Earth. The right members incorporate, for $\varepsilon_2 = 0$, the assumptions of constant weight and thrust directed along the path. $E \equiv h + V^2/2g$ is specific energy, χ heading angle, γ path angle to horizontal, and μ bank angle. The symbol V for velocity should be regarded as merely convenient shorthand for $V \equiv [2g(E - h)]^{\frac{1}{2}}$. T is full-throttle engine thrust, L lift, D drag, W weight, α angle of attack, η throttle control variable, $0 \leq \eta \leq 1$. The ε dependence of the right members represents little more than a convenience; it imbeds some approximations in the reduced-order system, in the manner of conventional perturbation theory.

The choice of specific energy $E = h + V^2/2g$ as a state variable reflects its relative freedom from phugoid-type motions, i.e., it is more slowly varying than either V or h. A synthesis of specific energy as a state variable via a transformation scheme in the same general spirit as that of the preceding section, but employing a further approximation of level-flight drag evaluation, is presented in Reference 23.

According to the considerations on choice of variables taken up earlier, there would seem to be an advantage in interchanging time t and some other variable as dependent and independent variables. This is the case if time does not appear explicitly, usually true in aircraft flight problems, and the new independent variable does. Of all the possibilities, E would appear most attractive, since it is once-differentiable sectionally; however, it is not monotonic, and hence does not suit the purpose. The best energy can perform as an independent variable is a rôle in the sublayer. This has not been investigated in the following, time being retained throughout, but it is thought worthwhile as a future research item.

Inequality constraints of the problem are

$$\beta_1 = h - h_T \geq 0 \qquad \text{(terrain limit)} \tag{83}$$

$$\beta_2 = \bar{q} - g\rho(E - h) \geq 0 \qquad \text{(dynamic pressure limit)} \tag{84}$$

$$\beta_3 = \bar{M} - M \geq 0 \qquad \text{(Mach limit)} \tag{85}$$

$$\beta_4 = \hat{C}_L(M) - C_L \geq 0 \qquad \text{(aerodynamic lift coefficient limit)} \tag{86}$$

$$\beta_5 = (\bar{n}W/qS) - C_L \geq 0 \qquad \text{(normal load factor limit on lift coefficient)} . \tag{87}$$

The first three constraints are depicted by the boundaries shown in the sketch of Figure 1.

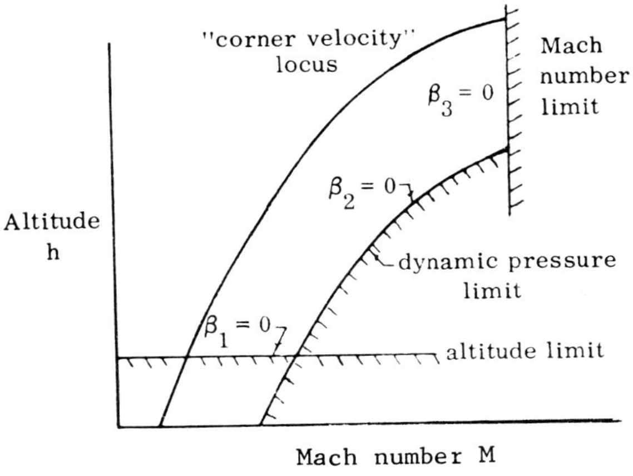

Fig. 1. Mach Number-Altitude Chart Showing Boundaries and "Corner Velocity" Locus.

The Euler equations are

$$\varepsilon_2 \dot{\lambda}_h = - \partial\tilde{H}/\partial h \tag{88}$$

$$\varepsilon_2 \dot{\lambda}_\gamma = - \partial\tilde{H}/\partial \gamma \tag{89}$$

$$\varepsilon_1 \dot{\lambda}_E = - \partial\tilde{H}/\partial E \tag{90}$$

$$\varepsilon_1 \dot{\lambda}_\chi = - \partial\tilde{H}/\partial \chi \tag{91}$$

$$\dot{\lambda}_x = - \partial\tilde{H}/\partial x \tag{92}$$

$$\dot{\lambda}_y = - \partial\tilde{H}/\partial y \tag{93}$$

$$\dot{\lambda}_W = - \partial\tilde{H}/\partial W \tag{94}$$

$$\partial H/\partial \alpha = 0 , \quad \partial H/\partial \mu = 0 \tag{95}$$

where the Hamiltonian function H is given by

$$H \equiv \sum_{j=1}^{n} \lambda_j f_j \tag{96}$$

the f_j being the right members of the state equations (76)-(82), and the augmented Hamiltonian $\tilde{H}$ defined by

$$\tilde{H} = H + \sum_{i=1}^{5} \lambda_i \beta_i \tag{97}$$

The terms of the summation correspond to treatment of the inequality constraints by the technique of Valentine.

The state system has been chosen to illustrate decoupling of motions on three time scales. With $\varepsilon_2 = 0$, the relatively fast kinetic-potential energy interchange is discarded in favor of a model featuring instantaneous interchange. When, in addition, $\varepsilon_1 = 0$, then even the slower processes of heading and total-energy change become instantaneous, yielding a highly-simplified rectilinear motion model. To satisfy the requirements of the Tihonov theory that $\varepsilon_2/\varepsilon_1 \to 0$ as $\varepsilon_1 \to 0$, a single interpolation parameter ε could be employed and ε_1 taken as ε, ε_2 as ε^2. If rigid-body short-period motion were to be included in the model, a third parameter $\varepsilon_3 = \varepsilon^3$ would be introduced.

In high-order systems, such as the aircraft model at hand, separation into more than two groups by the introduction of additional parameters will often be appropriate. The grouping adopted in the preceding equations seems fairly natural with, perhaps, a slight arbitrariness in the common time-scale choice for heading and total-energy changes. It might be appropriate for some situations, and/or as a possible gross simplification, to idealize total-energy changes as fast compared to heading changes, thus effecting still further mathematical separation; this was pursued in the study of Reference 8 and will be reviewed in the following.

B. Dash and Cruise Arcs as Reduced Solutions

Consider first the problem of minimum-time motion between end states fixed except for fuel expenditure open. One begins by solving the optimization problem for the reduced system, $\varepsilon_1 = \varepsilon_2 = 0$ (i.e., the rectilinear motion model), then performs corrections for the initial and terminal transients in boundary-layer approximation. One obtains

$$\sin \bar{\gamma} = 0, \quad \cos \bar{\gamma} = 1 \tag{98}$$

$$L = W_0 \tag{99}$$

$$\bar{T}\eta = D \tag{100}$$

$$\sin \bar{\mu} = 0, \quad \cos \bar{\mu} = 1 \tag{101}$$

$$\bar{x} = x_0 + t\,\bar{V} \sin \bar{\chi} \tag{102}$$

$$\bar{y} = y_0 + t\,\bar{V} \cos \bar{\chi} \,. \tag{103}$$

Here $\bar{T}$ is full-throttle thrust; superscribed bars generally denote the solution of the reduced state-Euler system. The multipliers $\bar{\lambda}_x$ and $\bar{\lambda}_y$ are constant and $\bar{\lambda}_W = 0$. $\bar{E}$, $\bar{h}$, $\bar{\chi}$ minimize H subject to Equations (98)-(101). This is an exercise in determining the steady-state level-flight performance envelope and in selecting the maximum level-velocity point. The multipliers $\bar{\lambda}_x$ and $\bar{\lambda}_y$ are chosen so that $\bar{\chi}$ is the proper constant heading to reach the specified terminal point; they are scaled in magnitude according to the transversality condition $\bar{H} = -1$.

Other problems such as minimum fuel, maximum range, etc., exhibiting rectilinear motion over a central portion of the path, can be treated similarly. Attention is now directed to the transitions at either end.

C. Boundary-Layer System: Energy-Heading Transient

Consider transition to and from the rectilinear motion in boundary-layer approximation. Near the initial point, a stretched time scale is introduced via the adoption of a time variable $\tau = t/\varepsilon_1$. For small τ, one obtains a reduction in the order of the state-Euler system for $\varepsilon_1 = 0$ by virtue of $dx/d\tau = dy/d\tau = dW/d\tau = d\lambda_x/d\tau = d\lambda_y/d\tau = d\lambda_W/d\tau = 0$. One is thus led to an energy model for optimal transitional flight.

$$V \sin \gamma = 0 \tag{104}$$

$$(g/V)[L \cos \mu/W_0 - \cos \gamma] = 0 \tag{105}$$

$$dE/d\tau = V(T\eta - D^*)/W_0 \tag{106}$$

$$d\chi/d\tau = gL \sin \mu/VW_0 \tag{107}$$

$$\begin{aligned} \frac{d}{d\tau}\lambda_E = &-\frac{\partial}{\partial E}\left[\lambda_E \frac{V(T\eta - D^*)}{W_0} + \lambda_\chi \frac{gL^* \sin \mu}{VW_0}\right] - \sum_{i=1}^{5} \lambda_i \frac{\partial \beta_i}{\partial E} \\ &-\frac{\partial}{\partial E}\{\overline{\lambda}_x V \sin \chi + \overline{\lambda}_y V \cos \chi - \overline{\lambda}_W \eta Q\} \end{aligned} \tag{108}$$

$$(d/d\tau)\lambda_\chi = -(\partial/\partial\chi)\{\overline{\lambda}_x V \sin \chi + \overline{\lambda}_y V \cos \chi\} \tag{109}$$

$$\partial H^*/\partial\mu = 0 \tag{110}$$

$$\partial H^*/\partial h = 0 \tag{111}$$

where

$$\begin{aligned} H^* \equiv {} & \lambda_E(T\eta - D^*)V/W_0 + \lambda_\chi gL^* \sin \mu/VW_0 + \sum_{i=1}^{5} \lambda_i\beta_i \\ & + \{\overline{\lambda}_x V \sin \chi + \overline{\lambda}_y V \cos \chi - \overline{\lambda}_W \eta Q\}\,. \end{aligned} \tag{112}$$

For simplicity, it has been elected to account for Equations (104) and (105) in the remaining equations by substitution of $\gamma = 0$ and evaluating drag D for $L = L^* = W_0/\cos \mu$, designating it D^*, rather than to employ multipliers λ_h and λ_γ.

The multipliers $\overline{\lambda}_x$, $\overline{\lambda}_y$, $\overline{\lambda}_W$ are known constants from the optimal steady-flight solution on the reduced system. When they are not constant, as would be the case, for example, if winds aloft were included in the reduced-system model, the initial values would appear in the boundary-layer state-Euler system. These are, nonetheless, merely known constants for purposes of the boundary-layer calculation.

The terms set off in braces, $\{\cdot\}$, for ready identification, correspond to a performance index

$$\int_0^\infty \{\overline{\lambda}_x V \sin \chi + \overline{\lambda}_y V \cos \chi - \overline{\lambda}_W \eta Q\} d\tau . \qquad (113)$$

The reduced system furnishes the form of this index, and its numerical solution the required initial values of the barred quantities appearing in it. Variables satisfying the boundary-layer equations will be denoted by superscript $(*)$.

Initial values of multipliers not determined by transversality conditions must be chosen to suppress unstable transients, i.e., to attain values of E and χ for large τ equal to the initial values of the reduced system (the dash values) as these make the inhomogeneous terms in the differential equations for the multipliers λ_E and λ_χ vanish.

Drag is approximated as quadratic in C_L:

$$D = qS\{C_{D_0} + C_{D_B}(1-\eta) + C_{D_{C_L^2}}[C_L^2\eta + \overline{C}_L^2(1-\eta)]\} \qquad (114)$$

in which the C_{D_B} term accounts for speed-brake incremental drag. The $\overline{C}_L$ term arises from "relaxation." That is, it makes convex the "hodograph figure" traced out by the controls in the space $\dot{E}$, $\dot{\gamma}$, $\dot{\chi}$, admitting high drag equal to that approachable via "chattering" between C_L bounds with sectionally-continuous control histories [4]. While the variable η, previously introduced as engine throttle control variable, plays a part in this, its use is fortuitous in the sense that an appropriate similar variable would be introduced solely for relaxation purposes even for flight of an unpowered vehicle (e.g., a re-entry glider); a separate variable will generally be needed in minimum-fuel or bounded-fuel studies. The lift-coefficient bounds are determined for the boundary-layer system by the forms

$$\beta_4 = \cos\mu - \frac{W}{qS\hat{C}_L} \geq 0 \tag{115}$$

$$\beta_5 = \cos\mu - \cos\bar{\mu} \geq 0 \tag{116}$$

taken by the applicable constraints in conjunction with $L = W_0/\cos\mu$. It is worth noting that the lift-coefficient bound $\bar{C}_L$ depends differently on E and h, depending which of the two inequalities determines it and, particularly, that the first partial derivatives of the relaxed drag exhibit jump discontinuities across the boundary of transition in E, h space.

The control variable μ is given for positive η by the smallest in magnitude of the three quantities

$$\arctan\frac{\lambda_\chi gSq}{2\lambda_E WV^2\eta C_{D_{C_L^2}}}, \quad \arccos\frac{W}{qS\bar{C}_L}, \quad \bar{\mu}. \tag{117}$$

The first furnishes a minimum of H in the absence of constraints [4], while the second and third are threshhold values for the constraints β_4 and β_5. For $\eta = 0$, only the second and third candidates must be considered, as the relaxed drag model exhibits no interior minimum.

For the case of unspecified fuel expenditure, $\overline{\lambda}_W = 0$, the multiplier, λ_E, assumes the rôle of throttle "switching function," i.e., the operation min H yields

$$\eta = 0 \quad \text{if} \quad \lambda_E > 0 \tag{118}$$

$$\eta = 1 \quad \text{if} \quad \lambda_E < 0 . \tag{119}$$

The case $\lambda_E = 0$ is singular and occurs at switchings between throttle bounds. Singular arcs along which $\lambda_E = 0$ over a nonzero time interval appear with some system models but not with others. For example, they appear in the constant-altitude minimum-fuel case of Reference 20 but neither in the constant-altitude minimum-time case of Reference 21 nor in the variable-altitude minimum-time case of Reference 22 in which normal load factor limit was not included. Singular arcs corresponding to intermediate throttle operation may best be determined by noting that the state system is already in the canonical form of Reference 23, discarding the $\dot{E}$ state equation and finding the extremals of the resulting state system with E treated as a control variable. With the present model, this leads to steady maximum-lift turning at the intersection of the constraints β_1, β_4, and β_5. Unusually high thrust is required for equilibrium in such a turn and, in fact, the hypothetical Mach 3 aircraft of our example does not possess it.

The altitude that minimizes H subject only to the constraints $\beta_4 \geq 0$ and $\beta_5 \geq 0$ is found by computational one-dimensional search, e.g., the routine of Reference 24. The values of h corresponding to the vanishing of β_1, β_2, and β_3 are termed h_1, h_2, and h_3. They are compared to each other and to the value furnishing an interior minimum of H, if one exists, which is called h_{min}; the largest of the four values is then selected.

When the minimizing h makes β_1, β_2, or β_3 vanish, but not β_4 or β_5, the corresponding multiplier is calculated according to the Valentine procedure (from $\partial\tilde{H}/\partial h = 0$) as

$$\lambda_i = -\frac{\dfrac{\partial H}{\partial h}}{\dfrac{\partial \beta_i}{\partial h}} . \tag{120}$$

The other two multipliers will be zero except at transition points characterized by two of the β_i, $i = 1,2,$ or 3, vanishing simultaneously.

Similar considerations apply to β_4 and β_5, which depend explicitly on the control μ. From $\partial\tilde{H}/\partial\mu = 0$,

$$\lambda_4 = -\frac{\dfrac{\partial H}{\partial \mu}}{\dfrac{\partial \beta_4}{\partial \mu}} \tag{121}$$

if $\beta_4 = 0$ and $\lambda_5 = 0$. If, however, $\beta_5 = 0$ applies simultaneously, then

$$\lambda_4 = -\frac{\dfrac{\partial H}{\partial h}}{\dfrac{\partial \beta_4}{\partial h}} \tag{122}$$

and $\lambda_5 \neq 0$. Since λ_5 does not enter the Euler system, it need not be calculated. Altitude-airspeed combinations at which maximum C_L ($\beta_4 = 0$) and maximum normal load factor ($\beta_5 = 0$) coincide define a "corner velocity locus" as sketched in Figure 1 [25,26]. The term is not of variational origin, but rather refers to the discontinuity in slope appearing in maximum normal load factor as a function of velocity at the coincidence of the aerodynamic and structural limits, as shown in the sketch of Figure 2.

It is worth noting that the minimum of H as a function

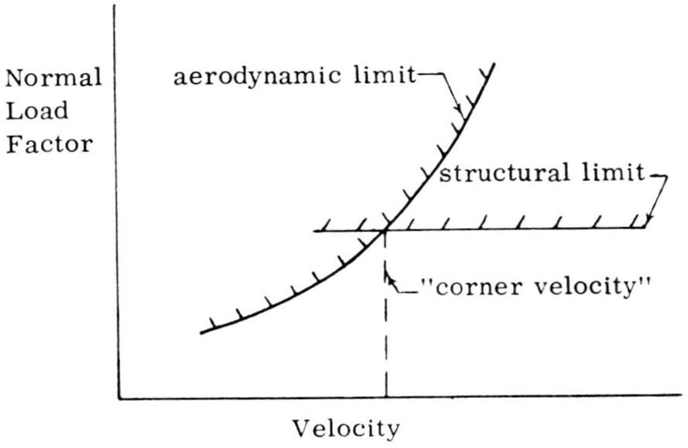

Fig. 2. Aerodynamic and Structural Limits.

of h to be located by numerical search will often be V-shaped, i.e., the slope dH/dh negative to the left and positive to the right, when it occurs on the corner velocity locus (see sketch of Figure 3). The "total" derivative symbol for this

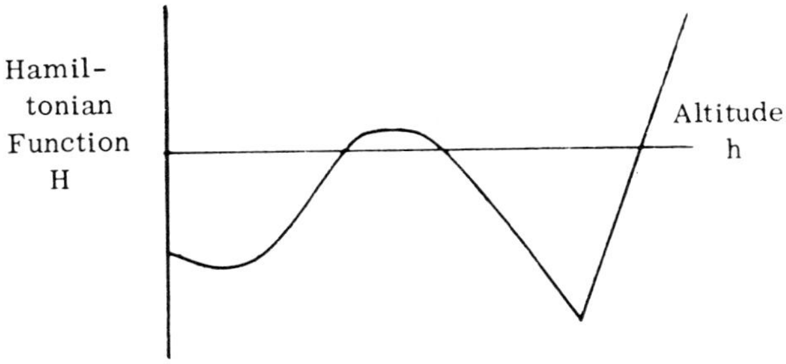

Fig. 3. Hamiltonian Versus Altitude.

slope is adopted for the purpose of the moment to mean:

$$\frac{dH}{dh} = \frac{\partial H}{\partial h} + \frac{\partial H}{\partial \mu}\frac{\partial \mu}{\partial h} \tag{123}$$

and the jump of interest at the value of h of the corner velocity locus is occasioned by a shift in the $\partial\mu/\partial h$ factor of the second member between the values appropriate to the aerodynamic and load factor limits.

It has been found helpful to evaluate computationally the left-hand and right-hand derivatives at $\beta_4 = \beta_5 = 0$ as an aid in locating the minimum. Note also that the sometime appearance of two minima of $H(h)$ denotes nonconvexity, a need for further relaxation, and an outside chance of h-control chattering singular arcs (!); this has not been pursued. Since the one-dimensional search routine locates only one of the minima, it ought to be biased by choice of initial guess of a low altitude, away from the V-shaped minimum; this minimum can be separately examined by evaluation of H and its derivatives on the corner velocity locus.

D. Sublayer System: Energy Interchange

Because the left members of Equations (76), (77), (88), and (89) never entered the generation of the system Equations (104)-(112), there is a mismatch of that system's solution with specified initial values. Transitions in sub-boundary-layer approximation are required. These are described by a further stretching of scale, $T = \varepsilon_1\tau/\varepsilon_2$, which leads, for $\varepsilon_2 = 0$, to $dE/dT = d\chi/dT = d\lambda_E/dT = d\lambda_\chi/dT = 0$, and to the system

$$dh/dT = V^* \sin\gamma \tag{124}$$

$$d\gamma/dT = (g/V^*)[(L\cos\mu/W_0) - \cos\gamma] \tag{125}$$

$$\partial\hat{H}/\partial C_L = 0 \tag{126}$$

$$\partial\hat{H}/\partial\mu = 0\,. \tag{127}$$

Here

$$V^* \equiv [2g(E_0^* - h)]^{\frac{1}{2}} \tag{128}$$

and

$$\hat{H} \equiv \left[\lambda_h V^* \sin\gamma + \lambda_\gamma \frac{g}{V^*} \left(\frac{L \cos\mu}{W_0} - \cos\gamma \right) \right] + \sum_{i=1}^{5} \lambda_i \beta_i$$

$$+ \left\{ \lambda^*_{E_0} \frac{V^*(T\eta - D)}{W_0} + \lambda^*_{\chi_0} \frac{gL \sin\mu}{V^* W_0 \cos\gamma} \right\}. \qquad (129)$$

The terms set off in braces, $\{\cdot\}$, again for ready identification, correspond to a performance index

$$\int_0^\infty \left\{ \lambda^*_{E_0} \frac{V^*(T\eta - D)}{W_0} + \lambda^*_{\chi_0} \frac{gL \sin\mu}{V^* W_0 \cos\gamma} \right\} dT \qquad (130)$$

for the sublayer optimization, the quantities subscripted zero being the initial values of the boundary-layer solution.

The inequality constraint functions β_i appearing in the summation are those in the original form (83)-(87). Since the first three inequalities are essentially state inequalities for the boundary-layer problem, the extended form of the Valentine device ("super-Valentine") due to Jacobson and Lele [27] may be appropriate, especially if an "exact" solution is needed; if not, an integral penalty function treatment [28] offers an alternative.

Bank angle μ is determined by

$$\tan\mu = -\frac{\lambda^*_{\chi_0}}{\lambda_\gamma \cos\gamma} \qquad (131)$$

and lift coefficient C_L by the least of

$$C_{L_{min}} = \frac{g\left(\lambda^*_{\chi_0} \dfrac{\sin\mu}{\cos\gamma} + \lambda_\gamma \cos\mu\right)}{2\eta^* \lambda^*_{E_0} V^{*2} C_{D_{C_L^2}}} \qquad (132)$$

and the two bounds $\hat{C}_L$ and $\frac{\overline{n}W}{qS}$. Should $\eta^* = 0$, only the bounds remain as candidates and the lesser of the two obtains.

Where the indirect method is to be used for computational solution of the sublayer system, the values of λ_h and λ_γ taken on at the initial point of the boundary-layer solution are useful as first estimates for an iteration process, and these are given by $\lambda^*_{h_0} = 0$ and

$$\lambda^*_{\gamma_0} = \frac{2W_0 \lambda^*_{E_0} \eta^* V^{*2} C_{D_{C_L^2}}}{gq^*S \cos^2\mu^*} - \lambda^*_\chi \tan \mu^* . \tag{133}$$

E. Instantaneously-Variable-Speed Model

In the preceding development, heading changes and energy changes were presumed to take place on the same time scale in the second of three groupings. It is of interest, as a gross approximation, to idealize energy changes as fast compared to heading changes; this results in a reduced-system model intermediate in complexity between the energy-state model and the popular constant-speed model. The introduction of interpolation parameters must be done differently and, in the context of the preceding development, there might be three time scales assumed, with heading change on the same scale as horizontal displacements; the ε_1 factor on the left member of Equation (79) would then simply be missing. The consequences of this last modelling assumption will be pursued briefly in the present section, at least as far as the characteristics of the reduced system are concerned. The reduced system is:

$$0 = \frac{V(T\eta - D)}{W} \tag{134}$$

$$\dot{\chi} = \frac{gL \sin \mu}{VW} \tag{135}$$

$$\dot{x} = V \sin \chi \tag{136}$$

$$\dot{y} = V \cos \chi . \tag{137}$$

The condition (134) represents a constraint on the four control variables h, μ, η, E. Note that the variable E, and hence speed V, can now jump! Two control-variable combinations are of particular interest: the maximum steady-speed point and the point of greatest-level-turning rate at which energy can be maintained. This last will usually be attained in high-lift-coefficient, low-altitude flight. It may correspond to maximum lift coefficient or maximum load factor, or it may be an "interior" extremum, depending upon the configuration, propulsion system, etc.; for a turbojet, this best turning circle will generally occur at sea level subsonically or transonically.

Consider the model given by:

$$\dot{\chi} = \bar{\omega}\sigma \quad ; \quad -1 \leq \sigma \leq 1 \tag{138}$$

$$\dot{x} = V(\sigma) \sin \chi \tag{139}$$

$$\dot{y} = V(\sigma) \cos \chi . \tag{140}$$

If $V(\sigma)$ is represented by

$$V(\sigma) = \bar{V} - (\bar{V} - v)|\sigma| \tag{141}$$

this corresponds to linear interpolation with respect to turn rate between maximum level-flight speed $\bar{V}$ and the speed v at maximum turning rate $\bar{\omega}$. In some cases, it might roughly approximate part of the convex hull of a "relaxed" variational problem, but perhaps it might best be regarded merely as a sweeping assumption, not too inconsistent with the already sweeping assumption of instantaneous jumps in speed.

The V, ω hodograph figure for the Mach 3 aircraft of Reference 29 is illustrated in Figure 4. It happens that, for this particular aircraft/propulsion system combination, turning

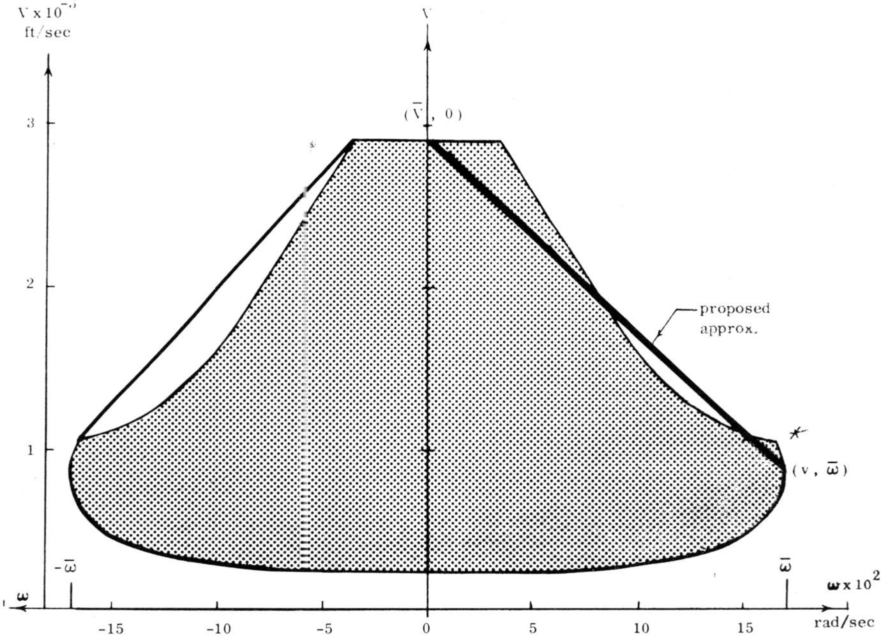

Fig. 4. Velocity-Turn Rate Hodograph Figure.

rate is limited by the power available to maintain energy at almost all airspeed-altitude combinations rather than by normal load factor or lift-coefficient limits. The over-pressure constraint limits altitude choice at speeds above the point marked with an asterisk. The vee-shaped approximation of Equation (141) (right side of figure) is seen to bear some resemblance to the convex hull (left side). The tailoring of the approximation neglects the possibility of flight at low speeds $V < v$ which is rarely of interest in mission-shaping applications. The nonconvexity of the approximation itself also rules out its use where low-speed arcs enter the solution.

A somewhat more satisfactory model featuring two control variables, σ and V, is given by

$$\dot{\chi} = \omega(V)\sigma, \qquad -1 \leq \sigma \leq 1 \tag{142}$$

$$\dot{x} = V \sin \chi \tag{143}$$

$$\dot{y} = V \cos \chi . \tag{144}$$

With appropriate bounds on V and the characteristic defined by the boundary of the shaded region for $\omega(V)$ adopted, a model faithful at least in a certain respect may be obtained; with <u>linear</u> $\omega(V)$ and bounds $\overline{V}$ and v on V, a simple model exhibiting convexity is available. It can be shown that, with linear $\omega(V)$, the solution of the Mayer problem for the single-vehicle third-order vehicular-motion model contains no arcs with V intermediate between bounds or turn rate intermediate in the sense $0 < |\sigma| < 1$.

Sectionally linear $V(\sigma)$ models offer obvious advantages for mission analyses: simple performance characterization in terms of three turn-and-dash parameters, plus paths represented by a neat, predictable sequence of circular-arc turns and straight-line dashes. The attendant drawbacks are the unrealistic velocity jumps, and their appearance suggests that the model is quite an approximate one — of interest mainly as a more realistic alternative to a constant-speed model for analyses in which it is important to hold the order to three on account of other complicating factors.

IV. COMPUTATIONAL RESULTS

The configuration, aerodynamic, and propulsion system data employed for the computations are those of the hypothetical Mach 3 aircraft of Reference 29 and the preceding chapter. These data are generally fairly realistic; however, the drag is taken quadratic with angle of attack, speed-brake drag increment a constant fraction of zero-lift drag, and maximum C_L assumed independent of Mach number, all in the interest of simplifying the example computations somewhat.

Computations were performed of a family of "energy turns" of this aircraft, i.e., minimum-time turns from specified initial heading and specific energy to various specified terminal headings and terminal energies, with horizontal position components open $(\overline{\lambda}_x = \overline{\lambda}_y = 0)$. The initial energy for the family was taken as 170,000 ft., the initial heading angle and time as zero. The value of the ratio of multiplier initial values is the single parameter of the family, which yields to computational survey treatment. The system is solved simultaneously by fourth-order Runge-Kutta integration proceeding forward in time. Minimization of the Hamiltonian H with respect to bank angle μ and throttle variable η is carried out in closed form, as in the preceding analysis, while the minimization with respect to altitude h is done by a numerical search routine.

Figures 5, 6, and 7 present time histories of energy, heading angle, and altitude, while Figure 8 shows the same family of optimal maneuvers in the Mach number-altitude chart. Also shown in Figure 8 is the symmetric "energy climb" path computed without regard to the dynamic pressure limitation $\beta_2 \geq 0$; it is seen to violate it grossly due to high wing loading and other system characteristics. With the multiplier λ_E initially negative, the trajectories are full-throttle turns of varying sharpness, the milder maneuvers taking place entirely at the minimum altitude allowable by the constraint β_i, $i = 1,2$, or 3. The members of the full-throttle portion of the family that develop maximum normal load factor initially spend brief periods at higher altitudes but quickly come to the minimum-altitude boundary. The more extreme of these ride down the corner-velocity locus briefly [cf. 25]. Two full-throttle turn histories have been computed with the dynamic pressure limitation $\beta_2 \geq 0$ removed (Figure 8), even though

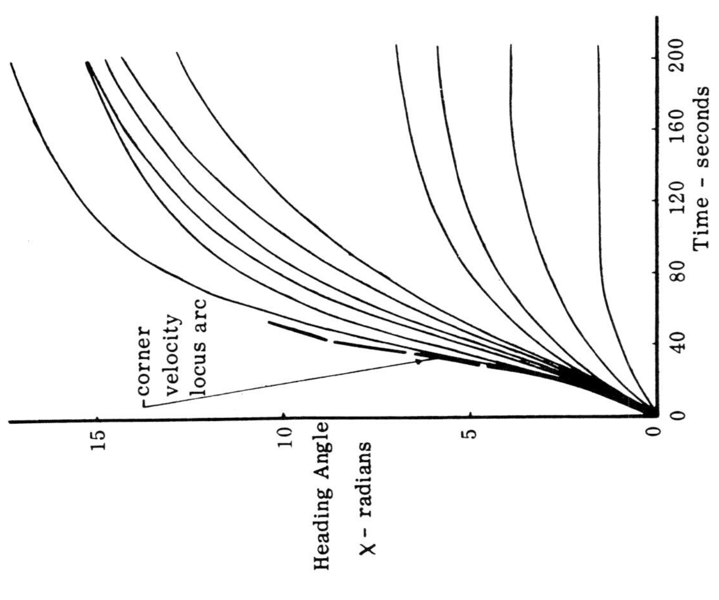

Fig. 6. Heading Histories in Optimal Turns.

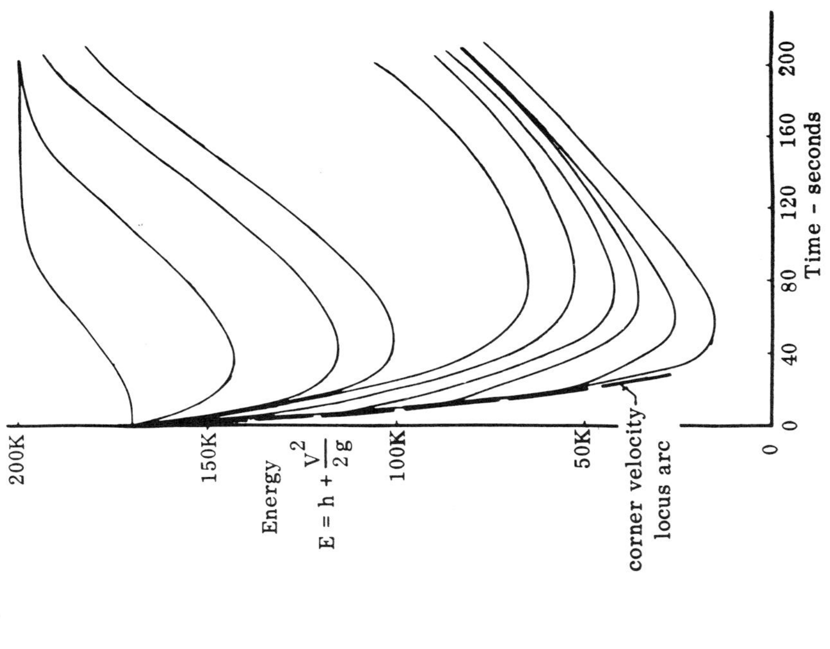

Fig. 5. Energy Histories in Optimal Turns.

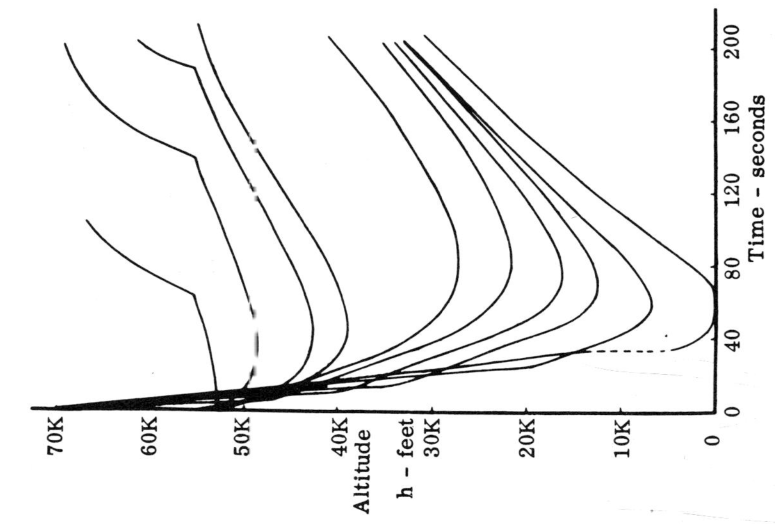

Fig. 7. Altitude Histories in Optimal Turns.

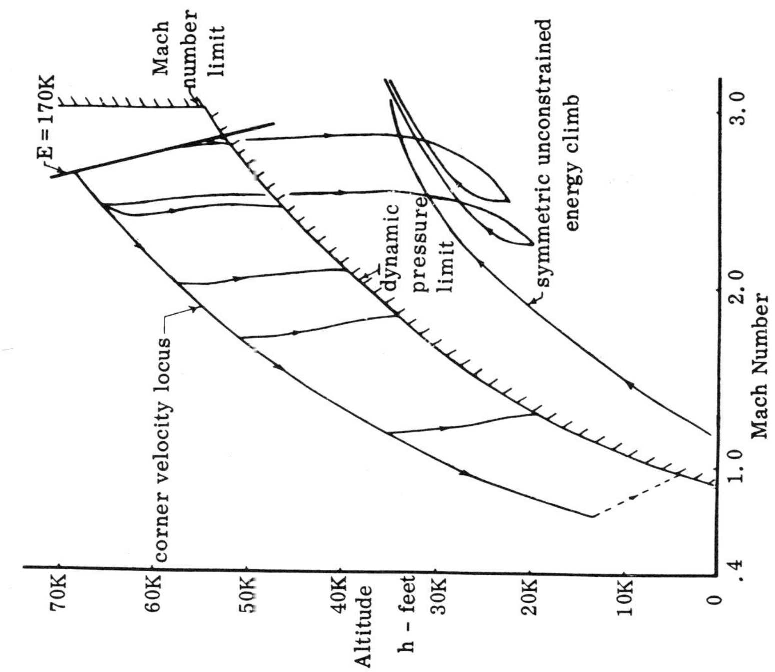

Fig. 8. Optimal Turns in the Mach-Altitude Chart.

the pressures met are unrealistically high for design purposes — 3000 lb/ft^2! Except for the initial transient, these optimal maneuvers fall below the symmetric energy climb path; this behavior is predicted by the Boyd/Christie instantaneous E versus $\dot{\chi}$ trade-off analysis [30]. More accessible flight regimes exhibit such patterns for other aircraft designs, and this "maximum-maneuver corridor" [30] then assumes the central rôle played by the dynamic-pressure placard boundary in the present example.

With λ_E initially positive, all trajectories start with throttle closed, $\eta = 0$, and speed brakes extended, and they ride the corner-velocity locus for various periods. Departure from the corner-velocity locus is preceded by switching of the throttle to open, $\eta = 1$, as a result of λ_E sign change. Speed brakes have been assumed actuated by the throttle in this study. The trajectories subsequently behave like those of the full-throttle part of the family. Deliberate, if temporary, reduction in energy so as to attain higher turn-rates is a hallmark of optimal turn maneuvers; the tendency will no doubt be accentuated with future high thrust/weight designs which permit fast recovery of energy.

The smooth, if rapid, transition from corner velocity to dynamic-pressure-limit characteristic of the supersonic range becomes an instantaneous constant-energy transition in altitude in the transonic range This is a variational "corner" not unlike a similar jump occurring in symmetric energy climbs when two branches are present. This is depicted in Figure 8 by a dashed constant-energy arc.

Optimal turns of extended duration tend to develop a central portion that approaches turning at the highest rate steadily sustainable within the flight envelope. For the example aircraft, this is approximately 0.15 rad/sec at Mach

0.9 and sea level, E approximately 15,000 ft. One turn history remaining in this Mach-altitude range for a brief central portion of its history is shown in the figures. Extended turning maneuvers are of particular interest in connection with the classical circling encounter, in which turn capability and execution decides which aircraft becomes the pursuer and which the evader.

A. Sublayer Computations

The sublayer differential equations presented in a previous section were solved numerically in the spirit of the indirect method, i.e., using difference quotient (so-called "secant") approximation partial derivatives of various terminal quantities with respect to initial multiplier values and an iteration scheme to meet the desired terminal conditions at a chosen (finite) final time. The iteration consisted of minimization of the integral (130) plus terminal altitude and path angle quadratic penalty terms by a modification of the Davidon-Fletcher-Powell algorithm described in Reference 31. While this would seem a reasonable approach on account of the low dimension of the problem, the well-known ill-conditioning of the Euler system for lifting flight offered the usual obstacles and the computations, while successful, were not altogether straightforward.

Solutions shown in Figure 9 correspond to initial transitions to one of the energy-turn trajectories from five thousand feet above and from five thousand feet below. The trajectory chosen is a full-throttle example beginning on the corner-velocity locus at E = 170K and remaining on it for some time, with bank angle constant at the value for limit load-factor.

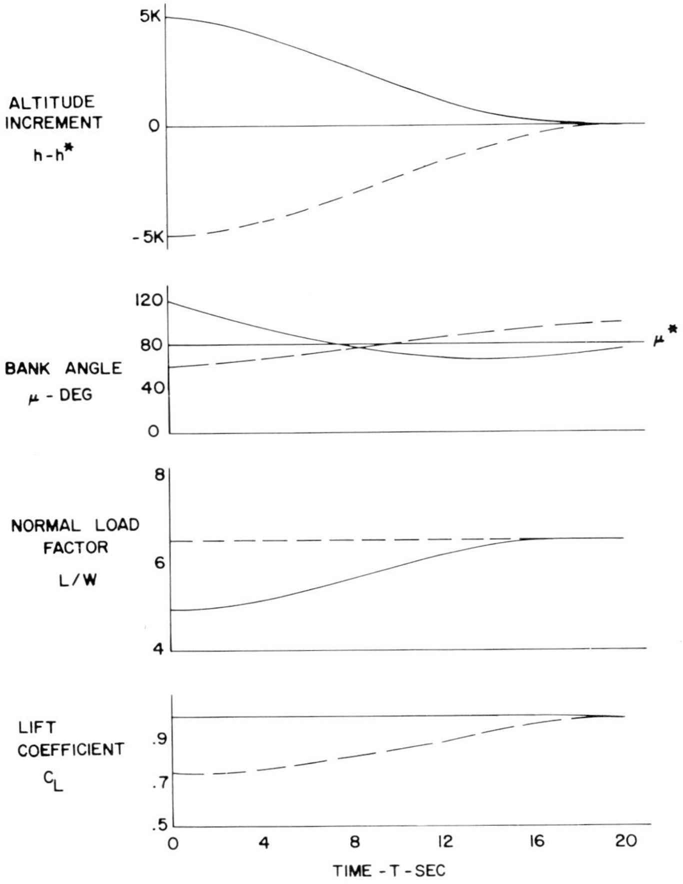

Fig. 9. Initial Transition to Full-Throttle Corner-Velocity Extremal in Sublayer Approximation.

V. CONCLUDING REMARKS

The presently reported effort has focused on singular perturbations of the state-Euler system of differential equations and has concentrated on the optimal aircraft flight application. Both choices tend to accentuate the positive aspects, which may be just as well, in order to draw attention to the possibilities and to encourage further research. On the other hand, the rôle of the various other necessary conditions in a singularly perturbed problem deserves attention if a usable theory and appropriately-tailored computational techniques are to be developed for use in general optimal control and differential game problems. Also, criteria are needed for decision on whether or not a given problem is a good candidate for singular perturbations. Such matters are the stuff of theses and there is obviously much to be done.

For example, the transition between branches in the two-dimensional energy-climb problem takes place in the preceding, as an act of faith, according to the Weierstrass condition for the reduced problem. Whether a jump in altitude at constant specific energy takes place according to this rule or some other, the discontinuity requires an interior boundary-layer transition as a patch. Since certain of the "control variables" of reduced solutions are not the genuine article, weak as well as strong minima should be investigated and so should the consequent multiple covering arising when the Hamiltonian has more than one minimum.

The computation of higher-order corrections in the expansion parameter should be investigated and actually carried out for suitable examples, as a matter of research interest. The complexity of the computations would seem roughly comparable to that of second-order variational methods, but the problems of ill-conditioning and numerical-error propagation ought to be mitigated by the time-scale separation. A higher order correction process for turn-dash modelled aircraft flight is examined in [32].

Use of the various vehicle models examined here in differential gaming is of considerable future interest. Steps in this direction are reported in [33] and [34].

VI. ACKNOWLEDGMENTS

The assistance of Theodore Edelbaum, Andrew Jazwinski, Leon Lefton, Agnes Michalowski, Frank Mummolo, Ann Muzyka,

Samuel Pines, and Saul Serben is gratefully acknowledged, as is the advice and stimulation of Col. John R. Boyd of USAF Systems Command Headquarters, Andrews AFB, in the course of many discussions.

VII. REFERENCES

1. F. KAISER, Der Steigflug mit Strahlflugzeugen-Teil I, Bahngeschwindigkeit besten Steigens, Vefsuchsbericht 262-02-L44, Messerschmitt A. G., Augsburg, April 1944. Translated as Ministry of Supply RTP/TIB, Translation GDC/15/148T.
2. G. LEITMANN, On Optimal Control of Lifting Vehicles, Proc. Fourth U. S. Nat. Cong. Appl. Mech. 1, 243 (1962).
3. W. WASOW, "Asymptotic Expansions for Ordinary Differential Equations," Interscience, New York, 1965.
4. H. J. KELLEY and T. N. EDELBAUM, Energy Climbs, Energy Turns, and Asymptotic Expansions, Jnl. of Aircraft 7, 93 (1970).
5. H. J. KELLEY, Singular Perturbations for a Mayer Variational Problem, AIAA Jnl. 8, 1177 (1970).
6. H. J. KELLEY, Boundary-Layer Approximation to Powered-Flight Attitude Transients, Jnl. of Spacecraft and Rockets 7, 879 (1970).
7. H. J. KELLEY, Flight Path Optimization with Multiple Time Scales, Jnl. of Aircraft 8, 238 (1971).
8. H. J. KELLEY, Reduced-Order Modeling in Aircraft Mission Analysis, AIAA Jnl. 9, 349 (1971).
9. H. J. KELLEY and L. LEFTON, Supersonic Aircraft Energy Turns, presented at Fifth IFAC Congress, Paris, France, June 12-17, 1972; also Automatica, 8, 575 (1972).
10. A. N. TIHONOV, Systems of Differential Equations Containing Small Parameters in the Derivatives, Matematichiskii

Sbornik 31, (1952). (English translation by A. Muzyka, Dept. of Transportation Systems Center, Cambridge, Mass.)

11. R. E. O'MALLEY, Singular Perturbation of the Time-Invariant Linear State Regulator Problem, J. Differential Equations, (1972).

12. H. J. KELLEY, State Variable Selection and Singular Perturbations, AACC invited session, 1972 Joint Automatic Control Conference, Stanford Univ., Palo Alto, Calif., August 16-18, 1972; in "Singular Perturbations: Order Reduction in Control System Design," American Society of Mechanical Engineers, New York, 1972.

13. H. ASHLEY, Multiple Scaling in Flight Vehicle Dynamic Analysis — A Preliminary Look, AIAA Guidance, Control and Flight Dynamics Conference, Huntsville, Ala., 1967.

14. K. J. LUSH, A Review of the Problem of Choosing a Climb Technique with Proposals for a New Climb Technique for High Performance Aircraft, Aeronautical Research Council Rept. Memo. 2557, 1951.

15. E. S. RUTOWSKI, Energy Approach to the General Aircraft Performance Problem, Jnl. Aerospace Sciences 21, 187 (1954).

16. A. MIELE, Optimum Climbing Technique for a Rocket-Powered Aircraft, Jet Propulsion 25, (1955).

17. H. HEERMANN and P. KRETSINGER, The Minimum Time Problem, Jnl. Astronautical Sciences 11, 93 (1964).

18. J. R. BOYD, T. P. CHRISTIE, and J. E. GIBSON, Energy Maneuverability, Eglin Air Force Base, Fla., 1966 and subsequent.

19. A. E. BRYSON, M. N. DESAI, and W. C. HOFFMAN, The Energy-State Approximation in Performance Optimization of Supersonic Aircraft, AIAA Paper 68-877, Pasadena, Calif., August 1968.

20. A. E. BRYSON and M. M. LELE, Minimum Fuel Lateral Turns at Constant Altitude, *AIAA Jnl. 7*, 559 (1969).

21. J. K. HEDRICK and A. E. BRYSON, Minimum Time Turns for a Supersonic Airplane at Constant Altitude, *Jnl. of Aircraft 8*, 182 (1971).

22. J. K. HEDRICK and A. E. BRYSON, Three-Dimensional Minimum-Time Turns for Supersonic Aircraft, AIAA Third Aircraft Design and Operations Meeting, Seattle, Wash., July 12-14, 1971.

23. H. J. KELLEY, R. E. KOPP, and H. G. MOYER, Singular Extremals, Chapter 3 of "Topics in Optimization" (G. Leitmann, ed.), Academic Press, New York, 1967.

24. I. L. JOHNSON and G. E. MYERS, One-Dimensional Minimization Using Search by Golden Section and Cubic Fit Methods, NASA Manned Spacecraft Center Internal Note 67-FM-172, 1967.

25. A. E. PREYSS and R. E. WILLES, Air Combat Maneuvering, USAF Academy notes, 1969; also Vol. II of "Advanced Aircraft Propulsion/Engagement Study," USAF Academy Technical Report 71-7, September 1971.

26. W. R. LAIDLAW, Fighter Aircraft Criteria, *Technical Review*, Society of Experimental Test Pilots, *9*, 37 (1969).

27. D. H. JACOBSON and M. M. LELE, A Transformation Technique for Optimal Control Problems with a State Variable Inequality Constraint, *IEEE Trans. Automatic Control AC-14*, (1969).

28. H. J. KELLEY, Method of Gradients, Chapter 6 of "Optimization Techniques" (G. Leitmann, ed.), Academic Press, New York, 1962.

29. H. J. KELLEY, M. FALCO, and D. J. BALL, Air Vehicle Trajectory Optimization, SIAM Symposium on Multivariable System Theory, Cambridge, Mass., November 1-3, 1962.

30. J. R. BOYD, Maximum Maneuver Concept, informal briefing, USAF Systems Command Headquarters, Andrews AFB, Md., July 1971; also briefing notes by J. R. Boyd, T. P. Christie, and R. E. Drabant, Eglin AFB, Fla., August 1971.

31. H. J. KELLEY, G. E. MYERS, and I. L. JOHNSON, An Improved Conjugate Direction Minimization Procedure, AIAA Jnl. 8, 2091 (1970).

32. H. J. KELLEY, 3-D Energy Management for Supersonic Aircraft, presented at Fifth IFAC Symposium on Automatic Control in Space, Genoa, Italy, June 4-8, 1973.

33. H. J. KELLEY and L. LEFTON, Differential Turns, AIAA Atmospheric Flight Mechanics Specialists Conference, Palo Alto, Calif., Sept. 11-13, 1972; to appear in AIAA Jnl.

34. H. J. KELLEY, Some Aspects of Two-on-One Pursuit/Evasion, AIAA Aerospace Sciences Meeting, Jan. 10-12, 1973; Automatica 9, (1973).

Differential Dynamic Programming—A Unified Approach to the Optimization of Dynamic Systems*

DAVID Q. MAYNE
Department of Computing and Control, Imperial College, London SW7 2BT, England

*This work was done during the author's visit to the Division of Engineering and Applied Physics, Harvard University, and was supported by the U.S. Army Research Office, the U.S. Air Force Office of Scientific Rearch and the U.S. Office of Naval Research under the Joint Services Electronics Program by Contracts N00014-67-A-0298-0006, 0005. and 0008.

I. INTRODUCTION

The purpose of this chapter is to present certain exact expressions for the change, ΔV, in cost due to a change in control, and to indicate, by example, the unifying role these expressions can play. On the one hand, the expressions are useful for obtaining conditions of optimality, particularly sufficient conditions, and, on the other hand, for obtaining optimization algorithms, including the powerful differential dynamic programming (D.D.P.) algorithms [1-8]. The expressions enable two arbitrary controls to be compared, thus permitting the consideration of strong variations in control. The fact that the two controls may be arbitrary should facilitate the construction of new algorithms; many of the estimates of ΔV available in the literature are more restricted; commonly, one of the controls must be optimal.

Exact expressions for the change in cost have, in fact, a long history. A common technique is to add to the integrand of the cost function a perfect differential in order to transform the original problem into a simpler problem (see, for example, Caratheodory [9]); in this way Weierstrass' excess function can be obtained, leading to sufficient conditions for the optimality of a curve [9,10]. Recently, Brockett [11] has used this technique to good effect in his treatment of linear systems. The global sufficiency theorems based on the existence of a function satisfying the Hamilton-Jacobi-Bellman partial differential equation are similar. In fact, the main use of the technique in the classical literature has been to prove sufficiency theorems. The exact expression for ΔV obtained by Weierstrass was said, by Young [12], to have revolutionized the calculus of variations. In contrast, exact expressions for ΔV do not appear to have played a central role in modern control theory, although they do appear as tools in the derivation of optimization algorithms [5,8]. An explicit derivation of these expressions, valid for arbitrary changes in control, appears in [13-15].

II. ASSUMPTIONS AND PRELIMINARY RESULTS

A. System Description

The system is described by:

$$\dot{x}(t) = f(x(t), u(t), t) . \tag{1}$$

The solution of (1) due to initial condition (x_1, t_1) and control u is denoted by $x(t; x_1, t_1, u)$. The solution due to initial condition (x_1, t_1) and policy k:

$$u(t) = k(x(t), t) \tag{2}$$

is denoted by $x(t; x_1, t_1, k)$ i.e., $x(t; x_1, t_1, k)$ is the solution of:

$$\dot{x}(t) = f(x(t), k(x(t), t), t) \tag{3}$$

with initial condition (x_1, t_1).

$V^u(x_1, t_1)$ denotes the cost due to control u and initial condition (x_1, t_1):

$$V^u(x_1,t_1) \triangleq \int_{t_1}^{t_\delta} L(x(t;x_1,t_1,u),\ u(t),t)dt$$

$$+ F(x(t_\delta;x_1,t_1,u)\ . \qquad (4)$$

Similarly, $V^k(x_1,t_1)$ is the cost due to policy k and initial condition $(x_1 t_1)$:

$$V^k(x_1,t_1) \triangleq \int_{t_1}^{t_\delta} L(x(t;x_1,t_1,k),\ k(x(t;x_1,t_1,k),t),t)dt$$

$$+ F(x(t_\delta;x_1,t_1,k)\ . \qquad (5)$$

B. Basic Assumptions

The following basic assumptions are made. The bracketted statements are the extra assumptions required when a policy k rather than a control u is being considered. S denotes the set $\{x,u,t \mid x \in E_n,\ u \in \Omega, t \in T\}$ where Ω is a bounded subset of E_m, $T \triangleq [t_0,t_\delta]$.

H1. $f(x,u,t)$, $L(x,u,t)$ and their partial derivatives with respect to x (with respect to (x,u)) up to order s exist and are continuous in (x,u,t) for all $(x,u,t) \in S$. ($k(x,t)$ and its partial derivatives with respect to x up to order s exist and are continuous in (x,u,t) for all $(x,u,t) \in S$, except at a finite number of times where they have left and right limits).

H2. $\|f(x,u,t)\| \leq M(\|x\| + 1)$, $M < \infty$ for all $(x,u,t) \in S$. ($\|f(x,k(x,t),t)\| \leq M(\|x\| + 1)$, $M < \infty$, for all $x \in E_n$, $t \in T$.)

H3. The partial derivatives $f_t(x,u,t)$, $L_t(x,u,t)$ exist and are continuous in (x,u,t) for all $(x,u,t) \in S$.

When H1, H2 include the bracketted statements they will, where necessary, be referred to as H1A, H2A.

C. Properties of Trajectories

Let G denote the class of piecewise continuous functions from T into Ω. Let $\Theta(u)$ denote the set of points in (t_0, t_δ) at which u is continuous.

Assertion 1. Let H1, H2 be satisfied, $s = 1$. For all $u \in G$ there exists a unique solution $x(t;x_0,t_0,u)$ to Equation (1) satisfying $x(t_0;x_0,t_0,u) = x_0$, ($x(t;x_0,t_0,u)$ is a continuous function of t, differentiable over $\Theta(u)$). Moreover, the functions $x(t;x_0,t_0,u)$ are uniformly bounded over T for all $u \in G$.

This is a standard result. H2 prevents the possibility of a finite escape time.

We will often be concerned with comparing the effect of two controls $u, \bar{u} \in G$. We define the 'distance' $d(u,\bar{u})$ between $u, \bar{u}$ as follows:

$$d(u,\bar{u}) \triangleq \int_{t_0}^{t_\delta} \|u(t) - \bar{u}(t)\| dt . \tag{6}$$

Note that $d(u,\bar{u}) \leq \varepsilon$ permits strong variations. Also:

$$\left[\int_{t_0}^{t_\delta} \|u(t) - \bar{u}(t)\|^2 dt/(t_\delta - t_0) \right]^{\frac{1}{2}} \geq d(u,\bar{u})/(t_\delta - t_0) .$$

An alternative definition of distance, employed by Halkin [16], is the following. Let E denote the subset of T such that for $t \in E$, $u(t) \neq \bar{u}(t)$, and let $\mu(E)$ denote the length of the set E. Then,

$$d_1(u,\bar{u}) \triangleq \mu(E) . \tag{7}$$

Let $x(t)$ denote $x(t;x_0,t_0,u)$, $\bar{x}(t)$ denote $x(t;x_0,t_0,\bar{u})$ and $\delta x(t)$ denote $x(t) - \bar{x}(t)$.

<u>Assertion</u> 2. Let H1, H2 be satisfied, $s = 1$. Let $\bar{u}$, $u \in G$. If either: (i) $f_u(x,u,t)$ exists and is continuous in (x,u,t) for all $(x,u,t) \in S$ and $d(u,\bar{u}) \leq \varepsilon$, or (ii) $d_1(u,\bar{u}) \leq \varepsilon$, then $\|\delta x(t))\| \leq c\varepsilon$, $c < \infty$, for all $t \in T$.

<u>Proof</u>: (i) From Assertion 1, $\|\delta x(t)\| \leq c_1$, $c_1 < \infty$, $\|u(t) - \bar{u}(t)\| \leq c_2 < \infty$ for all $t \in T$. Since f_x, f_u exist and are continuous, there exists constants $c_3, c_4 < \infty$ such that $\|f(x(t),u(t),t) - f(\bar{x}(t),u(t),t)\| \leq c_3\|\delta x(t)\|$, $c_3 < \infty$, $\|f(x(t),u(t),t) - f(x(t),\bar{u}(t),t)\| \leq c_4\|u(t) - \bar{u}(t)\|$, $c_4 < \infty$ for all $t \in T$.

$$\delta x(t) = \int_{t_0}^{t} [f(x(\tau),u(\tau),\tau) - f(\bar{x}(\tau),u(\tau),\tau)$$
$$+ f(\bar{x}(\tau),u(\tau),\tau) - f(\bar{x}(\tau),\bar{u}(\tau),\tau)]d\tau$$

$$\|\delta x(t)\| \leq \int_{t_0}^{t_\delta}[c_3\|\delta x(\tau)\| + c_4\|u(\tau) - \bar{u}(\tau)\|]d\tau$$
$$\leq c_4 e^{c_3(t_\delta - t_0)} d(u,\bar{u}) .$$

(ii)
$$\|\delta x(t)\| \leq \int_{t_0}^{t_\delta} [c_3\|\delta x(\tau)\| + \|\Delta f(\tau)\|]d\tau$$

where $\Delta f(t) \triangleq f(\bar{x}(t),u(t),t) - f(\bar{x}(t),\bar{u}(t),t)$. (8)

Clearly $\|\Delta f(t)\| \leq c_5$, $c_5 < \infty$, $t \in E$ and $\Delta f(t) = 0$, $t \notin E$, so that $\|\delta x(t)\| \leq c_5 e^{c_3(t_\delta - t_0)} d_1(u,\bar{u})$.

A useful approximation to $\delta x(t)$ is given by $\widehat{\delta x}(t)$, the solution of:

$$\delta\dot{\hat{x}}(t) = f_x(\bar{x}(t),\bar{u}(t),t)\delta\hat{x}(t) + \Delta f(t) \tag{9}$$

$$\delta\hat{x}(t_0) = 0\ . \tag{10}$$

<u>Assertion</u> 3. Let H1, H2 be satisfied, $s = 2$. Let $u, \bar{u} \in G$. If either (i) $f_u(x,u,t)$; $f_{xu}(x,u,t)$ exist and are continuous in (x,u,t) for all $(x,u,t) \in S$ and $d(u,\bar{u}) \leq \varepsilon$, or, (ii) $d_1(u,\bar{u}) \leq \varepsilon$, then $\|\delta x(t) - \delta\hat{x}(t)\| \leq c\varepsilon^2$, $c < \infty$, for all $t \in T$.

<u>Proof</u>: (i) Let $z(t) \triangleq \delta x(t) - \delta\hat{x}(t)$.

Then $z(t_0) = 0$ and

$$\begin{aligned}\dot{z}(t) &= f(x(t),u(t),t) - f(\bar{x}(t),u(t),t) \\ &\qquad - f_x(\bar{x}(t),\bar{u}(t),t)\delta\hat{x}(t) \\ &= f_x(\bar{x}(t),u(t),t)\delta x(t) - f_x(\bar{x}(t),\bar{u}(t),t)\delta\hat{x}(t) + r(t)\end{aligned}$$

where $\|r(t)\| \leq c_1\|\delta x(t)\|^2 \leq c_2\varepsilon^2$, c_1, $c_2 < \infty$, for all $t \in T$.

Hence:

$$\begin{aligned}\dot{z}(t) &= f_x(\bar{x}(t),\bar{u}(t),t)z(t) + r(t) \\ &\qquad + [f_x(\bar{x}(t),u(t),t) - f_x(\bar{x}(t),\bar{u}(t),t)]\delta x(t)\end{aligned}$$

$$\|z(t)\| \leq \int_{t_0}^{t_\delta} c_3\|z(\tau)\| + \|r(\tau)\| + c_4\|u(\tau) - \bar{u}(\tau)\|\,\|\delta x(\tau)\|d\tau$$

where

$$\|\delta x(\tau)\| \leq c_5\ \varepsilon, \quad c_3,c_4,c_5 < \infty$$

that is,

$$z(t) \leq e^{c_3(t_\delta - t_0)}\ [c_2 + c_4c_5]\varepsilon^2$$

(ii) $$\|z(t)\| \leq \int_{t_0}^{t_\delta} c_3\|z(\tau)\| + \|r(\tau)\| + \|\Delta f_x(\tau)\|\,\|\delta x(\tau)\|d\tau$$

where

$$\Delta f_x(t) = f_x(\overline{x}(t),u(t),t) - f_x(\overline{x}(t),\overline{u}(t),t) .$$

Clearly $\|\Delta f_x(t)\| \leq c_6$, $c_6 < \infty$, $t \in E$ and $\Delta f_x(t) = 0$, $t \notin E$.

The result follows:

<u>Assertion</u> 4. If either:

(i) H1, H2 are satisfied, $s = 2$, f is linear in u, and $d(u,\overline{u}) \leq \varepsilon$, or $d_1(u,\overline{u}) \leq \varepsilon$, or,

(ii) H1A, H2A are satisfied, $s = 2$, and $\|u(t) - \overline{u}(t)\| \leq \varepsilon$ for all $t \in T$.

Then: $\|\delta x(t) - \delta\hat{x}(t)\| \leq c\varepsilon^2$, $c < \infty$, for all $t \in T$, where $\delta\hat{x}(t)$ is the solution of:

$$\delta\dot{\hat{x}}(t) = f_x(\overline{x}(t),\overline{u}(t),t)\delta\hat{x}(t) + f_u(\overline{x}(t),\overline{u}(t),t)\Delta u(t) \tag{11}$$

$$\delta\hat{x}(t_0) = 0 \tag{12}$$

where

$$\Delta u(t) \triangleq u(t) - \overline{u}(t) . \tag{13}$$

<u>Proof</u>: (i) $\Delta f(t) = f_u(\overline{x}(t),\overline{u}(t),t)\,\Delta u(t)$ and the result follows from Assertion 3.

(ii) $$\Delta f(t) = f_u(\overline{x}(t),\overline{u}(t),t)\,\Delta u(t) + r_1(t)$$

where: $\|r_1(t)\| \leq c_1\varepsilon^2$, $c_1 < \infty$. Therefore, if $\Delta f(t)$ in the proof of Assertion 3 is replaced by $f_u(\overline{x}(t),\overline{u}(t),t)\,\Delta u(t)$, on extra error of norm $\leq c_1\varepsilon^2$ is introduced; this does not alter the conclusion of Assertion 3.

Note, the hypotheses in Assertions 2(i) (3(i)) can be replaced by: (i) H1A and H2A are satisfied, $s = 1$ ($s = 2$), and $d(u,\overline{u}) \leq \varepsilon$.

D. Properties of Cost Function

Assertion 5: Let H1, H2 be satisfied, $u \in G$. Then: (i) $V^u(x,t)$, $V^k(x,t)$ and their partial derivatives with respect to x up to order s exist and are continuous in (x,t).

(ii) $V_t^u(x,t)$, $V_t^k(x,t)$ and their partial derivatives with respect to x up to order $s - 1$ exist and are continuous in (x,t) except where u or k are discontinuous.

Proof: Consider $V^u(x_1,t_1)$ (with the extra assumptions on k, $\overline{f}(x,t) \triangleq f(x,k(x,t),t)$ satisfies the original assumptions on f. Hence, with these extra assumptions, the following discussion is applicable to V^k). Consider the following system:

$$\dot{z}(t) = g(z(t),t) \tag{14}$$

where $z \triangleq (x_0,x)$, $g \triangleq (L,f)$, i.e.

$$\dot{x}_0(t) = L(x(t),u(t),t) \,. \tag{15}$$

Clearly:

$$V^u(x_1,t_1,u) = H(z(t_\delta;z_1,t_1)) \tag{16}$$

where

$$z_1 \triangleq (0,x_1) \tag{17}$$

and

$$H(x_0,x) \triangleq x_0 + F(x) \,.$$

H is s times continuously differentiable.

(i) (McShane [17], Theorem 69.4). Let $z(t;\, z_1,t_1)$ denote the solution of Eq. (11) with initial condition (z_1,t_1). (The dependence on u is omitted for convenience) $z(t;z_1,t_1)$ and its partial derivatives with respect to z_1 of all orders less than or equal to s exist and are continuous in (t,z_1,t_1) for all $t,t_1 \in T$, all $z_1 \in E_{n+1}$. Setting $t = t_\delta$ and using Eq. (16), yields (i).

(ii) Let $t,t_1 \in \Theta(u)$. There exists an $\varepsilon > 0$ such that if $|\delta t_1| < \varepsilon$, u is continuous on $[t_1,t_1 + \delta t_1]$. Let z^i denote ith component of z, etc. For $i = 1 \ldots n$:

$$\varphi^i(\delta t_1) \triangleq [z^i(t;z_1,t_1 + \delta t_1) - z^i(t;z_1,t_1)]/\delta t_1$$

$$z^i(t;z_1,t_1 + \delta t_1) = z^i(t;y,t_1),$$

where:

$$\begin{aligned} y^i &\triangleq z^i(t_1;z_1,t_1 + \delta t_1) \\ &= z_1^i - g^i(z(t_i^*;z_1,t_1 + \delta t_1),t_i^*)\delta t_1 \end{aligned}$$

where

$$t_i^* \in [t_1,t_1 + \delta t_1]$$

$$z^i(t;y,t_1) - z^i(t;z_1,t_1) = \sum_{j=1}^{n} (\partial z^i/\partial z_i^j)(t;z_i^*,t_1)(y^j - z_i^j)$$

where $z_i^* = \theta z_1 + (1-\theta)(y - z_1)$ for some $0 \leq \theta \leq 1$.
Hence:

$$\varphi^i(\delta t_1) = \sum_{j=1}^{n} (\partial z^i/\partial z_1^j)(t;z_i^*,t_1)[-g^j(z(t_j^*;z_1,t_1 + \delta t_1),t_j^*)] .$$

As $\delta t_1 \to 0$, $t_i^* \to t_1$, $y \to z_1$, $z_i^* \to z_1$. Hence from the continuity of z_{z_1} and g, z_{t_1} exists for all t, $t_1 \in \Theta(u)$ and is given by:

$$z_{t_1}(t;z_1,t_1) = - z_{z_1}(t;z_1,t_1)\, g(z_1,t_1) . \tag{18}$$

The right hand side of Eq. (15), and its partial derivatives with respect to z, up to order $s-1$, exist and are continuous in (z_1,t_1) for all $t,t_1 \in \Theta(u)$, possessing left and/or right limits at the remaining points. Allowing $t \to t_\delta$ from the left, and using Eq. (16) yields (ii).

III. EXACT EXPRESSIONS FOR ΔV

We come now to the central results of this chapter, a series of exact expressions for the change of cost resulting from the adoption of a new control (or policy). In the sequel, $\bar{u},u \in G$; $\bar{x}(t)$, $x(t)$ denote, respectively, $x(t;x_0,t_0,\bar{u})$, $x(t;x_0,t_0,u)$. $\bar{k}$ and k denote policies which, with initial condition $(x_0 t_0)$ generate, respectively, $(\bar{x},\bar{u})$ and (x,u), i.e.

$$x(t;x_0,t_0,\bar{k}) \equiv \bar{x}(t)$$

$$\bar{k}(\bar{x}(t),t) \equiv \bar{u}(t) .$$

Similarly for k. Hence:

$$V^{\bar{k}}(x_0,t_0) = V^{\bar{u}}(x_0,t_0)$$

$$V^{k}(x_0,t_0) = V^{u}(x_0,t_0) .$$

Let ΔV denote $V^{u}(x_0,t_0) - V^{\bar{u}}(x_0,t_0)$.

A. The expressions for ΔV

Assertion 6. Let H1, H2 (H1A, H2A for (iii), (iv)) be satisfied, $s = 1$. Then:

$$\text{(i)}\quad \Delta V = \int_{t_0}^{t_\delta} [H(\bar{x}(t),u(t),V_x^{u}(\bar{x}(t),t),t) - H(\bar{x}(t),\bar{u}(t),V_x^{u}(\bar{x}(t),t),t)]dt$$

$$\text{(ii)}\quad \Delta V = \int_{t_0}^{t_\delta} [H(x(t),u(t),V_x^{\bar{u}}(x(t),t),t) - H(x(t),\bar{u}(t),V_x^{\bar{u}}(x(t),t),t)]dt$$

(iii) $\Delta V = \int_{t_0}^{t_\delta} [H(\bar{x}(t),k(\bar{x}(t),t),V_x^k(\bar{x}(t),t),t)$

$- H(\bar{x}(t),\bar{u}(t),V_x^k(\bar{x}(t),t),t)]dt$

(iv) $\Delta V = \int_{t_0}^{t_\delta} [H(x(t),u(t),V_x^{\bar{k}}(x(t),t),t)$

$- H(x(t),\bar{k}(x(t),t),V_x^{\bar{k}}(x(t),t),t)]dt$

where

$$H(x,u,\lambda,t) \triangleq L(x,u,t) + \lambda^T f(x,u,t) . \tag{19}$$

Proof: (i) It follows from Assertion 5 that:
For all $t \in \Theta(\bar{u})$:

$$\frac{d}{dt} V^{\bar{u}}(x(t),t) = - L(\bar{x}(t),\bar{u}(t),t)$$
$$= V_t^{\bar{u}}(\bar{x}(t),t) + [V_x^{\bar{u}}(\bar{x}(t),t)]^T f(\bar{x}(t),\bar{u}(t),t) . \tag{20}$$

For all $t \in \Theta(\bar{u}) \cap \Theta(u)$

$$\frac{d}{dt} V^u(\bar{x}(t),t) = V_t^u(\bar{x}(t),t) + [V_x^u(\bar{x}(t),t)]^T f(\bar{x}(t),\bar{u}(t),t)$$

$$= V_t^u(\bar{x}(t),t) + [V_x^u(\bar{x}(t),t)]^T f(\bar{x}(t),u(t),t)$$

$$- [V_x^u(\bar{x}(t),t)]^T [f(\bar{x}(t),u(t),t) - f(\bar{x}(t),\bar{u}(t),t)]$$

$$= - L(\bar{x}(t),u(t),t) - [V_x^u(\bar{x}(t),t)]^T$$

$$[f(\bar{x}(t),u(t),t) - f(\bar{x}(t),\bar{u}(t),t)] .$$

Hence, for all $t \in \Theta(\bar{u}) \cap \Theta(u)$

$$- \frac{d}{dt} [V^u(\bar{x}(t),t) - V^{\bar{u}}(\bar{x}(t),t)] = [H(\bar{x}(t),u(t),V_x^u(\bar{x}(t),t),t)$$
$$- H(\bar{x}(t),\bar{u}(t),V_x^u(\bar{x}(t),t),t)] \tag{21}$$

which proves (i), since $T - \{\theta(\bar{u}) \cap \theta(u)\}$ consists of, at most, a finite set of points.

(ii) Interchange u with $\bar{u}$ and x with $\bar{x}$ in (i).

(iii) Replace V^u by V^k and $u(t)$ by $k(\bar{x}(t),t)$ in (i).

(iv) Replace $\bar{u}$ by u, $\bar{x}$ by x, k by $\bar{k}$ in (iii)

Usually in the control literature an approximate version of (i) is presented, where $V_x^u(\bar{x}(t),t)$ is replaced by the solution $\bar{\lambda}(t)$ of the usual adjoint differential equation. Although (i) cannot be directly employed, since $V_x^u(\bar{x}(t),t)$ is generally not known, it seems preferable to use these exact expressions, which are in a useful form, as a starting point, both for the derivation of algorithms and of conditions of optimality. Before doing so, we present a few more useful results.

Assertion 7. (i) Let H1, H2 be satisfied. For all $t \in \theta(u)$,

$$V_t^u(x,t) = -H(x,u(t),V_x^u(x,t),t) .$$

For all $t \in T - \theta(u)$, the left and/or right limits of V_t^u are given by the same expression with $u(t)$ having the corresponding left and/or right limit.

(ii) Let H1A, H2A be satisfied. For all $t \in \theta(k)$,

$$V_t^k(x,t) = -H(x,k(x,t),V_x^k(x,t),t)$$

having left and/or right limits for all $t \in T - \theta(k)$. $\theta(k)$ is the subset of T in which k is continuous.

Proof: The result follows from Eq. (20), with $\bar{u}$ replaced by u or k.

In Assertion 6 the cost V^u is compared with cost $V^{\bar{u}}$

of a nominal control $\bar{u}$, or the cost $V^{\bar{k}}$ of a policy $\bar{k}$. Alternative expressions for V^u can be obtained, however, using an arbitrary differentiable function V^*.

Assertion 8. Let H1, H2 be satisfied, $s = 1$. Let $V^*(x,t)$ and its partial derivatives $V^*_x(x,t), V^*_t(x,t)$ exist and be continuous in (x,t). Then:

$$V^u(x_0,t_0) - V^*(x_0,t_0) = \int_{t_0}^{t_\delta} [H(x(t),u(t),V^*_x(x(t),t),t) + V^*_t(x(t),t)]dt + V^u(x(t_\delta),t_\delta) - V^*(x(t_\delta),t_\delta)$$

where:

$$V^u(x(t_\delta),t_\delta) = F(x(t_\delta)) \; .$$

Proof:

$$V^u(x_0,t_0) = \int_{t_0}^{t_\delta} [L(x(t),u(t),t) + \dot{V}^*(x(t),t)]dt + F(x(t_f)) + V^*(x_0,t_0) - V^*(x(t_f),t_f) \; .$$

If V^* is, in fact, the cost due to $\bar{u}$, then Assertion 8 reduces to Assertion 6(ii).

In the following two subsections, we obtain exact and approximate differential equations for calculating $V^u_x(\bar{x}(t),t)$, $V^u_{xx}(\bar{x}(t),t)$, $V^k_x(\bar{x}(t),t)$, $V^k_{xx}(\bar{x}(t),t)$.

B. Exact Differential Equation for $V_x(\bar{x}(t),t)$, $V_{xx}(\bar{x}(t),t)$

Let $\bar{\lambda}(t)$, $\lambda(t)$, $\bar{P}(t)$, $P(t)$ be defined as follows:

$$\bar{\lambda}(t) \triangleq V^{\bar{u}}_x(\bar{x}(t),t) \tag{22}$$

$$\lambda(t) \triangleq V_x(\bar{x}(t),t) \tag{23}$$

$$\bar{P}(t) \triangleq V^{\bar{u}}_{xx}(\bar{x}(t),t) \tag{24}$$

$$P(t) \triangleq V_{xx}(\bar{x}(t),t) \tag{25}$$

where V_x, V_{xx} denote V^u_x or V^k_x and V^u_{xx} or V^k_{xx}, according to the context.

Assertion 9. (i) Let H1, H2 be satisfied, $s = 2$. Then $\bar{\lambda}(t)$ is the solution of:

$$-\dot{\bar{\lambda}}(t) = H_x(x(t),\bar{u}(t),\bar{\lambda}(t),t) \tag{26}$$

$$\bar{\lambda}(t_\delta) = F_x(\bar{x}(t_\delta)) . \tag{27}$$

(ii) Let H1, H2 be satisfied, $s = 3$. Then $\bar{P}(t)$ is the solution of:

$$-\dot{\bar{P}}(t) = H_{xx}(\bar{x}(t),\bar{u}(t),\bar{\lambda}(t),t)$$

$$+ f^T_x(\bar{x}(t),\bar{u}(t),t)\bar{P}(t) + \bar{P}(t)f_x(\bar{x},(t),\bar{u}(t),t) \tag{28}$$

$$\bar{P}(t_\delta) = F_{xx}(\bar{x}(t_\delta)) . \tag{29}$$

Proof: (i)

$$\frac{d}{dt} V^{\bar{u}}_x(\bar{x},(t),t) = V^{\bar{u}}_{xt}(\bar{x}(t),t) + V^{\bar{u}}_{xx}(\bar{x}(t),t)f(\bar{x}(t),\bar{u}(t),t). \tag{30}$$

Since

$$-V^{\bar{u}}_t(\bar{x}(t),t) = H(\bar{x}(t),\bar{u}(t),\bar{\lambda}(t),t) \tag{31}$$

$$-V^{\bar{u}}_{xt}(\bar{x}(t),t) = H_x(\bar{x}(t),\bar{u}(t),\bar{\lambda}(t),t)$$

$$+ V^{\bar{u}}_{xx}(\bar{x}(t),t)f(\bar{x}(t),\bar{u}(t),t) . \tag{32}$$

The result follows.

(ii) $$\frac{d}{dt} V^{\bar{u}}_{xx}(\bar{x}(t),t) = V^{\bar{u}}_{xxt}(\bar{x}(t),t) + \sum_{i=1}^{n} V^{\bar{u}}_{xxx_i}(\bar{x}(t),t) f_i(\bar{x}(t),\bar{u}(t),t) .$$

V_{xxt} is evaluated using Eq. (32) leading to the result.

Assertion 10. (i) Let H1, H2 be satisfied, $s = 2$. Then $V^u_x(\bar{x}(t),t)$ is the solution of:

$$-\dot{\lambda}(t) = H_x(\bar{x}(t),u(t),\lambda(t),t) + V^u_{xx}(\bar{x}(t),t)\Delta f(t) \tag{33}$$

$$\lambda(t_\delta) = F_x(\bar{x}(t_\delta)) \tag{34}$$

where $\Delta f(t)$ is defined in Eq. (8).

(ii) Let H1, H2 be satisfied, $s = 3$. Then $V^u_{xx}(\bar{x}(t),t)$ is the solution of:

$$-\dot{P}(t) = H_{xx}(\bar{x}(t),u(t),\lambda(t),t) + f^T_x(\bar{x}(t),u(t),t)P(t) + P(t)f_x(\bar{x}(t),u(t),t) + \sum_{i=1}^{n} V^u_{xxx_i}(\bar{x}(t),t)\Delta f_i(t) \tag{35}$$

$$P(t_\delta) = F_{xx}(x(t_\delta)) . \tag{36}$$

Proof: (i)

$$\frac{d}{dt} V^u_x(\bar{x}(t),t) = V^u_{xt}(\bar{x}(t),t) + V^u_{xx}(\bar{x}(t),t)f(\bar{x}(t),\bar{u}(t),t) . \tag{37}$$

But, now

$$-V^u_t(\bar{x}(t),t) = H(\bar{x}(t),u(t),\lambda(t),t) \tag{38}$$

$$-V^u_{xt}(\bar{x}(t),t) = H_x(\bar{x}(t),u(t),\lambda(t),t) + V_{xx}(\bar{x}(t),t)f(\bar{x}(t),u(t),t) . \tag{39}$$

The result follows, the presence of the term involving $\Delta f(t)$ being due to the presence of $u(t)$, not $\bar{u}(t)$ in Eq. (31) and Eq. (32).

(ii) The proof of (ii) is similar.

Assertion 11. (i) Let H1A, H2A be satisfied, $s = 2$. Then $V_x^k(\bar{x}(t),t)$ is the solution of:

$$
\begin{aligned}
-\dot{\lambda}(t) = {} & H_x(\bar{x}(t),u^*(t),\lambda(t),t) \\
& + [k_x(\bar{x}(t),t)]^T H_u(\bar{x}(t),u^*(t),\bar{\lambda}(t),t) \\
& + V_{xx}^k(\bar{x}(t),t)\Delta f(t)
\end{aligned} \tag{40}
$$

$$
\lambda(t_\delta) = F_x(x(t_\delta)) . \tag{41}
$$

(ii) Let H1A, H2A be satisfied, $s = 3$. Then $V_{xx}^k(\bar{x}(t),t)$ is the solution of:

$$
\begin{aligned}
-\dot{P}(t) = {} & H_{xx}(\bar{x}(t),u^*(t),\lambda(t),t) + f_x^T(\bar{x}(t),u^*(t),t)P(t) \\
& + P(t)f_x(\bar{x}(t),u^*(t),t) \\
& + K^T(t)[H_{ux}(\bar{x}(t),u^*(t),\lambda(t),t) + f_u^T(\bar{x}(t),u^*(t),t)P(t)] \\
& + [H_{ux}(\bar{x}(t),u^*(t),\lambda(t),t) + f_u^T(\bar{x}(t),u^*(t),t)P(t)]^T K(t) \\
& + K^T(t)H_{uu}(\bar{x}(t),u^*(t),\lambda(t),t)K(t) \\
& + \sum_{i=1}^{n} H_u^i(\bar{x}(t),u^*(t),\lambda(t),t)\gamma_i(t) \\
& + \sum_{i=1}^{n} V_{xxx_i}^k(\bar{x}(t),t)\Delta f_i(t)
\end{aligned} \tag{42}
$$

$$
P(t_\delta) = F_{xx}(\bar{x}(t),t) \tag{43}
$$

where

$$u^*(t) \triangleq k(\bar{x}(t),t) \tag{44}$$

$$K(t) \triangleq k_x(\bar{x}(t),t) \tag{45}$$

$$\gamma_i(t) \triangleq k^i_{xx}(\bar{x}(t),t) \tag{46}$$

$$\Delta f(t) \triangleq f(\bar{x}(t),u^*(t),t) - f(\bar{x}(t),\bar{u}(t),t) \tag{47}$$

and H^i_u is the ith component of H_u, k^i is the ith component of k, Δf_i is the ith component of Δf etc.

<u>Proof</u>: The results follow from Assertion 10 if the dependence of u on x is taken into account.

C. <u>Approximate Differential Equations for</u> $V_x(\bar{x}(t),t), V_{xx}(\bar{x}(t),t)$

The differential equation for $V^u_x(\bar{x}(t),t)$ has a term involving $V^u_{xx}(\bar{x}(t),t)$. This term arises because we are interested in the rate of change of V^u_x not along x, the trajectory generated by u, but along $\bar{x}$. The differential equation for $V^u_x(x(t),t)$ would be similar to Eq. (26), i.e., it would not involve V^u_{xx}. Because of this involvement of V^u_{xx} in the differential equation for V^u_x (and of V^u_{xxx} in the differential equation for V^u_{xx}), Assertion 10 (and Assertion 11) cannot be directly employed for computation. In this subsection we estimate the error arising from the omission of unknown quantities in the relevant differential equations. Firstly, we define an extra term:

$$a(t) \triangleq V(\bar{x}(t),t) - V^{\bar{u}}(\bar{x}(t),t) \tag{48}$$

where V denotes V^u or V^k according to the context. Clearly:

$$\Delta V = a(t_0) \,. \tag{49}$$

<u>Assertion</u> 12. (i) Let H1, H2 be satisfied, $s = 2$. If

either, (a) $f_u(x,u,t)$ exists and is continuous in (x,u,t) and $d(u,\bar{u}) \leq \varepsilon$, or, (b) $d_1(u,\bar{u}) \leq \varepsilon$, then $\hat{a}(t)$, and $\hat{\lambda}(t)$, where:

$$-\dot{\hat{a}}(t) = H(\bar{x}(t),u(t),\hat{\lambda}(t),t) - H(\bar{x}(t),\bar{u}(t),\hat{\lambda}(t),t) \tag{50}$$

$$-\dot{\hat{\lambda}}(t) = H_x(x(t),u(t),\hat{\lambda}(t),t) \tag{51}$$

with the usual boundary conditions

$$\hat{a}(t_\delta) = 0 \tag{52}$$

$$\hat{\lambda}(t_\delta) = F_x(\bar{x}(t_\delta)) \tag{53}$$

are estimates of $a(t) = V^u(\bar{x}(t),t) - V^{\bar{u}}(\bar{x}(t),t)$ and $\lambda(t) = V_x^u(\bar{x}(t),t)$ such that:

$$\|a(t) - \hat{a}(t)\| \leq c_1\varepsilon^2, \quad c_1 < \infty \tag{54}$$

and

$$\|\lambda(t) - \hat{\lambda}(t)\| \leq c_2\varepsilon, \quad c_2 < \infty\,. \tag{55}$$

(ii) Let H1A, H2A be satisfied, $s = 2$. If either (a) $d(u^*,\bar{u}) \leq \varepsilon$, or (b) $d_1(u^*,\bar{u}) \leq \varepsilon$, then $\hat{a}(t)$, $\hat{\lambda}(t)$, where $\hat{a}(t)$ satisfies Eq. (50) and Eq. (52) and $\hat{\lambda}(t)$ satisfies:

$$-\dot{\hat{\lambda}}(t) = H_x(\bar{x}(t),u^*(t),\hat{\lambda}(t),t) + K^T(t)H_u(\bar{x}(t),u^*(t),\hat{\lambda}(t),t)$$

and Eq. (53), are estimates of $a(t) = V^k(\bar{x}(t),t) - V^{\bar{u}}(\bar{x}(t),t)$ and $\lambda(t) = V_x^k(\bar{x}(t),t)$ such that Eq. (54) and Eq. (55) are satisfied.

<u>Proof</u>: (i)

$$-\frac{d}{dt}(a(t) - \hat{a}(t)) = [\lambda(t) - \hat{\lambda}(t)]^T \Delta f(t) \tag{57}$$

$$-\frac{d}{dt}\,(\lambda(t) - \hat{\lambda}(t)) = f_x^T(\bar{x}(t),u(t),t)(\lambda(t) - \hat{\lambda}(t))$$
$$+ P(t)\Delta f(t) \tag{58}$$

$$a(t_\delta) - \hat{a}(t_\delta) = 0;\ \lambda(t_\delta) - \hat{\lambda}(t_\delta) = 0 \tag{59}$$

(a) $\|\Delta f(t)\| \leq c_3\|\bar{u}(t) - u(t)\|,\ c_3 < \infty$

(b) $\|\Delta f(t)\| \leq c_4,\ c_4 < \infty,\ t \in E;\ \Delta f(t) = 0,\ t \notin E.$

Hence $\|\lambda(t) - \hat{\lambda}(t)\| \leq c_1\varepsilon,\ c_1 < \infty$. Making use of this in Eq. (57) yields $\|\varepsilon(t) - \hat{a}(t)\| \leq c_2\varepsilon^2,\ c_2 < \infty$.

(ii) The proof is similar.

Assertion 13. (i) Let H1, H2 be satisfied, $s = 3$. If either (a) $f_u(x,u,t)$ exists and is continuous in (x,u,t) and $d(u,\bar{u}) \leq \varepsilon$ or, (b) $d_1(u,\bar{u}) \leq \varepsilon$, then $\hat{a}(t)$, $\hat{\lambda}(t)$, $\hat{P}(t)$ where $\hat{a}(t)$ satisfies Eq. (50) and Eq. (52), $\hat{\lambda}(t)$ satisfies Eq. (53) and:

$$-\dot{\hat{\lambda}}(t) = H_x(\bar{x}(t),u(t),\hat{\lambda}(t),t) + \hat{P}(t)\,\Delta f(t) \tag{60}$$

and $P(t)$ satisfies:

$$-\dot{\hat{P}}(t) = H_{xx}(\bar{x}(t),u(t),\hat{\lambda}(t),t) + f_x^T(\bar{x}(t),u(t),t)\hat{P}(t)$$
$$+ \hat{P}(t)f_x(\bar{x}(t),u(t),t) \tag{61}$$

$$\hat{P}(t_\delta) = F_{xx}(\bar{x}(t_\delta)) \tag{62}$$

are estimates of $a(t) = V^u(\bar{x}(t),t) - V^{\bar{u}}(\bar{x}(t),t), \lambda(t) = V_x^u(\bar{x}(t),t), P(t) = V_{xx}^u(\bar{x}(t),t)$, satisfying:

$$\|a(t) - \hat{a}(t)\| \leq c_1\varepsilon^3,\ c_1 < \infty \tag{63}$$

$$\|\lambda(t) - \hat{\lambda}(t)\| \leq c_2\varepsilon^2,\ c_2 < \infty \tag{64}$$

$$\|P(t) - \hat{P}(t)\| \leq c_3\varepsilon, \quad c_3 < \infty . \tag{65}$$

(ii) Let H1A, H2A be satisfied, $s = 3$. If either, (a) $d(u,\bar{u}) \leq \varepsilon$, or, (b) $d_1(u,u) \leq \varepsilon$, then $\hat{a}(t)$, $\hat{\lambda}(t)$, $\hat{P}(t)$, where $\hat{a}(t)$ satisfies Eq. (50) and Eq. (52), $\hat{\lambda}(t)$ satisfies Eq. (53) and:

$$-\dot{\hat{\lambda}}(t) = H_x(\bar{x}(t),u(t),\hat{\lambda}(t),t) + K^T(t)\, H_u(\bar{x}(t),u(t),\hat{\lambda}(t),t) + \hat{P}(t).\ \Delta f(t) \tag{66}$$

and $\hat{P}(t)$ satisfies Eq. (62) and:

$$\begin{aligned}-\dot{\hat{P}}(t) = {} & H_{xx}(\bar{x}(t),u*(t),\hat{\lambda}(t),t) + f_x^T(\bar{x}(t),u*(t),t)\hat{P}(t) \\ & + \hat{P}(t)\, f_x(\bar{x}(t),u*(t),t) \\ & + K^T(t)\,[H_{ux}(\bar{x}(t),u*(t),\hat{\lambda}(t),t) + f_u^T(\bar{x}(t),u*(t),t)\,\hat{P}(t)] \\ & + [H_{ux}(\bar{x}(t),u*(t),\hat{\lambda}(t),t) + f_u^T(\bar{x}(t),u*(t),t)\hat{P}(t)]^T K(t) \\ & + K^T(t)\, H_{uu}(\bar{x}(t),u^*(t),\hat{\lambda}(t),t)K(t) \\ & + \sum_{i=1}^{n} H_u^i(x(t),u*(t),\hat{\lambda}(t),t)\gamma_i(t)\end{aligned} \tag{67}$$

are estimates of $a(t) = \Delta V(t), \lambda(t) = V_x^k(\bar{x}(t),t)$ and $P(t) = V_{xx}^k(\bar{x}(t),t)$ such that Eqs. (63)-(65) are satisfied.

<u>Proof</u>:

$$-\frac{d}{dt}\,(a(t) - \hat{a}(t)) = [\lambda(t) - \hat{\lambda}(t)]^T.\ \Delta f(t) \tag{68}$$

$$\begin{aligned}-\frac{d}{dt}\,(\lambda(t) - \hat{\lambda}(t)) = {} & F^T(t)(\lambda(t) - \hat{\lambda}(t)) \\ & + [P(t) - \hat{P}(t)]\ \Delta f(t)\end{aligned} \tag{69}$$

$$\begin{aligned}-\frac{d}{dt}\,(P(t) - \hat{P}(t)) = {} & F^T(t)[P(t) - \hat{P}(t)] \\ & + [P(t) - \hat{P}(t)]F(t)\end{aligned}$$

$$+ \sum_{i=1}^{n} (\lambda_i(t) - \hat{\lambda}_i(t))G_i(t)$$

$$+ \sum_{i=1}^{n} V^k_{xxx_i}(\bar{x}(t),t)\,\Delta f_i(t) \tag{70}$$

and

$$F(t) \triangleq f_x(\bar{x}(t),u^*(t),t) + f_u(\bar{x}(t),u^*(t),t)K(t) \tag{71}$$

and $G_i(t)$, $i = 1,\ldots,n$, are piecewise continuous functions. Eqs. (68)-(70) have terminal conditions of zero at time t_δ. Let $z(t)$ denote a vector whose components are the elements of $\lambda(t) - \hat{\lambda}(t)$ and $P(t) - \hat{P}(t)$. Hence, for some piecewise continuous matrix functions $\bar{F}(t),\bar{G}(t)$:

$$-\frac{d}{dt} z(t) = \bar{F}(t)z(t) + \bar{G}(t)\,\Delta f(t)\,.$$

Hence, using the same techniques as before:

$$\|z(t)\| \leq c_4\varepsilon,\ c_4 < \infty\,.$$

Hence Eq. (65) is satisfied. Using the result in Eq. (69) yields Eq. (64). Using Eq. (64) in Eq. (68) yields Eq. (63).

None of the differential equations obtained so far correspond to the classical results. One difference is the appearance of $u(t)$ or $u^*(t)$ in place of $\bar{u}(t)$ on the right hand side of the various differential equations.

Assertion 14. If either:
(i) H1, H2 are satisfied, $s = 2$, and $d_1(u,\bar{u}) \leq \varepsilon$ or
(ii) H1A, H2A are satisfied, $s = 2$, and $d(u,\bar{u}) \leq \varepsilon$ then Assertion 12(i) holds with $u(t)$ replaced by $\bar{u}(t)$ in Eq. (51) (i.e. $\hat{\lambda}(t)$ replaced by $\bar{\lambda}(t)$ in Eq. (50). Also, if H1A, H2A are satisfied, $s = 3$, and $d(u,\bar{u})$ or $d_1(u,\bar{u}) \leq \varepsilon$, then Assertion 13(ii) holds with $u^*(t)$ replaced by $\bar{u}(t)$

in Eq. (67).

IV. CONDITIONS OF OPTIMALITY

We now have the tools for determining the actual change of cost ΔV, or an estimate $\Delta\hat{V}$ thereof, for arbitrary changes of control or policy; the tedious work has been done, and the expressions obtained can be used fairly directly to obtain conditions of optimality and optimization algorithms. The results presented in this section are, mostly, well known. Our purpose in presenting them is two fold: firstly, they are easily obtained, using the expression for ΔV already obtained, thus showing the central role these expressions can play — it is not necessary to invent a new procedure for each situation; secondly, since conditions of optimality are naturally closely entwined with numerical optimization, the results provide some useful hints for the algorithms discussed later.

A. Necessary Conditions of Optimality

The major difficulty in obtaining necessary conditions of optimality, to which the present discussion contributes nothing, is proving [16] the existence of a separating hyperplane between the reachable set of the linearized system (Eqs. (9) and (10)) and the set which is the intersection of the set of improved costs and the set of permissible final states (we are assuming here a problem with terminal cost only, i.e. $L = 0$, and terminal constraints on the state). However, for the free end-point problem, or for the terminal constrained problems, given the existence of the separating hyperplane, (a new problem is defined in which the cost is $c^T x(t_\delta)$, c being the normal to the separating hyperplane) necessity of the minimum principle follows directly. For, if there exists a $u_1 \in \Omega$, $t_1 \in \Theta(\bar{u})$ such that $H(\bar{x}(t_1), u_1, \bar{\lambda}(t_1), t_1) <$ $H(x(t_1), \bar{u}(t_1), \bar{\lambda}(t_1), t_1)$, we can construct a new control defined by:

$$u(t) = \bar{u}(t) \quad \text{for all} \quad t \notin (t_1 - \varepsilon, t_1)$$

$$u(t) = u_1 \quad \text{for} \quad t \in [t_1 - \varepsilon, t_1] .$$

Clearly, $V_x^u(\bar{x}(t_1), t_1) = \bar{\lambda}(t_1)$. From the continuity of the integrand in Assertion 6(i), there exists an $\varepsilon > 0$ such that $\Delta V < 0$, contradicting the optimality of $\bar{u}$.

B. Sufficient Conditions

The exact expressions for ΔV are much more useful for proving the sufficiency of the minimum principle. Several examples illustrate this claim.

1. f, L, F linear in x.

Consider the system defined by:

$$f(x,u,t) = A(t)x + \varphi(u,t)$$

$$L(x,u,t) = m^T(t)x + \theta(u,t)$$

$$F(x) = n^T x$$

where A, φ, m, θ are continuous. Note that:

$$H_x(x,u,\lambda,t) = m(t) + A^T(t)\lambda$$

is independent of x. It follows that, for all $t \in T$, for all $x \in E_n$

$$V_x^{\bar{u}}(x,t) = V_x^{\bar{u}}(\bar{x}(t),t) \triangleq \bar{\lambda}(t) .$$

If we now use Assertion 6(ii), with $\bar{\lambda}(t)$ replacing $V_x^u(x(t),t)$, we see that satisfaction of the minimum principle by $(\bar{x},\bar{u})$ implies $\Delta V \geq 0$ for all $u \in G$.

2. f linear in x, L, F convex in x.

Let f be defined as in Section IV.B.1 and L and F by:

$$L(x,u,t) = \eta(x,t) + \theta(u,t)$$

$$F(x) = \xi(x)$$

such that η and ξ are convex in x and L, F satisfy H1, H2, $s = 2$. Then:

$$\hat{L}(x,u,t) \overset{\Delta}{=} \eta(\bar{x}(t),t) + \eta_x^T(\bar{x}(t),t)(x - \bar{x}(t)) + \theta(u,t)$$

$$L(x,u,t) \geq \hat{L}(x,u,t)$$

$$\hat{F}(x) \overset{\Delta}{=} \hat{F}(\bar{x}(t_\delta))(x - \bar{x}(t))$$

$$F(x) \geq \hat{F}(x) .$$

Let $\hat{V}$ denote the cost with L replaced $\hat{L}$, F by $\hat{F}$, and $\Delta\hat{V}$ denote $\hat{V}^u(x_0,t_0) - \hat{V}^u(x_0,t_0)$. Clearly $\Delta V \geq \Delta\hat{V}$. But, from Section IV.B.1, satisfaction of the minimum principle by $(\bar{x},\bar{u})$ implies $\Delta\hat{V} \geq 0$, and, hence, $\Delta V \geq 0$, for all $u \in G$.

3. Terminal Constraints.

Consider the problem immediately above, with the added terminal constraint:

$$x(t_\delta) = x_\delta .$$

Assume that $(\bar{x},\bar{u})$ satisfy the minimum principle for the modified problem with terminal cost:

$$\hat{F}(x) = F(x) + c^T(x - x_0)$$

and no constraint, i.e. with:

$$\bar{\lambda}(t_\delta) = F_x(x(t_\delta)) + c .$$

Since, for all $u \in G$ such that $x(t_\delta) = x_\delta$ the total cost for the original and modified problems are the same, satisfaction of the minimum principle implies, as before, that $\Delta\hat{V} \geq 0$ and $\Delta V \geq 0$ for the <u>original</u> problem.

Assume now that we have a terminal inequality constraint:

$$\alpha^T x(t_\delta) - b \leq 0 .$$

Assume that $(\bar{x},\bar{u})$, where

$$\alpha^T \bar{x}(t_\delta) = b$$

satisfy the minimum principle for the modified problem with terminal cost:

$$\hat{F}(x) = F(x) + c[\alpha^T x - b]$$

for some scalar $c > 0$, i.e., with

$$\bar{\lambda}(t_\delta) = F_x(\bar{x}(t_\delta)) + c\alpha .$$

Since for all $u \in G$ satisfying the terminal constraint, $c[\alpha^T x(t_\delta) - b] = c[\alpha^T(x(t_\delta) - \bar{x}(t_\delta))] \leq 0$, we have that the change of cost for the original problem is not less than the change of cost for the modified problem. Hence, optimality of $(\bar{x},\bar{u})$ for the original problem follows.

4. f Linear; L:F Quadratic.

Consider the system:

$$\dot{x}(t) = A(t)\, x(t) + B(t)\, u(t)$$

with:

$$L(x,u,t) \triangleq \tfrac{1}{2}\, x^T\, Q(t)x + \tfrac{1}{2}\, u^T u$$

$$F(x) \triangleq \tfrac{1}{2}\, x^T\, Q_f x$$

where A, B, Q are piecewise continuous, $Q, Q_f \geq 0$. Consider the policy $\bar{k}$ defined by:

$$\bar{k}(x,t) = K(t)x$$

$$K(t) = -B^T(t)\, P(t)$$

where $P(t)$ is the solution of the normal Riccati equation, assumed to exist in $[t_0, t_\delta]$. It is easily shown that $V^{\bar{k}}$ is quadratic in x.

$$V^{\bar{k}}(x,t) = \tfrac{1}{2} x^T P(t)x$$

so that

$$V_x^{\bar{k}}(x,t) = P(t)x$$

$$V_{xx}^{\bar{k}}(x,t) = P(t) .$$

In fact $P(t)$ is the solution of Eq. (42), along any $(\bar{x},\bar{u})$, since V_{xxx} is zero. Using Assertion 6(iv), the cost $V^u(x_0,t_0) - V^{\bar{k}}(x_0,t_0)$ for any $u \in G$ is given by:

$$\begin{aligned}\Delta V = \int_{t_0}^{t_\delta} [\tfrac{1}{2} x^T(t)Q(t)x(t) &+ \tfrac{1}{2} u^T(t)u(t) \\ &+ x^T(t)P(t) \{A(t)\dot{x}(t) + B(t)u(t)\} \\ - \tfrac{1}{2}x^T(t)Q(t)x(t) - \tfrac{1}{2} x^T(t)K^T(t)K(t)x(t) & \\ - x^T(t)P(t) &\{A(t) + B(t)K(t)\}x(t)]dt \\ = \int_{t_0}^{t_\delta} [u(t) - \bar{K}(t)x(t)]^T[u(t) - \bar{K}(t)x(t)]dt \geq 0 .\end{aligned}$$

Consider now the addition of a terminal constraint:

$$x(t_\delta) = x_\delta .$$

We deal with this by considering a modified problem with terminal cost $\hat{F}(x) \triangleq F(x) + c^T x$. Consider the policy $\bar{k}$ defined by:

$$\bar{k}(x,t) = u^*(t) + K(t)x(t)$$

$$u^*(t) = - B^T(t)\lambda(t)$$

$$K(t) = - B^T(t)P(t)$$

where

$$\lambda(t) = V_x^{\bar{k}}(0,t)$$

$$P(t) = V_{xx}^{\bar{k}}(0,t) \, .$$

Using Eqs. (33) and (35), with $\bar{x}(t) = 0$, $\bar{u}(t) = 0$,

$$-\dot{\lambda}(t) = [A(t) + B(t)K(t)]\lambda(t) + [P(t)B(t) + K^T(t)R(t)] \, u^*(t)$$

$$\lambda(t_\delta) = c$$

and $P(t)$ is the solution of the normal Riccati equation, assumed to exist on T. Hence:

$$\dot{x}(t) = A_1(t)x(t) + B(t)u^*(t), \; x(t_0) = x_0$$

$$u^*(t) = -B^T(t)\lambda(t), \quad \text{and} \quad A_1(t) \leq A(t) + B(t)K(t)$$

$$-\dot{\lambda}(t) = A_1^T(t)\lambda(t), \; \lambda(t_\delta) = c$$

so that $x(t_\delta) = -W(t_\delta, t_0)c$ where W is the controllability matrix corresponding to $(A_1(t), B(t))$. From Assertion 6(iv), $\bar{k}$ is optimal for the modified problem (i.e., $\Delta V \geq 0$ for all $u \in G$). Therefore, $\bar{k}$ is optimal for the original problem with terminal constraint:

$$x(t_\delta) = -W(t_\delta, t_0)c = x_\delta \, .$$

Obviously, if the system is completely controllable, x_δ may be arbitrary. For $c = c_1$, say, the resultant policy $\bar{k}$ satisfies $V^u(x_0, t_0) \geq V^{\bar{k}}(x_0, t_0)$, for all $u \in G$ such that $x(t_\delta; x_0, t_0, u) = -W(t_\delta, t_0)c_1$.

5. Hamilton-Jacobi-Bellman Sufficiency Result.

From Assertion 6(iv), if there exists a $\bar{k}$, satisfying H1A, H2A, and

$$H(x,u,V_x^{\bar{k}}(x,t),t) \geq H(x,\bar{k}(x,t),V_x^{\bar{k}}(x,t),t)$$

for all $(x,u,t) \in S$, then $V^u(x,t) \geq V^{\bar{k}}(x,t)$ for all $u \in G$, all $x \in E_n$, all $t \in T$. From Assertion 7:

$$-V_t^{\bar{k}}(x,t) = H(x,\bar{k}(x,t),V_x^{\bar{k}}(x,t),t) .$$

6. The Weierstrass Excess Function.

For the calculus of variation problem of choosing an optimal curve, the system dynamics are:

$$\dot{x}(t) = u(t)$$

i.e.,

$$H(x,u,\lambda,t) = L(x,u,t) + \lambda^T u .$$

Let $\bar{k}$ be a policy which satisfies our assumptions and also:

$$H_u(x,\bar{k}(x,t), V_x^{\bar{k}}(x,t),t) = 0$$

for all $x \in E_n$, $t \in T$. Hence:

$$V_x^{\bar{k}}(x,t) = - L_u(x,\bar{k}(x,t),t)$$

and, from Assertion 6(iv):

$$\Delta V = \int_{t_0}^{t_\delta} E(x(t),u(t),\bar{k}(x(t),t),t)dt$$

where

$$E(x,u,w,t) \triangleq L(x,u,t) - L(x,w,t) - L_u(x,w,t)(u - w)$$

is the Weierstrass Excess Function. Clearly, the non-negativity of E for $w = \bar{k}(x,t)$, for all $x \in E_n$, $t \in T$, is a sufficient condition for the global optimality of $\bar{k}$.

7. A Local Sufficiency Theorem for Strong Variations in Control.

The sufficiency results in the previous six subsections have been global. For more general systems only local sufficiency can be established. The usual procedure in the literature to prove local sufficiency employs an expression for the second variation valid only for weak variations in control. Use of Assertion 6(iv) enables a result for strong variations in control to be obtained.

Assertion 15. Let H1A, H2A be satisfied $s = 3$ and let $\bar{u}(t) \in$ interior of Ω for all $t \in T$. If for all $t \in T$.

(i) $H_u(\bar{x}(t),\bar{u}(t),\bar{\lambda}(t),t) = 0$

(ii) $H_{uu}(\bar{x}(t),\bar{u}(t),\bar{\lambda}(t),t) > 0$

(iii) $H(\bar{x}(t),u,\bar{\lambda}(t),t) > H(\bar{x}(t),\bar{u}(t),\bar{\lambda}(t),t)$ for all $u \in \Omega$ such that $u \neq \bar{u}(t)$.

(iv) The matrix Riccati differential equation:

$$-\dot{\bar{P}}(t) = H_{xx}(\bar{x}(t),\bar{u}(t),\bar{\lambda}(t),t) + f_x^T(\bar{x}(t),\bar{u}(t),t)\bar{P}(t) + \bar{P}(t)f_x(\bar{x}(t),\bar{u}(t),t) - \bar{K}(t)H_{uu}(\bar{x}(t),\bar{u}(t),\bar{\lambda}(t),t)\bar{K}(t) \tag{72}$$

$$\bar{P}(t_\delta) = F_{xx}(x(t_\delta)) \tag{73}$$

(where

$$\bar{K}(t) \triangleq - H_{uu}^{-1}(\bar{x}(t),\bar{u}(t),\bar{\lambda}(t),t) \; [H_{ux}(\bar{x}(t),\bar{u}(t),\bar{\lambda}(t),t) + f_u^T(\bar{x}(t),\bar{u}(t),t)\bar{P}(t)] \tag{74}$$

and $\bar{\lambda}(t)$ is the solution of Eqs. (26) and (27)) is bounded (has no conjugate points) in T, then, $\bar{u}$ is locally optimal in the sense that $V^u(x_0,t_0) > V^{\bar{u}}(x_0,t_0)$ for all $u \in G$, such that $u \neq \bar{u}$ and $d(u,\bar{u}) \leq \varepsilon$ for some $\varepsilon > 0$.

Proof: Consider the policy $\bar{k}$ defined by:

$$\bar{k}(x,t) = \bar{u}(t) + \bar{K}(t) \; [x - \bar{x}(t)] .$$

From (iv), the solution $\overline{P}(t)$ exists for all $t \in T$, so that $\overline{k}$ is well defined and satisfies our assumptions. From Assertion 15(ii), since $u^* = \overline{u}$, and $H_u(\overline{x}(t),\overline{u}(t),\overline{\lambda}(t),t) = 0$ it follows that

$$V_k^{\overline{k}}(\overline{x}(t),t) = \overline{\lambda}(t)$$

$$V_{xx}^{\overline{k}}(\overline{x}(t),t) = \overline{P}(t) .$$

Also, the control u that minimizes the second order expansion with respect to (x,u) of $H[x,u,V_x^{\overline{k}}(x,t),t]$ about $(\overline{x}(t),\overline{u}(t))$ is $\overline{u}(t) + \overline{K}(t)\,[x - \overline{x}(t)]$. It follows, as is shown in the Appendix, that there exists an $\varepsilon > 0$, and an $\alpha > 0$ such that if $d(u,\overline{u}) \leq \varepsilon$:

$$H[\overline{x}(t) + \delta x,\ u(t),\ V_x^{\overline{k}}(\overline{x}(t) + \delta x,t),t)]$$

$$-H[\overline{x}(t) + \delta x,\ \overline{u}(t) + \overline{K}(t)\delta x,\ V_x^{\overline{k}}(\overline{x}(t) + \delta x,t),t)]$$

$$\geq \tfrac{1}{2}\,\alpha w^T(t)w(t) + r(t)$$

where, for all $t \in T$:

$$|r(t)| \leq c[d(u,\overline{u})]^3$$

so that:

$$\Delta V \geq \tfrac{1}{2}\,\alpha \int_{t_0}^{t_\delta} w^T(t)w(t)dt + \int_{t_0}^{t_\delta} r(t)dt$$

$$\geq \tfrac{1}{2}\,\alpha\overline{c}[d(u,\overline{u})]^2 + \overline{r} \qquad \text{(see Appendix)}$$

where $|\overline{r}| \leq c[t_\delta - t_0][d(u,\overline{u})]^3$.

Hence, there exists an $\overline{\varepsilon} \in (0,\varepsilon)$ such that $d(u,\overline{u}) \leq \overline{\varepsilon}$ implies:

$$\Delta V \geq \tfrac{1}{4}\,\alpha\overline{c}[d(u,\overline{u})]^2$$

which proves the assertion.

COROLLARY. There exists $\varepsilon > 0$ such that

$V^u(x_0 + \delta x, t_0) > V^{\bar{u}}(x_0 + \delta x, t_0)$, for all $u \in G$, such that $u \neq \bar{u}$, $d(u,\bar{u}) \leq \varepsilon$, for all δx such that $\|\delta x\| \leq \varepsilon$.

V. FURTHER SECOND ORDER ESTIMATES OF ΔV

Assertion 16. Let $u, \bar{u} \in G$.

If either (i) H1A, H2A are satisfied, $s = 3$ and $d(u,\bar{u}) \leq \varepsilon$ or (ii) H1, H2 are satisfied, $s = 3$ and $d_1(u,\bar{u}) \leq \varepsilon$

$$\Delta V = \int_{t_0}^{t_f} [\Delta H(t) + \{\Delta H_x(t) + \bar{P}(t)\Delta f(t)\}^T \delta x(t)]dt + e_1$$

where $|e_1| \leq c\varepsilon^3$, $c < \infty$

$$\delta x(t) \triangleq x(t) - \bar{x}(t)$$

$$\Delta H(t) \triangleq H(\bar{x}(t),u(t),\bar{\lambda}(t),t) - H(\bar{x}(t),\bar{u}(t),\bar{\lambda}(t),t)$$

$$\Delta H_x(t) \triangleq H_x(\bar{x}(t),u(t),\bar{\lambda}(t),t) - H_x(\bar{x}(t),\bar{u}(t),\bar{\lambda}(t),t)$$

$$\Delta f(t) \triangleq f(\bar{x}(t),u(t),t) - f(\bar{x}(t),\bar{u}(t),t)$$

and $\bar{\lambda}(t)$, $\bar{P}(t)$ are the solutions of Eqs. (26)-(29).

COROLLARY. Assertion (16) holds if $\delta x(t)$ is replaced by the approximation $\delta\hat{x}(t)$ given by Eqs. (11) and (12).

Proof:

$$\bar{\lambda}(t) = V_x^{\bar{u}}(\bar{x},(t),t)$$

$$\bar{P}(t) = V_{xx}^{\bar{u}}(\bar{x}(t),t)$$

for all $t \in T$. Expanding the integrand of Assertion 6(ii) about $\bar{x}(t)$ with respect to x up to second order and neglecting the second order terms yields the integrand of Assertion 16. The neglected terms are:

$$\tfrac{1}{2}\,\delta x^T(t)[H_{xx}(x*(t),u(t),\ V_x^{\bar{u}}(x*(t),t),t)$$
$$-\ H_{xx}(x*(t),\ V_x^{\bar{u}}(x*(t),t),t)$$
$$+\ \{f_x(x*(t),u(t),t)\ -\ f_x(x*(t),\bar{u}(t),t)\}^T V_{xx}^{\bar{u}}(x*(t),t)$$
$$+\ V_{xx}^{\bar{u}}(x*(t),t)\ \{f_x(x*(t),u(t),t)\ -\ f_x(x*(t),\bar{u}(t),t),t)$$
$$+\ \sum_{i=1}^{n}\ V_{xxx_i}^{\bar{u}}\ (x*(t),t),\ \Delta f_i(t)]\delta x(t)$$

where $x*(t) = \bar{x}(t) + \theta\delta x(t),\ 0 \le \theta \le 1.$

Since $\|\delta x(t)\| \le c_1\varepsilon,\ c_1 < \infty,$ for all $t \in T,$ and the terms inside the bracket and all of the form $\varphi(u(t)) - \varphi(\bar{u}(t)),$ the result follows.

If $\delta x(t)$ is replaced by $\delta\hat{x}(t)$ in the integrand of Assertion 16, then a further error of the form

$$\int_{t_0}^{t_\delta} [\theta(u(t)) - \theta(\bar{u}(t))]^T[\delta x(t) - \delta\hat{x}(t)]dt$$

is introduced. Since $\|\delta x(t) - \delta\hat{x}(t)\| \le c_2\varepsilon^2,\ c_2 < \infty,$ the error is not greater than $c_3\varepsilon^2,\ c_3 < \infty.$

This result leads to a necessary condition of optimality for singular control problems (Section 8). A further development of this result lends to a strong version of the usual second variation formula.

Assertion 17. Let the hypotheses of Assertion 16 be satisfied. Then

$$\Delta V = \int_{t_0}^{t_\delta} [\Delta H(t) + \Delta H_x^T(t)\delta\hat{x}(t) + \tfrac{1}{2}\ \delta\hat{x}^T(t)Q(t)\delta\hat{x}(t)]dt$$
$$+\ \tfrac{1}{2}\ \delta\hat{x}^T(t_\delta)F_{xx}(x(t_\delta))\delta\hat{x}(t_\delta)\ +\ e_1$$

$$|e_1| \leq c\varepsilon^3, \quad c < \infty$$

$$Q(t) \triangleq H_{xx}(\bar{x}(t),\bar{u}(t),\bar{\lambda}(t),t)$$

and the remaining terms are as defined in Assertion 16.

Proof: $\Delta f(t)$ in Assertion 16 is replaced by $\delta\dot{\hat{x}}(t) - f_x(\bar{x}(t),\bar{u}(t),t)\,\delta\hat{x}(t)$ and the term $\int_{t_0}^{t_\delta} \dot{\bar{P}}(t)\delta\hat{x}(t)dt$ integrated by parts, yielding Assertion 17 with an extra error term:

$$\tfrac{1}{2}\int_{t_0}^{t_\delta} \delta\hat{x}^T(t)[\bar{P}(t)\Delta f_x(t) + \Delta f_x^T(t)\bar{P}(t)]\delta\hat{x}(t)dt\ ,$$

where

$$\Delta f_x(t) \triangleq f_x(\bar{x}(t),u(t),\bar{\lambda}(t),t) - f_x(\bar{x}(t),\bar{u}(t),\bar{\lambda}(t),t)\ .$$

Clearly $\delta\hat{x}(t) \leq c_1\varepsilon,\ c_1 < \infty$, for all $t \in T$. Hence, the magnitude of the extra error term is not greater than $c_2\varepsilon^3,\ c_2 < \infty$; the result follows.

The weak version of Assertion 17 is the usual first and second variation result:

Assertion 18. Let $\bar{u},u \in G$. If either:
(i) H1 and H2 are satisfied $s = 3$, f and L are linear in u and $d(u,\bar{u})$ or $d_1(u,\bar{u}) \leq \varepsilon$.
(ii) H1A and H2A are satisfied $s = 3$ and $\|u(t) - \bar{u}(t)\| \leq \varepsilon$ for all $t \in T$.

$$\Delta V = \int_{t_0}^{t_\delta} H_u^T(\bar{x}(t),\bar{u}(t),\bar{\lambda}(t),t)\ \Delta u(t)dt$$

$$+ \int_{t_0}^{t_\delta} [\tfrac{1}{2}\Delta u^T(t)R(t)\Delta u(t) + \Delta u^T(t)C(t)\delta\hat{x}(t) + \tfrac{1}{2}\delta\hat{x}^T(t)Q(t)\delta\hat{x}(t)]dt +$$

$$+ \tfrac{1}{2}\, \delta\hat{x}^T(t) F_{xx}(\bar{x}(t_\delta))\delta\hat{x}(t_\delta) + e_1$$

$|e_1| \le c\varepsilon^3, \; c < \infty$

$$\Delta u(t) \triangleq u(t) - \bar{u}(t)$$

$$R(t) \triangleq H_{uu}(\bar{x}(t),\bar{u}(t),\bar{\lambda}(t),t)$$

$$C(t) = H_{ux}(\bar{x}(t),\bar{u}(t),\bar{\lambda}(t),t) \,.$$

$\delta\hat{x}(t)$ is the solution of Eqs. (11) and (12) and the remaining terms are as defined in Assertions 16 and 17.

Proof: Expand the integrand of Assertion 17 with respect to u up to terms of second order. The resultant error is not greater than $c_1\varepsilon^3$, $c_1 \le \infty$.

In Assertions (16)-(18) we were concerned with comparing V^u with $V^{\bar{u}}$. In the sequel, we will compare V^u with $V^{\bar{k}}$, where $\bar{k}$ is the local linear policy defined by:

$$\bar{k}(x,t) = \bar{u}(t) + \bar{K}(t)[x - \bar{x}(t)] \tag{75}$$

where $\bar{u} \in G$, and $\bar{K}$ is piecewise continuous. Clearly $\bar{k}$ generates $(\bar{x},\bar{u})$ so that $V^{\bar{k}}(x_0,t_0) = V^{\bar{u}}(x_0,t_0)$.

Assertion 19. Let $u,\bar{u} \in G$, $\bar{K}$ be piecewise continuous. Let H1A, H2A be satisfied, $s = 3$ and $d(u,\bar{u}) \le \varepsilon$ or $d_1(u,u) \le \varepsilon$.

$$\Delta V = \int_{t_0}^{t_\delta} [\Delta H(t) + \{\Delta H_x(t) + \bar{P}(t)\Delta f(t)\}^T \,\delta\hat{x}(t)]dt + e_1$$

$|e_1| \le c\varepsilon^3, \; c < \infty$

$$\Delta H(t) \triangleq H(\bar{x}(t),u(t),\bar{\lambda}(t),t) - H(\bar{x}(t),\bar{u}(t) + \bar{K}(t)\delta\hat{x}(t),\bar{\lambda}(t),t)$$

$$\Delta H_x(t) \triangleq H_x(\bar{x}(t),u(t),\bar{\lambda}(t),t) - H_x(\bar{x}(t),\bar{u}(t) + \bar{K}(t)\delta\hat{x}(t),\bar{\lambda}(t),t)$$

$$\Delta f(t) \triangleq f(\bar{x}(t),u(t),t) - f(\bar{x}(t),\bar{u}(t) + \bar{K}(t)\delta\hat{x}(t),t)$$

and $\bar{\lambda}(t), \bar{P}(t)$ are the solutions of:

$$-\dot{\bar{\lambda}}(t) = H_x(\bar{x}(t),\bar{u}(t),\bar{\lambda}(t),t) + \bar{K}^T(t)H_u(\bar{x}(t),\bar{u}(t),\bar{\lambda}(t),t) \tag{76}$$

$$\begin{aligned}-\dot{\bar{P}}(t) = {} & H_{xx}(\bar{x}(t),\bar{u}(t),\bar{\lambda}(t),t) + \bar{P}(t)f_x(\bar{x}(t),\bar{u}(t),t) \\ & + f_x^T(\bar{x}(t),\bar{u}(t),t)\bar{P}(t) + \bar{K}^T(t)H_{uu}(\bar{x}(t),\bar{u}(t),\bar{\lambda}(t),t)\bar{K}(t) \\ & + \bar{K}^T(t)[H_{ux}(\bar{x}(t),\bar{u}(t),\bar{\lambda}(t),t) + f_u^T(\bar{x}(t),\bar{u}(t),t)\bar{P}(t)] \\ & + [H_{ux}(\bar{x}(t),\bar{u}(t),\bar{\lambda}(t),t) + f_u^T(\bar{x}(t),\bar{u}(t),t),\bar{P}(t)]^T\bar{K}(t)\end{aligned} \tag{77}$$

with the usual terminal conditions.

Proof:

$$\bar{\lambda}(t) = V_x^{\bar{k}}(\bar{x}(t),t)$$
$$\bar{P}(t) = V_{xx}^{\bar{k}}(x(t),t) .$$

The proof follows in much the same way as for Assertion 16, using Assertion 6(iv) instead of 6(ii).

Assertion 20. Let the hypotheses of Assertion 19 be atisfied.

$$\begin{aligned}\Delta V = \int_{t_0}^{t_\delta} & [\Delta H(t) + \Delta H_x^T(t)\delta\hat{x}(t) + \tfrac{1}{2}\,\delta\hat{x}^T(t)Q(t)\delta\hat{x}(t)]dt \\ & + \tfrac{1}{2}\,\delta\hat{x}^T(t_\delta)F_{xx}(\bar{x}(t_\delta)\delta\hat{x}(t_\delta) + e_1\end{aligned}$$

$|e_1| \leq c\varepsilon^3, \ c < \infty$

$$Q(t) \triangleq H_{xx}(\bar{x}(t),\bar{u}(t),\bar{\lambda}(t),t)$$

and the remaining terms are as defined in Assertion 19.

Assertion 21. Let $u,\bar{u} \in G$, $\bar{K}$ be piecewise continuous. Let H1A, H2A be satisfied, $s = 3$ and either (i) f, L are linear in u and $d(u,\bar{u}) \leq \varepsilon$ or $d_1(u,\bar{u}) \leq \varepsilon$ or (ii) $\|u(t) - \bar{u}(t)\| \leq \varepsilon$ for all $t \in T$. Then:

$$\Delta V = \int_{t_0}^{t_\delta} [\tfrac{1}{2} w^T(t)R(t)w(t) + w^T(t)\{R(t)\bar{K}(t) + C(t) + f_u^T(\bar{x}(t),\bar{u}(t),t)\bar{P}(t)\}\delta\hat{x}(t)]dt$$

$$+ \int_{t_0}^{t_\delta} H_u^T(\bar{x}(t),\bar{u}(t),\bar{\lambda}(t),t)w(t)dt + e_1$$

$|e_1| \leq c\varepsilon^3$, $c < \infty$

$$w(t) \triangleq u(t) - \bar{u}(t) - \bar{K}(t)\, \delta\hat{x}(t) .$$

R and C are defined in Assertion 18. $\delta\hat{x}(t)$ is the solution of Eqs. (11) and (12). $\bar{\lambda}(t)$ and $\bar{P}(t)$ are the solutions of Eqs. (75) and (76) with the usual boundary conditions.

COROLLARY. If, in addition:

$$H_u(\bar{x}(t),\bar{u}(t),\bar{\lambda}(t),t) = 0$$

$$R(t)\bar{K}(t) + C(t) + f_u^T(\bar{x}(t),\bar{u}(t),t)\bar{P}(t) = 0$$

for all $t \in T$, then,

$$\Delta V = \int_{t_0}^{t_\delta} \tfrac{1}{2} w^T(t)R(t)w(t)dt + e_1 .$$

Proof: The integrand of Assertion 19 is expanded with respect to u about $\bar{u}(t)$.

VI. OPTIMIZATION ALGORITHMS

The optimization algorithms are well described in [5]. Here we will content ourselves by pointing out how Assertion 6 motivates the algorithms and gives the conditions required to ensure $\Delta V < 0$ at each iteration. However, one of the purposes of this paper is to give estimates of ΔV which can, hopefully, be used to obtain improved algorithms. The expressions for ΔV hold for arbitrary $u \in G$; the earlier expositions of D.D.P. emphasized expressions for ΔV valid for the particular choice of u employed in the algorithms.

A. The First Order D.D.P. Algorithm

Assertion 6(i) is used, or rather its practical version, Assertion 12(i), where $V_x^u(\bar{x}(t),t)$ is approximated by $\hat{\lambda}(t)$, the solution of Eqs. (49) and (51). $u(t)$ is chosen to minimize $H(\bar{x}(t),u,\hat{\lambda}(t),t)$ with respect to $u \in \Omega$, for all $t \in E$. $\mu(E) = \varepsilon$, the length of the set E, is altered, according to various rules [5,7], until the error ($\leq c_1\varepsilon^2, c_1 < \infty$) in the estimate $\hat{a}(t_0)$ of ΔV is sufficiently small, compared with ΔV. Computationally, $\mu(E)$ is reduced until a sufficient reduction in ΔV is obtained (e.g. ($\Delta V < 0$ and $|\Delta V/\hat{a}(t_0)| \geq 0.5$).

B. The Second Order D.D.P. Algorithm

Assertion 6(iii) is used, or rather its practical version, Assertion 13(ii). $u^*(t)$ is chosen to minimize $H(\bar{x}(t),u,\hat{\lambda}(t),t)$ with respect to $u \in \Omega$ (assuming, for the moment, that $u^*(t) \in$ interior of Ω), and K to minimize $P(t)$, for all $t \in T$, with $\gamma_i(t) \equiv 0$, $i = 1,\ldots,n$, i.e.

$$K(t) = - H_{uu}^{-1}[H_{ux} + f_u^T \hat{P}]$$

where the omitted arguments are as in Eq. (67). Eq. (67) becomes a Matrix Riccati differential equation whose solution on T is assumed to exist. (From the usual theory of the Riccati equation, $\hat{P}(t)$ due to this K is not more positive definite than $\hat{P}(t)$ due to any other K). The new control is:

$$u(t) = \bar{u}(t) \quad \text{for all} \quad t \in T - E$$

$$u(t) = u^*(t) + K(t)[x(t) - \bar{x}(t)] \quad \text{for} \quad t \in E$$

where $t_1 \in T - E$, $t_2 \in E$ implies $t_1 < t_2$. Again $\mu(E)$ is chosen to ensure satisfactory cost reduction.

If $(\bar{x}, \bar{u})$ satisfy the conditions of Assertion (15), then the policy k produced by the algirithm is the same as policy $\bar{k}$ of the Assertion. Hence, from the Corollary, the policy k is locally optimal, in the sense that there exists an $\varepsilon > 0$ such that,

$$V^u(x(t) + \delta x, t) > V^k(\bar{x}(t) + \delta x, t)$$

for all $u \in G$ satisfying $u \neq \bar{u}$, $d(u,u) \leq \varepsilon$, for all δx satisfying $\|\delta x\| \leq \varepsilon$.

Since ΔV can be estimated for arbitrary u or k, within the limits of our assumptions, the algorithms can be easily modified to avoid difficulties, such as $\hat{P}(t)$ exploding. In this case, $K(t)$ can be set zero when $\|\hat{P}(t)\|$ reaches a certain norm. Since $u^*(t)$ minimize H, ΔH is still nonpositive, so that $a(t_0) \leq 0$, and the improved k satisfies our assumptions.

C. Control Constraints

The second order algorithm in Section VI. B is easily modified. If $u^*(t)$ lies on the boundary of Ω, then δu is modified so that $u^*(t) + \delta u$ "approximately" minimizes $H(\bar{x}(t) + \delta x, u^*(t) + \delta u, \hat{\lambda}(t) + \hat{P}(t)\delta x, t)$ subject to $u^*(t) + \delta u \in \Omega$. One interpretation of "approximately" is to minimize the second order expansion of H with respect to δx and δu subject to $N(t)\delta u = 0$, where $N(t)$ defines a linearized version of the boundary of Ω at $u^*(t)$. The solution to this is:

$$\delta u = K(t)\delta x$$

$$K(t) = -\Pi(t)[N^T(t)R(t)N(t)]^{-1}N^T(t)C(t)\delta x$$

where $R(t)$ and $C(t)$ are defined in Assertion 18, $u^*(t)$ replacing $\bar{u}(t)$, $\hat{\lambda}(t)$ replacing $\bar{\lambda}(t)$. A quadratic approximation to the boundary of Ω is used in [5].

Application of the improvement policy k yields, for suitable choice of $\mu(E)$, a reduction in V, but does not necessarily ensure satisfaction of the control constraint. An alternative procedure [5] on the "forward run" is to minimize $H(x(t),u,\hat{\lambda}(t) + \hat{P}(t)\delta x(t),t)$ with respect to $u \in \Omega$ to yield $u(t)$. Since then, the actual improvement policy does not correspond exactly to the predicted improvement policy k, further errors are introduced which have not yet been assessed.

Convergence of the algorithms has not yet been properly examined.† One class of algorithms maintains the control u in the feasible region. Use of the linear improvement policy defined above does not necessarily guarantee this.

D. Bang-Bang Control

In some problems the optimal control is bang-bang. For these problems the effective control parameter is the time t_1 at which the control changes from one value, v^- say, to another value v^+. To be more definite let $\bar{u}$ denote the nominal control (with resultant trajectory $x(t;x_0,t_0,\bar{u}) \triangleq \bar{x}(t)$), such that $\bar{u} \in G$ and:

$$\bar{u}(t) = v^-, \quad t \in [t_a,\bar{t}_1)$$
$$\bar{u}(t) = v^+, \quad t \in (\bar{t}_1,t_b],$$

and u has left and right limits of v^- and v^+,

†See, however, Mayne and Polak, JOTA (to appear).

respectively, at $\bar{t}_1$. Let u denote a control, identical to $\bar{u}$ on $T - [t_a, t_b]$, and otherwise defined as $\bar{u}$ with $\bar{t}_1$ replaced by t_1. Essentially, the switching time t_1 defines u, and we write V^{t_1} in place of V^u, $V^{\bar{t}_1}$ in place of $V^{\bar{u}}$ etc. We wish to determine how V^{t_1} depends on t_1. Previous analyses [5,18] are straightforward in principle, if tedious in detail.

Using Assertion 6 both the left and right derivatives can be calculated and shown to be equal.

Let $\hat{u}$ denote the control defined by:

$$\hat{u}(t) = v^-, \qquad t \in [t_a t_b]$$

$$\hat{u}(t) = \bar{u}(t), \quad t \in T - [t_a, t_b] .$$

Let

$$\hat{x}(t;x_1) \triangleq x(t;x_1,\bar{t}_1,\hat{u})$$

$$x^{t_1}(t;x_1) \triangleq x(t;\hat{x}(t_a;x_1),t_1,u)$$

$$\bar{x}_1 \triangleq \bar{x}(\bar{t}_1) .$$

These trajectories are shown in Figure 1. x^{t_1} 'peels away' from $\hat{x}$ at t_1. $\hat{x}(t;\bar{x}_1) = \bar{x}(t)$ for $t \leq \bar{t}_1$, $\hat{x}(t;x_1) = x^{t_1}(t;x_1)$ for $t \leq t_1$. Let

$$\theta(x_1,t_1) \triangleq V^{t_1}(\hat{x}(t_a;x_1),t_a) .$$

<u>Assertion</u> 22. Let H1, H2, H3 be satisfied, $s = 3$. Let $u,\bar{u} \in G$. Then:

(i) $$\theta^-_{t_1}(\bar{x}_1,\bar{t}_1) = \theta^+_{t_1}(\bar{x}_1,\bar{t}_1)$$

$$= H(\bar{x}_1,v^-,\bar{\lambda}(\bar{t}_1),\bar{t}_1) - H(\bar{x}_1,v^+,\bar{\lambda}(\bar{t}_1),\bar{t}_1) \qquad (78)$$

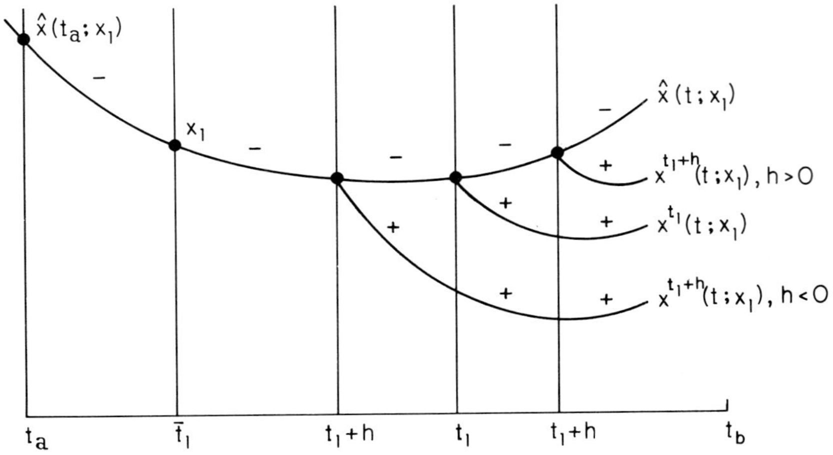

$$\text{(ii)} \quad \Theta^-_{x_1 t_1}(\bar{x}_1, \bar{t}_1) = \Theta^+_{x_1 t_1}(\bar{x}_1, \bar{t}_1)$$

$$= H_x(\bar{x}, v^-, \bar{\lambda}(\bar{t}_1), \bar{t}_1) - H_x(\bar{x}_1, v^+, \bar{\lambda}(\bar{t}_1), \bar{t}_1)$$

$$+ \bar{P}(\bar{t}_1)[f(\bar{x}_1, v^-, \bar{t}_1) - f(\bar{x}_1, v^+, \bar{t}_1] . \quad (79)$$

$$\text{(iii)} \quad \Theta^-_{t_1 t_1}(\bar{x}_1, \bar{t}_1) = \Theta^+_{t_1 t_1}(\bar{x}_1, \bar{t}_1))$$

$$= H_t(\bar{x}_1, v^-, \bar{\lambda}(\bar{t}_1), \bar{t}_1) - H_t(\bar{x}_1, v^+, \bar{\lambda}(\bar{t}_1), \bar{t}_1)$$

$$+ [H_x(\bar{x}_1, v^-, \bar{\lambda}(\bar{t}_1), \bar{t}_1)$$

$$- H_x(\bar{x}_1, v^+, \bar{\lambda}(\bar{t}_1), \bar{t}_1)]f(\bar{x}_1, v^-, \bar{t}_1)$$

$$+ [f(\bar{x}_1, v^-, \bar{t}_1)$$

$$- f(\bar{x}_1, v^+, \bar{t}_1)]^T[-H_x(\bar{x}_1, v^+, \bar{\lambda}(\bar{t}_1), \bar{t}_1)$$

$$+ \bar{P}(\bar{t}_1)\{f(\bar{x}_1, v^-, t_1) - f(\bar{x}_1, v^+, \bar{t}_1)\}] . \quad (80)$$

Where the superscripts $-$, $+$ denote the left and right limits, respectively and $\bar{\lambda}(t) \triangleq V_x^{\bar{t}_1}(\bar{x}(t), t), \bar{P}(t) \triangleq V_{xx}^{\bar{t}_1}(\bar{x}(t), \bar{t})$.

Proof: Let $\Delta\theta(h)$ denotes $\theta(x_1,t_1 + h) - \theta(x_1,t_1)$ where $t_1 \in (t_a,t_b)$.

(i) The left derivatives $(h < 0)$.

From Assertion 6(i), regarding x^{t_1} as the nominal, and x^{t_1+h} as the new trajectory:

$$\Delta\theta(h) = \int_{t_1+h}^{t_1} [H(\hat{x}(t;x_1),v^+,V_x^{t_1+h}(\hat{x}(t;x_1)t),t) - H(\hat{x}(t;x_1),v^-,V_x^{t_1+h}(\hat{x}(t;x_1),t)]dt$$

$$= \int_{t_1}^{t_1+h} [H(\hat{x}(t;x_1),v^-,V_x^{t_1+h}(\hat{x}(t;x_1),t)dt - H(\hat{x}(t;x_1),v^+,V_x^{t_1+h}(\hat{x}(t;x_1),t)]dt .$$

Since H, $\hat{x}$, V_x are all continuous functions, the limit $\Delta\theta(h)/h$ as h tends to zero from below exists and is given by:

$$\theta^-_{t_1}(x_1,t_1) = H(\hat{x}(t_1;x_1),v^-,V_x^{t_1^-}(\hat{x}(t_1;x_1),t_1) - H(\hat{x}(t_1;x_1),v^+,V_x^{t_1^-}(\hat{x}(t_1;x_1),t_1) . \quad (81)$$

($V_x^{t_1+h}$, $h < 0$, is associated with control v^+ in $(t_1 + h,t_1)$). Since $\hat{x}(t_1;x_1) = x_1$, Eq. (78) follows on setting $x_1 = \bar{x}(\bar{t}_1) = \bar{x}_1$, $t_1 = \bar{t}_1$. Also, since the necessary derivatives exist, Eq. (79) follows from the partial derivative of $\theta^-_{\bar{t}_1}(x_1,\bar{t}_1)$ with respect to x_1. Also, the <u>left</u> derivative of $\theta^-_{t_1}(x_1,\bar{t}_1)$ with respect to t_1 exists and is given by Eq. (80), since:

$$- \frac{d}{dt_1} V_x^{t_2}(\hat{x}(t_1;x_1),t_-) = H_x(\hat{x}(t_1;x_1),v^+,V_x^{t_2}(\hat{x}(t_1;x_1),t_1)$$
$$+ V_{xx}^{t_2}(\hat{x}(t_1;x_1),t_1)[f(\hat{x}(t_1;x_1),v^+,t_1) - f(\hat{x}(t_1;x_1),v^-,t_1)] \quad (82)$$

for $t_1 \geq t_2$.

(ii) The right derivatives $(h > 0)$.

From Assertion 6(ii),

$$\Delta\theta(h) = \int_{t_1}^{t_1+h} [H(\hat{x}(t;x_1),v^-,V_x^{t_1}(\hat{x}(t;x_1),t),t)$$
$$- H(\hat{x}(t;x_1),v^+,V_x^{t_1}(\hat{x}(t;x_1),t),t)]dt .$$

Hence:

$$\theta_{t_1}^+(x_1,t_1) = H(\hat{x}(t_1^+;x_1),v^-,V_x^{t_1}(\hat{x}(t_1^+;x_1),t_1^+),t_1^+)$$
$$- H(\hat{x}(t_1^+;x_1),v^+,V_x^{t_1}(\hat{x}(t_1^+,x_1),t_1^+),t_1^+) . \quad (83)$$

Eqs. (78) to (81) follow as before. Eq. (82) still holds since in (t_1,t_1+h), $V_x^{t_1}$ is associated with v^+, and $\hat{x}$ with v^-.

These results lead to necessary conditions of optimality: $\theta_{t_1} = 0$, $\theta_{t_1t_1} \geq 0$. They also lead to the improvement law at $\bar{t}_1$.

$$\delta t_1 = -\theta_{t_1t_1}^{-1}(\bar{x}_1,\bar{t}_1)[\theta_{t_1}(\bar{x}_1,\bar{t}_1) + \theta_{t_1x_1}(\bar{x}_1,\bar{t}_1)\{x - \bar{x}_1\}] \quad (84)$$

where x is the value of $x(\bar{t}_1)$ obtained using the improvement policy with $t_1 = \bar{t}_1$. If $\delta t_1 > 0$, then $x(t)$ is obtained, using initial condition $(x_1,\bar{t}_1)$, control v^- to $t_1 + \delta t_1$, control v^+ thereafter. If $\delta t_1 < 0$, control v^+ is applied from $t_1 + \delta t_1$ with initial condition $(x(\bar{t}_1 + \delta t_1),\bar{t}_1 + \delta t_1)$.

E. Free Terminal Time

If t_δ is a free parameter, then the cost may be expressed as $V^u(x,t;t_\delta)$, and the partial derivatives of V^u with respect to t_δ are of interest. F may now be a function, $F(x,t)$, of (x,t). Let H1A, H2A, H3 be satisfied, $s = 2$, $u \in G$ and, in addition, let F and its derivatives with respect to (x,t) up to order 2 be continuous in (x,t). As before let $x(t)$ denote $x(t;x_0,t_0,u)$. Then:

(i) $$V^u_{t_\delta}(x_0,t_0;t_\delta) = L(x(t_\delta),u(t_\delta) + \dot{F}(x(t_\delta),t_\delta)$$
$$= L(x(t_\delta),u(t_\delta),t_\delta) + F_t(x(t_\delta),t_\delta) + F_x^T(x(t_\delta),t_\delta)f(x(t_\delta),u(t_\delta),t_\delta)$$
$$= H(x(t_\delta),u(t_\delta),\lambda(t_\delta),t_\delta) + F_t(x(t_\delta),t_\delta) .$$

(ii) $$V^u_{xt_\delta}(x_0,t_0;t) = H_x(x(t_\delta),u(t_\delta),\lambda(t_\delta),t_\delta) + F_{xt}(x(t_\delta),t_\delta)$$

(iii) $$V^u_{t_\delta t_\delta}(x_0,t_0;t_\delta) = \dot{L}(x(t_\delta),u(t_\delta),t) + \ddot{F}(x(t_\delta),t_\delta)$$

where $\dot{L}$ denotes $L_t + L_x^T f + L_u^T \dot{u}(t)$ etc., $\lambda(t_\delta)$ denotes $F_x(x(t_\delta),t_\delta)$.

F. Terminal Constraints

Terminal Constraints are treated in [5] by a formal Lagrange multiplier technique. Consider, for simplicity, the control problem with inequality constraint:

$$a^T x(t_\delta) - b \leq 0 .$$

Suppose that the second order D.D.P. algorithm of Section VI.B, with the terminal value of λ modified from $F_x(x(t_\delta))$ to $ca + F_x(x(t_\delta))$, for some $c > 0$, yields $(\bar{x},\bar{u})$ satisfying Assertion 15'. $\bar{u}$ is then locally optimal for a modified problem, without terminal constraints, for which the terminal cost is $\hat{F}(x(t_\delta))$ where:

$$\hat{F}(x) \triangleq F(x) + c(a^T x - b_1)$$

and b_1 is arbitrary. Let $b_1 = \bar{x}(t_\delta)$, yielding:

$$F(x) = \hat{F}(x) - ca^T(x - \bar{x}(t_\delta)) .$$

Hence $F \geq \hat{F}$ for all x such that:

(i) $ca^T(x - \bar{x}(t_\delta)) \leq 0$. Hence, there exists an $\varepsilon > 0$ such that $V^u(x_0,t_0) \geq V^{\bar{\bar{u}}}(x_0,t_0)$ for all $u \in G$ such that (ii) $d(u,u) \leq \varepsilon$, (iii) $x(t_\delta;x_0t_0,u)$ satisfies (i). Can c be changed to make $a^T\bar{x}(t_\delta) = b$, in which case a locally optimal solution to the original problem will have been obtained?

Changing c to $c + \delta c$, $|\delta c| \leq \varepsilon$, will change $\bar{\lambda}(t)$ to $\bar{\lambda}(t) + \delta\lambda(t)$, changing $\bar{u}(t)$ to $\bar{u}(t) + \delta\, u(t)$, $\bar{x}(t)$ to $\bar{x}(t) + \delta\, u(t)$ etc. Approximations $\delta\hat{\lambda}$, $\delta\hat{u}$, $\delta\hat{x}$ to those quantities can be obtained as follows: (See also Section IV.B.4.)

$$-\delta\dot{\hat{\lambda}}(t) = A_1^T(t)\delta\hat{\lambda}(t) \tag{85}$$

$$\delta\hat{\lambda}(t_\delta) = \delta c\ a \tag{86}$$

$$\delta\hat{u}(t) = -R^{-1}(t)B^T(t)\delta\hat{\lambda}(t) \tag{87}$$

$$\delta\dot{\hat{x}}(t) = A_1(t)\delta x(t) + B^T(t)\delta\hat{u}(t) \tag{88}$$

$$\delta\hat{x}(t_0) = 0 \tag{89}$$

where

$$A_1(t) \triangleq f_x(\bar{x}(t),\bar{u}(t),t) + f_u(\bar{x}(t),\bar{u}(t),t)K(t) \tag{90}$$

$$B(t) \triangleq f_u(\bar{x}(t),\bar{u}(t),t) \tag{91}$$

$$R(t) \triangleq H_{uu}(\bar{x}(t),\bar{u}(t),\bar{\lambda}(t),t) . \tag{92}$$

Hence:

$$a^T\delta\hat{x}(t_\delta) = - a^T W(t_\delta,t_0)a, \quad \delta c \tag{93}$$

where

$$W(t_\delta,t_0) = \int_{t_0}^{t_\delta} \Phi(t_\delta,t)B(t)R^{-1}(t)B^T(t)\Phi^T(t_\delta,t)dt \tag{94}$$

and $\Phi(t_\delta,t)$ is the transition matrix corresponding to $A_1(t)$. It can be shown that $\|\delta x(t_\delta) - \delta\hat{x}(t_\delta)\| \leq c_1\varepsilon^2$, $c_1 < \infty$. Hence Eq. (85) can be used, if $a^T Wa \neq 0$, to choose a δc to reduce $a^T\bar{x}(t_\delta) - b$. The algorithm is identical to that described in [5]; the above discussion gives a different interpretation. Each choice of c yields a locally optimal solution satisfying the constraint equation with b replaced by $a^T\bar{x}(t_\delta)$. c is changed, using Eq. (85), to make $a^T\bar{x}(t_\delta) = b$. See [19] where the use of this procedure for a special system, f linear, L convex, leads to an algorithm with proven convergence.

In practice one would compute.

$$a^T W(t_\delta,t_0)a = \int_{t_0}^{t_\delta} \delta\tilde{\lambda}^T(t)B(t)R^{-1}(t)B^T(t)\delta\tilde{\lambda}(t)dt$$

where

$$\begin{aligned}\delta\tilde{\lambda}(t) &\triangleq \delta\lambda(t)/\delta c \\ &= \Phi^T(t_\delta,t)a\end{aligned}$$

is the solution of Eq. (85) with terminal condition (Eq. 86) replaced by:

$$\tilde{\lambda}(t_\delta) = a .$$

VII. STATE CONSTRAINED PROBLEMS

The purpose of this section is to derive some recent results due to Jacobson et al [20] on state constrained problems using, in part, the expressions for ΔV previously derived. This avoids the difficulty of treating the differential equation as an equality constraint.

The basic problem considered is the same as that previously considered, except that there is no integral cost $(L=0)$, only terminal Cost F, and that there is an added constraint of the form:

$$S(x(t)) \leq 0 \quad \text{for all} \quad t \in T . \tag{95}$$

where $S : E_n \rightarrow E_m$ and its derivatives up to order p are continuous, $f(x,u,t)$ is replaced by $f(x,u)$. Although our discussion holds for arbitrary $m \leq n$ (recall the control is also m-dimensional) some of our conditions will be rather artificial for $m \neq 1$, although they can be modified, at the expense of extra complexity, to avoid this artificiality.

We will be concerned with necessary conditions of optimality. Let $(\overline{x},\overline{u})$ denote the optimal solution, assumed to exist. Let $\overline{y}(t)$, $y(t)$ denote respectively, $S(\overline{x}(t))$, $S(x(t))$. Let $\delta y(t)$ denote $y(t) - \overline{y}(t)$. An approximation $\delta\hat{y}(t)$ to $\delta y(t)$ can be obtained as the solution of:

$$\delta\dot{\hat{x}}(t) = A(t)\delta x(t) + B(t)\delta u(t)$$

$$\delta\hat{y}(t) = C(t)\delta\hat{x}(t)$$

$$\delta\hat{x}(0) = 0$$

where:

$$A(t) \triangleq f_x(\overline{x}(t),\overline{u}(t)),\ B(t) \triangleq \delta_u(\overline{x}(t),\overline{u}(t))$$

$$C(t) \triangleq S_x(\overline{x}(t))$$

$$\delta u(t) \triangleq u(t) - \overline{u}(t) .$$

Let $\bar{\lambda}(t)$ denote the solution of Eqs. (26) and (27). From a slight extension of previous results, it follows, if H1A, H2A are satisfied, $s = 2$ and $\max_{t\in T} \|\delta u(t)\| \leq \varepsilon$, $u, \bar{u} \in G$, that:

$$\Delta V = \Delta\hat{V} + e_1$$

where

$$\Delta\hat{V} = \int_{t_0}^{t_\delta} w^T(t)\delta u(t)dt$$

$$w(t) \triangleq H_u(\bar{x}(t), \bar{u}(t), \bar{\lambda}(t), t)$$

$$|e_1| \leq c_1\varepsilon^2, \quad c_1 < \infty$$

$$\|\delta y(t)\| \leq c_2\varepsilon, \quad c_2 < \infty, \quad \text{for all} \quad t \in T$$

$$\|\delta y(t) - \delta\hat{y}(t)\| \leq c_3\varepsilon^2, \quad c_3 < \infty, \quad \text{for all} \quad t \in T.$$

$x, \delta x, \delta\hat{x}, y, \hat{y} \triangleq \bar{y} + \delta\hat{y}$ are all continuous functions. In particular $y \in C^m$, the space of all m-dimensional continuous functions on T with norm $\|y\|_c \triangleq \max_{t\in T} \|y(t)\|$. The dual space is B^m, the space of m-dimensional functions of bounded variation, continuous from the right and usual norm $\|z*\|_B$, so that $\langle z*, y\rangle$ has the representation:

$$\langle z*, y\rangle = \int_{t_0}^{t_\delta} y^T(t)d\eta(t)$$

for some $\eta \in B^m$, and Stieltjes integration is implied. In the sequel we assume that Ω is convex.

The following assertion follows almost immediately from Luenberger [21]:

Assertion 23. Let H1A, H2A be satisfied, $s = 2$, $\bar{u} \in G$. Let $(\bar{x},\bar{u})$ be optimal. Then there exists a $r^* \geq 0$, $z^* \geq \theta$, where $r^* \in E^1$, $z^* \in B^m$ and $|r^*| + \|z^*\|_B > 0$ such that:

(i) $r^*\Delta\hat{V} + \langle z^*,\hat{y}\rangle \geq 0$ for all $u \in G$

(ii) $\langle z^*,\bar{y}\rangle = 0$. (So that (i) becomes $r^*\Delta\hat{V} + \langle z^*,\delta\hat{y}\rangle \geq 0$.)

(iii) η, the representation of z^*, is nondecreasing and is constant when $\bar{y} < \theta$.

Proof: In the space $W = E^1 \times C^m$ define the sets:

$$A \triangleq \{(r,z) : r \geq \Delta\hat{V},\ z \geq \hat{y} \text{ for some } u \in G\}$$

$$B \triangleq \{(r,z) : r \leq 0,\ z \leq \theta\}$$

where θ denotes the null function. A and B are convex, B has interior points, and A does not contain any interior points of B. (If A does, then there exists for some u satisfying the hypothesis, a $\Delta\hat{V} \leq r < 0$, $\hat{y} \leq z < \theta$. This implies that there exists a $u = \bar{u} + \alpha(u_1 - \bar{u})$, $\alpha \in (0,1]$ such that $\Delta V < 0$, $y < \theta$, contradicting optimality). Hence, there exists a separating hyperplane between A and B, i.e. a r^*, z^*, δ such that:

$$r^*r + \langle z^*,z\rangle \geq \delta \quad \text{for all } (r,z) \in A$$

$$r^*r + \langle z^*,z\rangle \leq \delta \quad \text{for all } (r,z) \in B.$$

Since $(0,\theta) \in A,B$, $\delta = 0$. Putting $r = \Delta\hat{V}$, $z = \hat{y}$ yields (i). For $u = \bar{u}$, i.e., $\hat{y} = \bar{y}$, (i) implies $\langle z^*,\bar{y}\rangle \geq 0$. But $z^* \geq \theta$, $\bar{y} \leq \theta$ implies $\langle z^*,\bar{y}\rangle \leq 0$. Hence, $\langle z^*,\bar{y}\rangle = 0$. $\langle z^*,z\rangle \leq 0$ for all $z \leq \theta$ implies, via the representation of $\langle z^*,z\rangle$, that η is nondecreasing.

COROLLARY. If there exists a $u \in G$ such that $\hat{y} < \theta$ then $r^* > 0$ (and is usually set equal to unity).

Proof: If $z^* = \theta$, then $r^* > 0$. If $z^* > \theta$, then there exists a $u \in G$ such that $\hat{y} < \theta$ and $\langle z^*, \hat{y}\rangle < 0$. Then (i) implies that $r^* > 0$, proving the corollary.

$\overline{\lambda}(t)$ satisfies:

$$-\dot{\overline{\lambda}}(t) = A^T(t)\overline{\lambda}(t), \quad \overline{\lambda}(t_\delta) = F_x(\overline{x}(t_\delta)) .$$

If $\Phi(t, t_0)$ is the transition matrix of A, then:

$$\overline{\lambda}(t) = \Phi^T(t_\delta, t)\ \overline{\lambda}(t_\delta) .$$

Also, for $u \in G$:

$$\langle z^*, \delta y\rangle = \int_{t_0}^{t_\delta} \left[\int_{t_0}^{t} C(t)\Phi(t,\tau)B(\tau)\delta u(\tau)d\tau\right]^T d\eta(t)$$

$$= \int_{t_0}^{t_\delta} \left[\int_{\tau}^{t_\delta} \delta u^T(\tau)B^T(\tau)\Phi^T(t,\tau)C^T(t)d\eta(t)\right] d\tau$$

$$= \int_{t_0}^{t_\delta} \delta u^T(\tau)B^T(\tau)\lambda'(\tau)d\tau \tag{96}$$

where:

$$\lambda'(\tau) \triangleq \int_{\tau}^{t_\delta} \Phi^T(t,\tau)C^T(t)d\eta(t) . \tag{97}$$

If η were differentiable, then λ' would satisfy the differential equation:

$$-\dot{\lambda}'(t) = A^T(t)\lambda'(t) + C^T(t)\dot{\eta}(t)$$

but such a representation is not valid if η is bounded variation. However, by assuming that C, A and $B(B(t) \triangleq f_u(\overline{x}(t), \overline{u}(t))$ have certain differentiability properties we can obtain such a representation. Assume that

C, B are p times and A is p - 1 times continuously differentiable, and define $B_0, \ldots, B_p, C_0, \ldots, C_p$ as follows:

$$B_0(t) \triangleq B(t)$$

$$B_{r+1}(t) \triangleq -\dot{B}_r(t) + A(t)B_r(t) \tag{98}$$

$$C_0(t) \triangleq C(t)$$

$$C_{r+1}(t) \triangleq \dot{C}_r(t) + C_r(t)A(t) \,. \tag{99}$$

(These terms are used for obtaining controllability and observability conditions for linear time varying systems). From Assertion 23 and Eq. (96) (with $r^* = 1$):

$$\Delta\hat{V} + \langle z^*, \delta\hat{y}\rangle = \int_{t_0}^{t_\delta} \delta u^T(t) B^T(t) dt \geq 0 \tag{100}$$

for all δu such that $\bar{u} + \delta u \in G$, where:

$$\lambda(t) \triangleq \bar{\lambda}(t) + \lambda'(t), \quad \text{for all} \quad t \in T \,. \tag{101a}$$

In Eq. (100) we make use of the fact that $\Delta\hat{V} = \int H_u^T(\bar{x}(t), \bar{u}(t), \bar{\lambda}(t))\delta u(t)dt$, and the fact that, since $L \equiv 0$, $H_u(\bar{x}(t), \bar{u}(t), \bar{\lambda}(t)) \equiv B^T(t)\bar{\lambda}(t)$. Hence:

$$\int_{t_0}^{t_\delta} \delta u^T(t) B^T(t)\lambda(t) dt \geq 0 \tag{*}$$

for all δu such that $\bar{u} + \delta u \in G$. If we make the further assumption that $\bar{u}(t) \in$ interior of Ω for all $t \in T$, then (*) implies:

$$\int_{t_0}^{t_\delta} v^T(t) w(t) dt = 0 \tag{**}$$

for all piecewise continuous v, where

$$w(t) \triangleq B^T(t)\lambda(t) = H_u(\bar{x}(t),\bar{u}(t)\lambda(t)) \tag{101b}$$

has bounded variation. Note, if η were differentiable, we could regard w as being the output of the following system:

$$-\dot{\lambda}(t) = A^T(t)\lambda(t) + C^T(t)\dot{\eta}(t)$$
$$w(t) = B^T(t)\lambda(t) .$$

If w is smooth (e.g. $w(t) \equiv 0$) and this system invertible, we could deduce the smoothness of $\dot{\eta}$. This is, in effect, what we will do in the sequel.

Assume, for convenience, that $\bar{x}$ consists of one boundary and two interior arcs, with entry and exit times t_a and t_b respectively, $T_1 \triangleq (t_a,t_b)$. From the definitions of $\bar{\lambda}$, λ' (see Eq. (97)) and λ (Eq. (101)) we obtain:

$$w(t) = B^T(t)\Phi^T(t_b,t)\lambda(t_b^-) + \int_t^{t_b^-} B^T(t)\Phi^T(\tau,t)C^T(\tau)d\eta(\tau) . \quad (***)$$

We recall from Assertion 23(iii) that $\eta \in B^m$, is nondecreasing, and is constant on $[t_0,t_a)$ and $(t_b,t_\delta]$.

Using (**) and the fact that $w \in B^m$, and is therefore continuous from the right, we deduce that $w(t) \equiv 0$. For, if not, there exists a piecewise continuous v contradicting (**).

Hence, integrating (***) by parts yields:

$$0 = -B^T(t)C^T(t)\eta(t) + B^T(t)\Phi^T(t_b,t)[\lambda(t_b^-) + C^T(t_b^-)\eta(t_b^-)]$$
$$- \int_t^{t_b^-} B^T(t)\Phi^T(\tau,t)C_1^T(\tau)\eta(\tau)d\tau \tag{102a}$$

where C_1 is defined in Eq. (99). If we assume that CB is nonsingular on T_1, then Eq. (102a) shows that η is absolutely continuous on T_1 (recall that η is constant on the interior arcs, i.e. on $[t_0, t_a)$, $(t_b, t_\delta])$.

We now ignore the case when CB is singular and consider the case when $CB \equiv 0$ on T_1 (this is the 'artificiality' referred to in the introduction: the analysis is, of course, applicable to the scalar case $(m = 1)$ and could be extended to the general multivariable case when CB is singular only with considerable complexity. The complexity involved is comparable to that required in discussing invertibility of a linear time-varying system).

If $CB \equiv 0$ on T_1, both sides of Eq. (102a) can be differentiated, yielding:

$$0 = B^T(t)C_1^T(t)\eta(t) - B_1^T(t)\Phi^T(t_b, t)[\lambda(t_b^-) + C^T(t_b^-)\eta(t_b^-)] - \int_t^{t_b^-} B_1^T(t)\Phi^T(\tau, t)C_1^T(\tau)\eta(\tau)d\tau . \quad (102b)$$

In obtaining this result we have made use of the definitions in Eqs. (98) and (99) and the fact:

$$\frac{\partial}{\partial t}\Phi^T(\tau, t) = -A^T(t)\Phi^T(\tau, t) .$$

It is proven in the next Assertion (Assertion 25) that, if $CB \equiv 0$ on T_1, then $C_1B \equiv CB_1$ on T_1. It is clear from Eq. (102b), if $C_1B \equiv CB_1$ is nonsingular for all $t \in T_1$, that η is absolutely continuous on T_1. If, on the other hand, $C_1B \equiv CB_1 \equiv 0$ on T_1, then both sides of Eq. (102b) can be differentiated, yielding:

$$0 = -B_1^T(t)C_1^T(t)\eta(t) + B_2^T(t)\Phi^T(t_b,t)[\lambda(t_b^-) + C^T(t_b^-)\eta(t_b^-)]$$
$$- \int_t^{t_b^-} B_2^T(t)\Phi^T(\tau,t)C_1^T(\tau)\eta(\tau)d\tau \ . \qquad (102c)$$

It is shown below that if $C_1B \equiv CB_1$ on T_1, then $C_2B \equiv C_1B_1 \equiv CB_2$ on T_1. Hence if C_2B is nonsingular, for all $t \in T_1$, then Eq. (102c) shows that η is absolutely continuous on T_1.

Proceeding iteratively as above we find that if $C_rB \equiv 0$ on T_1 for $r = 0,1,2 \ldots p-1$, then η is absolutely continuous on T_1. Hence λ is the solution of:

$$-\dot{\lambda}(t) = A^T(t)\lambda(t) \qquad t \in [t_0,t_a) \cup (t_b,t_\delta]$$
$$-\dot{\lambda}(t) = A^T(t)\lambda(t) + C^T(t)\dot{\eta}(t) \qquad t \in (t_a,t_b) \qquad (103a)$$

with possible jumps of the form ($\bar{u}$ assumed continuous at t_α):

$$\lambda(t_\alpha^-) = \lambda(t_\alpha^+) + C^T(t_\alpha)\mu(t_\alpha), \ \alpha = a,b \qquad (103b)$$

at t_a, t_b. Before obtaining an explicit representation for $\dot{\eta}$ we state and prove Assertion 25, an extension of a result due to Skoog [22].

<u>Assertion 25</u>. If C_rB_0 is constant (zero) on T_1 for $r = 0,1,2 \ldots p-1$, (assuming C and B are p times and A is $p-1$ times continuously differentiable) then:

$$C_pB_0 \equiv C_{p-1}B_1 \equiv \cdots C_1B_{p-1} \equiv C_0B_p \quad \text{on} \quad T_1 \ .$$

<u>Proof</u>: C_0B_0 is constant by hypothesis. Assume C_iB_j is constant for $i + j = n < p - 1$. Since C_iB_j constant implies $\dot{C}_iB_j + C_i\dot{B}_j \equiv 0$, i.e., $C_{i+1}B_j \equiv C_iB_{j+1}$ we have:

C_nB_0 constant implies $C_{n+1}B_0 \equiv C_nB_1$

$C_{n-1}B_1$ constant implies $C_nB_1 \equiv C_{n-1}B_2$

"

"

"

"

"

"

C_0B_n constant implies $C_1B_n \equiv C_0B_{n+1}$

i.e., $C_{n+1}B_0 \equiv C_nB_1 \equiv \cdots \equiv C_nB_1$

i.e., C_iB_j is constant for $i + j = n + 1$.

Hence, by induction, C_iB_j is constant for all i, j such that $i + j \leq p - 1$. But C_iB_j constant for $i + j = p - 1$ implies.

$$C_pB_0 \equiv C_{p-1}B_i \equiv \cdots C_1B_{p-1} \equiv C_0B_p \,.$$

Now from Eq. (101b) and the fact that $w \equiv 0$:

$$0 = \frac{d}{dt}\,[B^T(t)\lambda(t)] = -B_1^T(t)\lambda(t) - [C(t)B(t)]^T\,\dot{\eta}(t)\,.$$

If $CB \equiv 0$ on T, $r = 0 \ldots p - 1$, differentiation yields:

$$0 = B_2^T\lambda(t) + [C(t)B_1(t)]^T\,\dot{\eta}(t)\,.$$

If $CB_r \equiv 0$ on T_1, $r = 0 \ldots p - 2$, repeated differentiation yields

$$0 = (-1)^p[B_p^T\lambda(t) + [C(t)B_{p-1}(t)]^T \dot{\eta}(t)] .$$

Hence, on T_1, if CB_{p-1} is nonsingular on T_1:

$$\dot{\eta}(t) = - [C(t)B_{p-1}(t)]^{-1} B_p^T(t)\lambda(t) . \tag{104}$$

On the interior arcs η is constant, i.e., $\dot{\eta}$ is zero, so we are left with the problem of finding the possible jumps in λ at t_a, t_b. Consider t_a first. Denote $H_u(\bar{x}(t_a), \bar{u}(t_a^-),\lambda(t_a^-))$ by $H_u(t_a^-)$, etc. We have:

$$\begin{aligned} 0 &= H_u(t_a^-) - H_u(t_a^+) \\ &= B^T(t_a^-)\lambda(t_a^-) - B^T(t_a^+)\lambda(t_a^+) \\ &= [B(t_a^-) - B(t_a^+)]^T \lambda(t_a^+) + B^T(t_a^-)[\lambda(t_a^-) - \lambda(t_a^+)] \end{aligned} \tag{105}$$

where (if $\bar{u}$ is continuous at t_a):

$$\lambda(t_a^-) - \lambda(t_a^+) = C^T(t_a)\mu(t_a),\mu(t_a) \geq 0 \tag{106}$$

and

$$\begin{aligned}[\dot{\lambda}(t_a^-) - \dot{\lambda}(t_a^+)] = &- A^T(t_a)[\lambda(t_a^-) - \lambda(t_a^+)] \\ &+ C^T(t_a) \dot{\eta}(t_a^+) . \end{aligned} \tag{107}$$

Eqs. (105), (106) yield $\mu(t_a)$ if CB is nonsingular at t_a^-. If in addition $\bar{u}$ is continuous across the junction, then there is no jump in λ. We assume $\bar{u}$ is continuous in the sequel. If $CB \equiv 0$ on T_1 we evaluate $\dot{H}_u(t_a^-) - \dot{H}_u(t_a^+)$:

$$\begin{aligned} 0 = &[\dot{B}(t_a^-) - \dot{B}(t_a^-)]^T\lambda(t_a^+) \\ &- B_1^T(t_a^-) [\lambda(t_a^-) - \lambda(t_a^+)] \end{aligned}$$

which yields $\mu(t_a)$ if $CB_1 \equiv C_1B$ is nonsingular at t_a^-. If

$C_1B = CB_1$ is zero on T_1, then since:

$$[\dot{B}(t_a^-) - \dot{B}(t_a^-)]^T \lambda(t_a^+) = H_{uu}(t_a^+)[\dot{\bar{u}}(t_a^-) - \dot{\bar{u}}(t_a^+)]$$

positive definiteness of $H_{uu}(t_a^+)$ implies that $\dot{u}$ is continuous across t_a.

Proceeding iteratively we obtain, if, on T_1, $\bar{u}$ is continuous, H_{uu} is positive definite and $C_rB \equiv 0$, $r = 0,\ldots,p-2$, that:

(i) $\left(\frac{r}{u}\right)(t) \triangleq (d/dt)^r\bar{u}(t)$ is continuous across t_a, $r = 1\ldots p-2$

(ii) $$0 = \left[\left(\frac{p-1}{B}\right)(t_a^-) - \left(\frac{p-1}{B}\right)(t_a^+)\right]\lambda(t_a^+) + (-1)^{p-1}[C(t_a)B_{p-1}(t_a^-)]^T \mu(t_a). \quad (108)$$

Also, using Assertions 25 and 27:

(iii) $C(t)B_{p-1}(t) = C_{p-1}(t)B(t) = \left(\frac{p}{S}\right)_u (\bar{x}(t),t)$ on T_1.

Nonsingularity of $\left(\frac{p}{S}\right)_u$ on T_1 implies existence of a u such that $\hat{y} < \theta$; hence $r^* = 1$. Similar results hold for the exit junction, t_b replacing t_a in Eq. (108).

Summarizing, we have the following result, due originally to Jacobson et al [20]:

Assertion 26. Let

(i) f and its derivatives with respect to (x,u) up to order $p-1$ exist and be continuous in (x,u); S be p times continuously differentiable; $\left(\frac{r}{u}\right)$, $r = 1,\ldots,p$ be continuous on T_1

(ii) $C_rB \equiv 0$ on T_1, $r = 0,\ldots,p-2$; $C_{p-1}B$ nonsingular on

$T_1\left(\left(\begin{smallmatrix} r \\ S \end{smallmatrix}\right)_u (\bar{x}(t),t) \equiv 0 \quad \text{on} \quad T_1,\ r = 1,\ldots,p-1,\ \left(\begin{smallmatrix} p \\ S \end{smallmatrix}\right)_u (\bar{x}(t),t)\right.$ nonsingular on T_1 - see Assertion 27 $\Big)$.

(iii) $\bar{u}(t) \in$ interior of Ω for all $t \in T$.

Then a necessary condition for optimality of $(\bar{x},\bar{u})$ is:

$$H_u(\bar{x}(t),\bar{u}(t),\lambda(t)) = 0$$

for all $t \in T$, where λ satisfies Eqs. (103a) - (103c), $\dot{\eta}$ satisfies Eq. (104) on T_1 and is zero elsewhere, and $\lambda(t_\delta) = F_x(\bar{x}(t_\delta))$. If also:

(iv) $\bar{u}$ is continuous across t_a, t_b, $H_{uu}(t_\alpha^+)$, $\alpha = a,b$ is positive definite, then $\left(\begin{smallmatrix} r \\ \bar{u} \end{smallmatrix}\right)$, $r = 0,\ldots,p-2$ is continuous across t_a, t_b and $\mu(t_\alpha)$, $\alpha = a,b$ is given by Eq. (108).

It only remains to relate $C_r B$ to properties of S.

Let:

$$(\dot{S}) \triangleq S_x(x)f(x,u,t)$$

i.e.,

$$(\dot{S})(\bar{x}(t),\bar{u}(t)) = \frac{d}{dt}\, S(\bar{x}(t)) .$$

Let

$$\left(\begin{matrix} 1 \\ S \end{matrix}\right) \triangleq (\dot{S})$$

$$\left(\begin{matrix} r \\ S \end{matrix}\right) \triangleq \left(\begin{matrix} \dot{r-1} \\ S \end{matrix}\right)$$

$$\left(\begin{matrix} r \\ S \end{matrix}\right)_x \triangleq \frac{\partial}{\partial x}\left(\left(\begin{matrix} r \\ S \end{matrix}\right)(x)\right) .$$

<u>Assertion</u> 27. If S is p times continuously differentiable, then, for $r = 0\ldots p$,

$$\left(\begin{matrix} r \\ S \end{matrix}\right)_x (\bar{x}(t),\bar{u}(t)) = C_r(t)$$

$$\left(\overset{r}{S}\right)_u(\bar{x}(t),\bar{u}(t)) = C_{r-1}(t)B_0(t) + \left(\left(\overset{r-1}{S}\right)_{xu} f\right)(\bar{x}(t),\bar{u}(t)) .$$

<u>Proof</u>: $(S_x)(\bar{x}(t),\bar{u}(t)) = C(t) = C_0(t)$.

Suppose: $$\left(\overset{r}{S}\right)_x(\bar{x}(t),\bar{u}(t)) = C_r(t) .$$

Then: $$\left(\overset{r-1}{S}\right)_x = \left(\left(\overset{r}{S}\right)_x f\right)_x$$

$$= \left(\left(\overset{r}{S}\right)_{xx} f\right) + \left(\overset{r}{S}\right)_x f_x$$

$$= \left(\dot{\left(\overset{r}{S}\right)_x}\right) + \left(\overset{r}{S}\right)_x f_x$$

$$\left(\overset{r+1}{S}\right)_x(\bar{x}(t),\bar{u}(t)) = \dot{C}_r(t) + C_r(t)A(t)$$
$$= C_{r+1}(t) .$$

Hence, by induction:

$$\left(\overset{r}{S}\right)_x(\bar{x}(t),\bar{u}(t)) = C_p(t), \qquad r = 0,1,\ldots,p .$$

Also:

$$\left(\overset{r}{S}\right)_u = \left(\left(\overset{r-1}{S}\right)_x f\right)_u$$

$$= \left(\overset{r-1}{S}\right)_{xu} f + \left(\overset{r-1}{S}\right)_x f_u .$$

An obvious corollary is the following:

If $\left(\overset{r}{S}\right)(\bar{x}(t),\bar{u}(t)$ is independent of u, $r = 0,\ldots,p-1$, then

$$\left(\overset{p}{S}\right)_u(\bar{x}(t),\bar{u}(t)) = C_{p-1}(t)B_0(t)$$
$$= C_0(t)B_{p-1}(t) .$$

VIII. SINGULAR PROBLEMS

Consider the control problem as originally defined. Let $(\bar{x},\bar{u})$ be optimal, and assume that $H(\bar{x}(t),u,\bar{\lambda}(t),t)$, where $\bar{\lambda}$ is the solution of Eqs. (26) and (28), is independent of u for all $t \in T$. Obviously, the estimate $\int_{t_0}^{t_\delta} \Delta H(t)dt$ of ΔV is zero, and therefore of no use. However, Assertion 16 can be employed, and leads directly to a strong version of a condition of optimality recently presented by Jacobson [23].

Assertion 28. Let the hypotheses of Assertion 16 be satisfied. If $(\bar{x},\bar{u})$ are optimal, and $H(x(t),u,\bar{\lambda}(t),t)$ is independent of u for all $t \in T$, then:

$$\varphi(t) \triangleq \Delta f^T(t)\Delta H_x(t) + \Delta f^T(t)\bar{P}(t)\Delta f(t) \geq 0$$

for all $u \in \Omega$, all $t \in T$, where all terms are defined as in Assertion 16, u replacing $u(t)$.

Proof: This follows directly from the Corollary to Assertion 16. If the inequality is violated at $t_1 \in \Theta(\bar{u})$ for $u = v$, the control $u(t)$ defined by:

$$u(t) = \bar{u}(t),\ t \notin T_\varepsilon \triangleq [t_1 - \varepsilon, b_1]$$

$$u(t) = \bar{u}(t) + v,\ t \in T_\varepsilon$$

is employed. Hence, for $t \in T_\varepsilon$

$$\delta\hat{x}(t) = \Delta f(t*),\ [t - t_1 - \varepsilon]$$

where

$$t*(t) \in [t_1 - \varepsilon, t] .$$

Since $\varphi(t_1) \leq c < 0$, there exists an $\varepsilon = \varepsilon_1 > 0$ such $\Delta f^T(t*(t))[\Delta H_x(t) + \bar{P}(t)\Delta f(t)] < c/2 < 0$ for all $t \in T_{\varepsilon_1}$.

Hence, from the Assertion, there exists an $\varepsilon \leq \varepsilon_1$ such that $\Delta V < 0$, contradicting optimality.

Unfortunately to proceed further, in the present state of knowledge, we have to impose the further restriction that f and L are linear in u or that weak variations only are permitted. Thus, under the conditions of Assertion 18 (and the independence of H with respect to u) we easily obtain a weak version of Assertion 16:

$$\Delta\hat{V} = \int_{t_0}^{t_\delta} v^T(t)[C(t) + B^T(t)\overline{P}(t)]z(t)dt \tag{109}$$

$$|\Delta V - \Delta\hat{V}| \geq c\varepsilon^3, \quad c < \infty$$

$$\dot{z}(t) = A(t)z(t) + B(t)v(t) . \tag{110}$$

C is defined in Assertion 18, A and B in Section VII.

$$z \text{ replaces } \delta\hat{x}, \text{ and } v \text{ replaces } \delta u .$$

Let $\overline{C}(t) \triangleq C(t) + B^T(t)\overline{P}(t)$, $y(t) \triangleq \overline{C}(t)z(t)$; the integrand becomes $v^T(t)y(t)$. One way of obtaining sufficient conditions of optimality has been demonstrated by Jacobson [24]. (See also [26-29]). Add to the integrand of Eq. (109):

$$\tfrac{1}{2} z^T(t)\tilde{P}(t)[A(t)z(t) + B(t)v(t) - \dot{z}(t)]$$

where

$$-\dot{\tilde{P}}(t) = A^T(t)\tilde{P}(t) + \tilde{P}(t)A(t) - \tilde{Q}(t) \tag{111}$$

$$\tilde{P}(t_\delta) = \tilde{P}_\delta$$

and integrate by parts, yielding:

$$\Delta\hat{V} = \int_{t_0}^{t_\delta} [\tfrac{1}{2} z^T(t)Q(t)z(t) + v^T(t)\bar{\bar{C}}(t)z(t)]dt$$
$$-\tfrac{1}{2} z^T(t_\delta)\tilde{P}_f z(t_\delta) + \tfrac{1}{2} z^T(t_0)\tilde{P}(t_0)z(t_0) \tag{112}$$

$$\bar{\bar{C}}(t) \triangleq \bar{C}(t) + B^T\tilde{P}(t) = C(t) + B^T(t)\ [P(t) + \tilde{P}(t)]$$

and $z(t_0) = 0$ in this case.

Obviously, if there exists a $\tilde{P}$ satisfying the differential equation and boundary condition for $P_\delta < 0$ and $Q(t) > 0$ on T such that $\bar{\bar{C}}(t) \equiv 0$ on T, then $\Delta\hat{V} > 0$ for all v such that $d(z,0) > 0$. This extra restriction, that v be 'significant' [28] arises because of the lack of any term of the form $v^T(t)R(t)v^T(t)$ in the integrand of $\Delta\hat{V}$, and ensures that $\Delta\hat{V} \geq c_1\varepsilon^2$, $c_1 > 0$.

Let $\eta(t)$ devote the adjoint variable for the linearized system (Eqs. (109)-(110)):

$$-\dot{\eta}(t) = A^T(t)\eta(t) + \bar{C}^T(t)v(t) \tag{113}$$

$$\eta(t_\delta) = 0\,. \tag{114}$$

From Assertion 10, if $u(\tau) \equiv \bar{u}(\tau)$ on (t,t_δ) and f and L are linear, then the differential equation for $\lambda(t) \triangleq V_x^u(\bar{x}(t),t)$ is:

$$-\dot{\lambda}(t) = A^T(t)\lambda(t) + [C^T(t) + P(t)B(t)]\Delta u(t)$$

so that:

$$\eta(t) \equiv \lambda(t) - \bar{\lambda}(t)\,.$$

We will restrict attention to the case where f is linear in u, and L is zero. Let $H(z,v,\eta,t) \triangleq v^T\bar{C}(t)z + \eta^T(A(t)z + B(t)v)$,

$$\begin{aligned} H_u(t) &\triangleq H_v(\bar{z}(t),v(t),\bar{\eta}(t),t) \\ &= \bar{C}(t)\bar{z}(t) + B^T(t)\bar{\eta}(t) \end{aligned}$$

where $\bar{z}(t)$ and $\bar{\eta}(t)$ are zero on T.

$$\dot{H}_u(t) = C_1(t)z(t) - B_1^T(t)\bar{\eta}(t) + [C_0(t)B_0(t) - [C_0(t)B_0(t)]^T]v(t)$$

where C_r, B_r, $r = 0,1,2,\ldots$ are defined as before, except that $C_0(t) \equiv \bar{C}(t)$ on T. If

$$C_0(t)B_0(t) - [C_0(t)B_0(t)]^T \equiv 0$$

$$\ddot{H}_u(t) = C_2(t)\bar{z}(t) + B_\varepsilon^T(t)\bar{\eta}(t) + [C_1(t)B_0(t) - [C_0(t)B_1(t)]^T v(t) .$$

Proceeding iteratively we obtain:

If $\binom{r}{C}(t), \binom{r}{B}(t)$ exist for all $t \in T$, $r = 0,\ldots,p-1$ and

$$[C_{r-1}(t)B_0(t) + (-1)^r [C_0(t)B_{r-1}(t)] \equiv 0, \quad r = 1,\ldots,p-1 .$$

Then:

$$\binom{p}{H_u}(t) = [C_{p-1}(t)B_C(t) + (-1)^p [C_0(t)B_{p-1}(t)]]^T v(t) . \qquad (115)$$

Consider now transforming the problem defined by Eqs. (109), (110) as follows [29]:

$$\dot{v}_1(t) \triangleq v(t), v_1(t_0) = 0$$

$$z_1(t) \triangleq z(t) - B(t)v_1(t) .$$

As a consequence, the system equation becomes:

$$\dot{z}_1(t) = A(t)z_1(t) + B_1(t)v_1(t) \qquad (116)$$

and $\Delta\hat{V}$ becomes:

$$\Delta\hat{V} = \Delta\hat{V}_1 \triangleq \int_{t_0}^{t_\delta} v_1^T[-C_1(t)]z_1(t) + \tfrac{1}{2} v_1^T(t)R_1(t)v_1(t)$$

$$+ \tfrac{1}{2} \dot{v}_1^T(t)S_1(t)v_1(t)]dt$$

$$+ v_1^T(t_\delta)C_0(t_\delta)z_1(t_\delta)$$

$$+ \tfrac{1}{2} v_1^T(t_\delta)C_0(t_\delta)B(t_\delta)v_1(t_\delta) \qquad (117)$$

where

$$R_r(t) \triangleq (-1)^r [C_{r-1}(t)B_r(t) + B_r^T(t)C_{r-1}^T(t) + \frac{d}{dt}[C_{r-1}(t)B_{r-1}(t)]]$$

$$= (-1)^r [C_r(t)B_{r-1}(t) + B_r^T(t)C_{r-1}^T(t)]$$

$$S_r(t) \triangleq (-1)^r [C_{r-1}(t)B_{r-1}(t) - B_{r-1}^T(t)C_{r-1}^T(t)] .$$

By repeatedly applying the transformation:

$$\dot{v}_r(t) \triangleq v_{r-1}(t), \quad v_r(t_0) = 0$$

$$z_r(t) \triangleq z_{r-1}(t) - B_{r-1}(t)v_r(t)$$

we obtain:

$$\dot{z}_p(t) = A(t)z_p(t) + B_p(t)v_p(t) . \tag{118}$$

If, in addition $R_r(t)$ and $S_r(t)$ are identically zero on T, $r = 1,\ldots,p-1$, then:

$$\begin{aligned}\Delta\hat{V} = \Delta\hat{V}_p \triangleq \int_{t_0}^{t_\delta} & [v_p^T(t)[(-1)^p C_p(t)]z_p(t) + \tfrac{1}{2} v_p^T(t)R_p(t)v_p(t) \\ & + \tfrac{1}{2} \dot{v}_p^T(t)S_p(t)v_p(t)]dt \\ & + \sum_{r=1}^{p} [v_r^T(t_\delta)C_{r-1}(t_\delta)z_r(t_\delta) \\ & + \tfrac{1}{2} v_r^T(t_\delta)C_{r-1}(t_\delta)B_{r-1}(t_\delta)v_r(t_\delta)](-1)^{r-1} .\end{aligned} \tag{119}$$

Define Φ_{ij} as follows:

$$\Phi_{ij}(t) \triangleq C_i(t)B_j(t) + (-1)^{i+j+1}B_i^T(t)C_j^T(t) .$$

Note:

$$\left[\binom{r}{H_u}(t)\right]_u = \Phi_{r-1,0}(t)$$

$$S_r(t) = (-1)^r \Phi_{r-1,r-1}(t)$$

$$S_r(t) = (-1)^r \Phi_{r-1,r-1}(t) .$$

Making use of the defining property of C_r, B_r, $r = 0,1,2,\ldots$, it follows that, if $\Phi_{ij}(t)$ is constant for all $t \in T$, then:

$$\Phi_{i+1,j}(t) \equiv \Phi_{i,j+1}(t) \quad \text{on} \quad T .$$

A similar property, with C_iB_j replacing Φ_{ij}, was used to obtain Assertion 25. Hence, we have:

<u>Assertion</u> 29. If $\Phi_{r,0}$ is constant on T, $r = 0,\ldots,p-1$, then:

$$\Phi_{p,0} \equiv \Phi_{p-1,1} \equiv \cdots \Phi_{1,p-1} \equiv \Phi_{0,p} \quad \text{on} \quad T .$$

As a direct consequence, we have,

<u>Assertion</u> 30. Let A, $\overline{C}$ and B be $2p$ times continuously differentiable.

(i) If $\left[\binom{r}{H_u}(t)\right]_u \equiv 0$ on T, $r = 0,\ldots,2p-2$ then:

$$S_p(t) \equiv (-1)^p \left[\binom{2p-1}{H_u}(t)\right]_u \quad \text{on} \quad T .$$

(ii) If $\left[\binom{r}{H_u}(t)\right]_u \equiv 0$ on T, $r = 1,\ldots,2p-1$ then

$$R_p(t) \equiv (-1)^p \left[\binom{2p}{H_u}(t)\right]_u \quad \text{on} \quad T .$$

<u>Proof</u>: (i) $\left[\binom{r}{H_u}\right]_u \equiv 0$, $r = 1,\ldots,2p-2$ implies that

$$\Phi_{r,0} \equiv 0, \; r = 0,\ldots,2p-3$$

$$S_p = (-1)^p \Phi_{p-1,p-1} = (-1)^p \Phi_{2p-2,0} = (-1)^p \left[\binom{2p-1}{H_u}\right]_u$$

(ii) is similarly proved.

Hence, if $\left[\left(\overset{r}{H_u}\right)(t)\right]_u \equiv 0$ on T, $v = 0,\ldots,2p-2$, we consider the control problem defined by Eqs. (114), (115). $S_p(t)$ is anti-symmetric, and if it is nonzero, a v_p can be found, $d(v_p,0) \leq \varepsilon$, such that $\Delta\hat{V} < 0$ and, hence, $\Delta V < 0$ for some $\varepsilon > 0$. Hence, $S_p(t) \equiv 0$ on T is a necessary condition of optimality. Given $S_p(t) \equiv 0$ on T, if $R_p(t_1) \not\geq 0$, then a control, nonzero except on $[t_1 - \varepsilon, t_1]$ can be found, such that $\Delta\hat{V} < 0$, again contradicting optimality for some $\varepsilon > 0$. These conditions,

$$\left[\left(\overset{2p-1}{H_u}\right)(t)\right] \equiv 0 \quad \text{on} \quad T, \quad (-1)^p \left[\left(\overset{2p}{H_u}\right)(t)\right] \geq 0 \quad \text{on} \quad T,$$

as well known necessary conditions of optimality. Note that the first condition implies that $C_i(t)B_j(t) = B_i^T(t)C_j^T$ for all i, j such that $i + j = 2p - 2$.

To illustrate the use of these results in obtaining sufficient conditions of optimality [25-28] consider the case when the singular control is of first order, and $R_1(t) > 0$, $S_1(t) = 0$ for all $t \in T$.

We will use the system and cost function defined by Eqs. (116) and (117) to obtain a control law, and assess the performance of this law using Eq. (112). Applying linear optimal control theory yields:

$$v_1(t) = -K(t)z_1(t) \tag{120}$$

$$K(t) = R_1^{-1}(t)\,[B_1^T(t)P_1(t) - C_1(t)] \tag{121}$$

where P_1 is the solution, assumed to exist on T, of the following Riccati equation:

$$-\dot{P}_1(t) = A^T(t)P_1(t) + P_1(t)A(t) - K^T(t)R_1(t)K(t) \tag{122}$$

making use of the definitions of C_1, B_1 we obtain:

$$-\frac{d}{dt}[P_1(t)B(t) + \overline{C}^T(t)]$$

$$= [A(t) - B_1K(t)]^T [P_1(t)B(t) + \overline{C}^T(t)] \qquad (123)$$

so that if $P_1(t_\delta)B(t_\delta) + \overline{C}^T(t_\delta) = 0$, then $P_1B + \overline{C}^T \triangleq \widetilde{P}$ is identically zero on T, satisfying the requirement on $\widetilde{P}$ to establish $\Delta V \geq 0$. If $P_1B + \overline{C}^T$ is zero on T, then:

$$K(t)B(t) = R_1^{-1}[\dot{B}_1^T(t)P(t)B(t) - C_1(t)B(t)]$$

$$= -R_1^{-1}(t)[B_1^T(t)C^T(t) + C_1(t)B(t)]$$

$$= R_1^{-1}(t)R_1(t)$$

$$= I, \quad \text{for all } t \in T. \qquad (124)$$

Our choice for $P_1(t_\delta)$ is motivated by Eq. (117), substituting $-K(t_\delta)z_1(t_\delta)$ for $v_1(t_\delta)$. This yields:

$$P_1(t_\delta) = -[K^T(t_\delta)\overline{C}(t_\delta) + \overline{C}^T(t_\delta)K(t_\delta)$$

$$- K^T(t_\delta)\overline{C}(t_\delta)B(t_\delta)K(t_\delta)] - \widetilde{K}^T \Gamma \widetilde{K} \qquad (125)$$

where the last term has been arbitrarily added. $\Gamma \geq 0$ and $\widetilde{K}$ is given by:

$$\begin{bmatrix} K(t_\delta) \\ \widetilde{K} \end{bmatrix} [B(t_\delta), \widetilde{B}] = I_n . \qquad (126)$$

$\widetilde{B}$ being any $(n-m) \times n$ matrix such that $[B(t_\delta), \widetilde{B}]$ is non-singular. $K(t_\delta)\,B = I_m$, and $\widetilde{K}\,B(t_\delta) = 0$, $\widetilde{K}\,\widetilde{B} = I_{n-m}$. Hence, as required:

$$P_1(t_\delta)B(t_\delta) = -\overline{C}^T(t_\delta)$$

satisfying the requirement for $P_1B + \overline{C}$ to be zero on T. Hence, Eq. (124) holds. Also:

$$- \begin{bmatrix} B^T(t_\delta) \\ \tilde{B}^T \end{bmatrix} [P_1(t_\delta)][B(t_\delta),\tilde{B}] = \begin{bmatrix} \overline{C}(t_\delta)B(t_\delta) & \overline{C}(t_\delta)\tilde{B} \\ \tilde{B}^T\ \overline{C}^T(t_\delta) & \Gamma \end{bmatrix} . \quad (127)$$

Sufficient conditions for the positive definiteness (semi-definiteness) of $-P_1(t_\delta)$ are:

(i) $\overline{C}(t_\delta)B(t_\delta) > 0$

(ii) $\Gamma - \tilde{B}^T\ \overline{C}^T(t_\delta)\ [\overline{C}(t_\delta)B(t_\delta)]^{-1}\ \overline{C}(t_\delta)\ \tilde{B} > 0\ (\geq 0)$.

$\overline{C}B \geq 0$ in T is a necessary condition of optimality. Hence, we have:

<u>Assertion</u> 30. Let the hypotheses of Assertion 16 be satisfied. Let f, L be linear in u. Let the solution $(\overline{x},\overline{u})$ be singular with order 1, i.e.:

$$S_1(t) = -[C_0(t)B(t) - B^T(t)C_0(t)] = 0,$$

and

$$R_1(t) = -[C_1(t)B(t) + B_1^T(t)C_0(t)] > 0 \text{ for all } t \in T : C_0(t) = \overline{C}(t)].$$

C and B are assumed to be differentiable. Then sufficient conditions for $\Delta\hat{V} \geq 0$ are:

(i) $\overline{C}(t_\delta)B(t_\delta) = B^T(t_\delta)P(t_\delta)B(t_\delta) + C(t_\delta)B(t_\delta) > 0$

(ii) $\Gamma - \tilde{B}^T\ \overline{C}^T(t_\delta)[C(t_\delta)B(t_\delta)]^{-1}\ C(t_\delta)\tilde{B}. \geq 0$

(iii) The Matrix Riccati differential equation, Eq. (123) with terminal condition Eq. (127) has no conjugate point in $[t_0,t_f]$.

To extend this result to local optimality of the original system, condition (ii) must be strengthened to positive definiteness and comparison control restricted to those which, with norm $d(v,0) = 1$, produce $\|z(t_\delta)\| \geq c > 0$.

An alternative approach to this problem is the transformation approach [25,30] which has been generalized in [28]. The transformation approach has the virtue of a lower dimension Riccati differential equation. It is, however, not easily extended to higher order singular problems.

Finally, we obtain the control law for v:

$$v_1(t) = -K(t)z_1(t)$$

$$v(t) = -K_1(t)z_1(t) - K(t)B_1(t)v_1(t)$$

where

$$K_1(t) \triangleq \dot{K}(t) + K(t)A(t) .$$

Since $K(t)B(t) = I$ for all t:

$$K_1(t)B(t) = K(t)B_1(t) \quad \text{for all} \quad t$$

$$\therefore \quad v(t) = -K_1(t)z(t) + K_1(t)B(t)v_1(t) - K(t)B_1(t)v_1(t)$$

$$= -K_1(t)z(t) .$$

With this control law:

$$\frac{d}{dt}[K(t)z(t)] = [\dot{K}(t) + K(t)A(t) - K(t)B(t)K_1(t)]z(t)$$

$$= 0 \quad \text{for all} \quad t \in T ,$$

so that $K(t)z(t) \equiv 0$ if $K(t_0)z(t_0) = 0$.

IX. CONCLUSION

The main purpose of this work has been to present some exact expressions for the change of cost due to arbitrary controls, and to exhibit the central role these expressions can play, both in control theory and numerical optimization, by illustrating their application to the derivation of algorithms and conditions of optimality. In doing so we believe that we have provided a better theoretical basis for the powerful differential dynamic programming algorithms. However, the aim has been more general than this. It is hoped that the results, particularly the second order estimates for strong variations in control, will be useful in deriving further or adapting existing algorithms. The existing differential dynamic programming algorithms need adaptation to ensure convergence in constrained problems.

It is obvious from the references and the text that much inspiration has been drawn from the work of Jacobson. A distinguishing feature of this work is its concentration on arbitrary controls. We have also benefitted immensely from discussions with Professors Y. C. Ho and R. Brockett and Dr. J. C. Allwright and Dr. J. M. C. Clark.

X. REFERENCES

1. D. Q. MAYNE, "A Gradient Method for Determining Optimal Control of Non-Linear Discrete-Time Systems," Proc. I.F.A.C. Symposium on Self-Adaptive Control Systems, Plenum Press, New York, 1966.
2. D. Q. MAYNE, "A Second-Order Gradient Method for Determining Optimal Control of Non-Linear Discrete-Time Systems," Int. J. Control, 3, 85, (1966).
3. J. H. WESTCOTT, D. Q. MAYNE, et al., "Optimal Techniques for On-Line Control," Proc. Third I.F.A.C. Congress, London, 1966.
4. D. H. JACOBSON, "Second Order and Second Variation Methods for Determining Optimal Control," Int. J. Control, 7, 175, (1968).

5. D. H. JACOBSON, D. Q. MAYNE, "Differential Dynamic Programming," Elsevier Press, New York, 1970.

6. D. H. JACOBSON, "Differential Dynamic Programming Methods for Solving Bang-Bang Control Problems," I.E.E.E. Trans. Auto. Control, AC 13, 661, (1968).

7. S. B. GERSHWIN, and D. H. JACOBSON, "A Discrete-Time Differential Dynamic Programming Algorithm with Application to Optimal Orbit Transfer," AIAA Journal, 8, 1616, (1970).

8. P. DYER, and S. R. MCREYNOLDS, "The Computation and Theory of Optimal Control," Academic Press, New York, 1970.

9. C. CARATHEODORY, "Calculus of Variations and Partial Differential Equations of First Order," Holden-Day, San Francisco, 1965.

10. I. M. GELFAND, and S. V. FOMIN, "Calculus of Variations," Prentice Hall, New Jersey, 1963.

11. R. W. BROCKETT, "Finite Dimensional Linear Systems," Wiley, New York, 1970.

12. L. C. YOUNG, "Calculus of Variations and Optimal Control Theory," Saunders, Philadelphia, 1969.

13. D. Q. MAYNE, "Properties of a Cost Function Employed in a Second-Order Optimisation Algorithm," Report 5/70, C.C.D., Imperial College, London, (1970) (to appear Journal Math. Anal. and Applic.).

14. D. Q. MAYNE, "Second Order Estimates of the Change in Cost Due to Strong Variations in Control," Report 17/70, C.C.D., Imperial College, London, (1970).

15. D. Q. MAYNE, 'On the Sensitivity of the Cost Function to Changes in Switching Time," Report 6/71, C.C.D., Imperial College, London, 1971.

16. H. HALKIN, "Mathematical Foundations of System Optimization," in, Topics in Optimization, Editor G. Leitmann, Academic Press, 1967.

17. E. J. MCSHANE, "Integration," Princeton University Press, Princeton, New Jersey, 1944.

18. D. H. JACOBSON, "A Note on Error Analysis in Differential Dynamic Programming," I.E.E.E. Trans. Auto. Control, AC-14, 197, (1969).

19. R. W. BROCKETT, "Related Problems in Approximation Theory and Optimal Control," Proc. of 2nd Princeton Conference on Information Sciences and Systems, 21, (1968).

20. D. H. JACOBSON, M. M. LELE, and J. L. SPEYER, "New Necessary Conditions of Optimality for Control Problems with State-Variable Inequality Constraints," J. Math. Anal. Appl., (1971).

21. D. G. LUENBERGER, "Optimization by Vector Space Methods," Wiley, New York, 1969.

22. R. A. SKOOG, Ph.D. Thesis, M.I.T., 1969.

23. D. H. JACOBSON, "A New Necessary Condition of Optimality for Singular Control Problems," SIAM Journal Control, 7, 578, (1969).

24. D. H. JACOBSON, "Sufficient Conditions for the Non-Negativity of the Second Variation in Singular and Non-Singular Control Problems," SIAM Journal Control, 8, (1970).

25. J. L. SPEYER, and D. H. JACOBSON, "Necessary and Sufficient Conditions for Optimality for Singular Control Problems: A Transformation Approach," Journal Math. Anal. and Applic., to appear.

26. D. H. JACOBSON, and J. L. SPEYER, "Necessary and Sufficient Conditions for Optimality for Singular Control Problems: A Limit Approach," Journal Math. Anal. and Applic.,

to appear.

27. D. H. JACOBSON. "Totally Singular Quadratic Minimization Problems," I.E.E.E. Trans. Auto. Control, to appear.

28. D. Q. MAYNE, "Sufficient Conditions for Optimality for Singular Control Problems," Report 18/70, C.C.D., Imperial College, London, (1970).

29. B. S. GOH, "The Second Variation for the Singular Bolza Problem," SIAM Journal Control, 4, 309, (1966).

30. H. J. Kelley, "A Transformation Approach to Singular Sub in Optimal Trajectory and Control Problems," SIAM Journal Control, 2, 234, (1964).

31. H. J. KELLEY, "A Second Variation Test for Singular Extremals," AIAA Journal, 2, 1380, (1964).

32. H. J. KELLEY, R. E. KOPP, and H. G. MOYER, "Singular Extremals," in Topics in Optimization (G. Leitmann, Ed.) Academic Press, New York, 1967.

33. L. W. NEUSTADT. "A General Theory of Extremals," Journal of Computer and Systems Sciences, 3, 57, (1969).

IX. APPENDIX

LEMMA. Let Z, X denote closed bounded subsets of E_n and E_m respectively. Let $\varphi : ZxWxT \rightarrow E$ and its partial derivatives with respect to (z,w) be continuous, except at a finite number of points in T, and satisfy, for some $c_1 < \infty$.

(i) $\varphi(z,0,t) = 0$ for all $(z,t) \in ZxT$.

(ii) $\|\varphi_w(z,0,t)\| \leq c_1 \|z\|^2$, for all $(z,t) \in ZxT$.

(iii) $\varphi_{ww}(0,0,t)$ is positive definite $(\varphi_{ww} > 0)$ for all $t \in T$.

(iv) $\varphi(0,w,t) > 0$ for all $(w,t) \in WxT$ such that $w \neq 0$.

Then, there exists an $\varepsilon_3 > 0$ and an $\alpha > 0$ such that

$$\varphi(z,w,t) \geq \tfrac{1}{2}\alpha w^T w + r$$

for all $(z,w,t) \in ZxWxT$ such that $\|z\| \leq \varepsilon_3$, where $|r| \leq c_1 \|z\|^2 \|w\|$.

Proof: From the piecewise continuity of $\varphi_{ww}(0,0,\cdot)$, there exists an $\varepsilon_1 > 0$, $\alpha_1 > 0$ such that $\varphi_{ww}(z,w,t) - \alpha_1 I \geq 0$ for all $(z,w,t) \in ZxWxT$ such that $\|z\| \leq \varepsilon_1$, $\|w\| \leq \varepsilon_1$. Hence (i):

$$\varphi(z,w,t) \geq \tfrac{1}{2}\alpha_1 w^T w + \varphi_w^T(z,0,t)w .$$

From (iv) and the continuity of $\varphi(\cdot)$, there exists an $\varepsilon_2 > 0$, $\alpha_2 > 0$ such that $\varphi(z,w,t) \geq \alpha_2$ for all $(z,w,t) \in ZxWxT$ such that $\|z\| \leq \varepsilon_2$, $\|w\| > \varepsilon_1$. Let d denote max $w^T w$ subject to $w \in W$. The lemma follows from (ii) $(\|\varphi_w^T(z,0,t)w\| \leq c\|z\|^2\|w\|)$ with $\varepsilon_3 = \min[\varepsilon_1,\varepsilon_2]$, $\alpha = \min[\alpha_1,\alpha_2/d]$.

Consider now the system:

$$\dot{x}(t) = g(x(t),w(t),t),\ x(t_0) = x_0$$

where $g(x,w,t) \triangleq f(x,w + \bar{u}(t) + \bar{K}(t)(x - \bar{x}(t),t)$. If $w(t) \equiv 0$ the corresponding trajectory is $\bar{x}(t)$; if $w(t) \not\equiv 0$, the corresponding trajectory is $x(t)$. Since $g(\cdot)$ satisfies the hypothesis of Assertion 2(ii), $\delta x(t) \triangleq x(t) - \bar{x}(t)$ satisfies, for some $c_2 < \infty$, $\|\delta x(t)\| \leq c_2 d(w,0)$, for all $t \in T$. Hence, for some $c_3 < \infty$:

$$d(u,\bar{u}) \leq c_3 d(w,0) .$$

Since $w(t) \equiv u(t) - \bar{u}(t) - \bar{K}(t)\delta x(t)$, and, for some $c_4 < \infty$, $\|\delta x(t)\| \leq c_4 d(u,\bar{u})$ for all $t \in T$, it follows that for some $c_5 < \infty$, for all $u \in G$:

$$d(w,0) \leq c_5 d(u,\bar{u})$$

i.e.

$$d(w,0) \in [d(u,\bar{u})/c_3,\ c_5 d(u,\bar{u})] .$$

There exists an $\varepsilon > 0$ such that $d(u,\bar{u}) \leq \varepsilon$ implies that $\|\delta x(t)\| \leq \varepsilon_2$ for all $t \in T$. If we now define $\varphi(\cdot)$ by:

$$\varphi(\delta x,w,t) \triangleq H(\bar{x}(t) + \delta x,\ w + \bar{u}(t) + \bar{K}(t)\delta x,\ V_x^k(\bar{x}(t) + \delta x,t),t)$$

$$- H(\bar{x}(t) + \delta x,\ \bar{u}(t) + \bar{K}(t)\delta x,\ V_x^k(\bar{x}(t) + \delta x,t),t)$$

it is easily shown that $\varphi(\cdot)$ satisfies the hypotheses of the lemma. Hence, if $d(u,\bar{u}) \leq \varepsilon$ (so that $\|\delta x(t)\| \leq \varepsilon_3$ for all $t \in T$), then:

$$\varphi(\delta x(t),w(t),t) \geq \tfrac{1}{2}\,\alpha w^T(t)w(t) + r(t)$$

where $|r(t)| \leq c_1\|\delta x(t)\|^2\,\|w(t)\|$. Hence there exists a $c < \infty$ such that, for all $t \in T$, for all $u \in G$ such that $d(u,\bar{u}) \leq \varepsilon$:

$$|r(t)| \leq c[d(u,\bar{u})]^3 .$$

Also, from the above and the relation following Eq. (6), there exists a $\bar{c} \in (0,\infty)$ such that:

$$\int_{t_0}^{t_\delta} w^T(t)w(t)dt \geq \bar{c}[d(u,\bar{u})]^2 .$$

Estimation of Uncertain Systems

JACK O. PEARSON
Aerospace Group,
Hughes Aircraft Company,
Culver City, California

I. INTRODUCTION

Theoretically, the Kalman-Bucy filter gives the unbiased, minimum variance estimate of the state vector of a linear dynamic system disturbed by additive white noise when measurements of the state vector are linear, but disturbed by white noise. Such performance is hardly ever realized in actual practice since the information required to construct the Kalman-Bucy filter is only approximately known. The noise parameters and models may be based upon only relatively few data points, computer round-off errors may be significant, and the system model may not be adequate. When it is impractical or impossible to arrive at accurate information upon which to base the filter design, suboptimal and adaptive techniques must be considered. This chapter discusses a technique that has been established to allow Kalman-Bucy filtering in an uncertain environment when the complexity of the estimation algorithm is constrained.

A. Problem Statement

The state estimation problem with parameter uncertainties to be considered in this chapter is described as follows. Consider a linear continuous system whose plant can be described by the vector differential equation

$$dx/dt = A(t,a_1)\ x(t) + B(t)\ u(t)$$
$$x(t_0) = x_0 \tag{1}$$

with noisy measurements

$$z(t) = C(t)\ x(t) + v(t); \qquad t_0 \leq t \leq T \tag{2}$$

where

$x(t)$ - nxl state vector with dependence upon a_i's suppressed.

$A(t,a_1)$ - nxn state matrix with time-varying uncertain parameters a_1.

$B(t)$ - nxq input matrix.

$z(t)$ - rxl measurement vector with dependence upon a_i's suppressed.

$C(t)$ - rxn measurement matrix.

$u(t)$ - qxl Gaussian white noise process.

$v(t)$ - rxl Gaussian white noise process.

The noise processes are zero mean with covariance matrices

$$E(v(t)v(t_1)^T) = R(a_3)\ \delta(t - t_1)$$
$$E(u(t)u(t_1)^T) = Q(a_2)\ \delta(t - t_1) \tag{3}$$
$$E(v(t)u(t_1)^T) = 0$$

where a_2 and a_3 are the time-varying uncertain parameters associated with the noise covariance matrices.

The initial condition for the plant is zero mean and statistically independent of the plant and measurement noises. Its covariance matrix is given by $E(x_0 x_0^T) = \gamma_0$.

The Kalman-Bucy filter problem is to find a continuous estimate of $x(t)$, denoted $\hat{x}(t)$, which is a linear function of the observations $z(t_2)$, $0 \leq t_2 \leq t$, minimizing

$$\text{Trace}(D(t)\gamma(t)) = E((x(t) - \hat{x}(t))^T D(t)(x(t) - \hat{x}(t))) \tag{4}$$

where $D(t)$ is symmetric and positive definite.

If the noise statistics are all known and the plant model is accurate, the solution of the above problem is well known [1]. The estimator is defined by the vector differential equation

$$d\hat{x}/dt = F(t)\,\hat{x}(t) + G(t)\,z(t) \tag{5}$$

where $F(t) = A(t) - G(t)C(t)$

$$G(t) = \gamma(t)C^{T}R^{-1}\,.$$

Furthermore, the mean square estimation error expected if the estimator is employed to estimate the state vector can be evaluated from the matrix differential equation

$$d\gamma/dt = A\gamma + \gamma A^{T} - \gamma C^{T}R^{-1}C\gamma + BQB^{T}$$
$$\gamma(t_0) = \gamma_0\,. \tag{6}$$

The basic problem treated in this chapter is the establishment of a method of Kalman-Bucy filter design when the noise covariance matrices and the state matrix are uncertain, but bounded with known bounds. The direction in which the design method developed was motivated by the following two design goals.

In several applications of Kalman filtering the estimator must be designed subject to stringent computation speed and computer storage constraints. The processing of data for landing displays of all-weather aircraft is an example of such an application [2]. Adaptive estimators cannot be used to minimize the effects of uncertainty for such applications due to the tremendous storage requirements of its identification unit.

A very appealing property of the Kalman-Bucy filter is that the performance expected can be evaluated before the filter is actually employed. It is desired that a Kalman-Bucy filter designed for an uncertain system also possess this property in the sense that bounds on the performance expected may be evaluated before the filter is utilized. Adaptive estimators do not possess this property since the filter gain usually must be evaluated on-line, or via a complex system simulation.

The formulation of the Kalman-Bucy filter problem for a system with uncertain parameters that is considered in this chapter incorporates the two desired design goals as follows. The filter form is constrained to that given by Eq. (5). The filter matrices $F(t)$ and $G(t)$, which define the estimator, are chosen to minimize the secondary performance measure that defines the sensitivity of the primary performance measure, Eq. (4), to the parameter uncertainties relative to design choices $F(t)$ and $G(t)$. Since the secondary performance measures will be chosen to be functions of the bounds on the overall performance of each design choice, the optimization problem will be linear, and the performance bounds of the optimum filter can be evaluated off-line.

B. Chapter Outline

A presentation of some of the significant research toward the design of Kalman-Bucy filters for uncertain systems is given in Section II. This section is included for completeness and is intended to introduce the reader to the basic approaches toward the design of Kalman-Bucy filters for uncertain systems.

A general theory for the optimum design of uncertain systems is developed in Section III. The special case of convex systems is emphasized. Secondary performance criteria that measure the overall performance of each filter design are introduced.

Kalman-Bucy filters that must estimate the state vector

of systems with uncertain noise covariance matrices are considered in Section IV. Equations for the estimator that is optimum with respect to the secondary performance measure are derived. The bounding performance sets, defined by the minimax and minimin filters, are also evaluated.

Section V duplicates the results of Section IV for a system with uncertainties in the state matrix, $A(t,a_1)$. An example is given to illustrate the filter design.

II. A SURVEY OF KALMAN FILTERING WITH UNCERTAINTY

A. Introduction

The research in Kalman-Bucy filtering with uncertainty is primarily for discrete systems and can be divided into two major areas; analysis and synthesis. Analysis is confined to evaluating the actual error covariance matrix, and its stability and sensitivity properties. The goal of synthesis is improvement in the performance of the Kalman filter in an uncertain environment. Adaptive estimators, bounding estimators, and estimators derived from secondary sensitivity performance measures are examples of synthesis.

B. The Discrete Estimation Problem

All the research presented in this section considered versions of the following problem, which is listed here for convenience, or the continuous Kalman-Bucy problem given in Section I, A.

Consider the linear discrete system

$$x(k+1) = \Theta(k,\underline{a}) \, x(k) + B(k) \, u(k); \qquad x(0) = x_0$$

$$z(k) = C(k) \, x(k) + v(k)$$

where

$x(k)$ - nxl state vector with dependence upon a_i's suppressed.

$\theta(k,a_1)$ - nxn state transition matrix with uncertain parameters a_1.

$B(k)$ - nxq input matrix.

$z(k)$ - rx1 measurement vector with dependence upon a_i's suppressed.

$C(k)$ - rxn output matrix.

$u(k)$ - qx1 Gaussian white noise sequence.

$v(k)$ - rx1 Gaussian white noise sequence.

The noise sequences are zero mean with covariance matrices

$$E(v(k)v(j)^T) = R(k,a_3)\,\Delta(k-j)$$

$$E(u(k)u(j)^T) = Q(k,a_2)\,\Delta(k-j)$$

$$E(v(k)u(j)^T) = 0$$

where a_2 and a_3 are the uncertain parameters of the noise covariance matrices, and $\Delta(k-j)$ equals one if $k=j$ and equals zero otherwise.

The initial condition for the plant is zero mean and statistically independent of the plant and measurement noises. Its covariance matrix is given by $E(x_0x_0^T) = P_0$.

The Kalman-Bucy filter problem is to find a recursive estimate of $x(k)$, denoted $\hat{x}(k/k)$, that is a linear function of the observations $z(0),\ldots,z(k)$ minimizing

$$E((x(k) - \hat{x}(k/k))^T D(k)\,(x(k) - \hat{x}(k/k)))$$

where $D(k)$ is a symmetric, positive-definite matrix.

C. Analysis of the Effect of Parameter Uncertainty

The primary techniques for analyzing the effect of parameter uncertainties upon the Kalman-Bucy filter are with increasing complexity; determination of the increase in estimation error, analysis of the filter sensitivity, and statistical model testing.

Although parameter uncertainties may effect all statistics of the estimation error, the primary measure of the increase in estimation error is the actual error covariance matrix, defined by

$$P_a(k) = E((x(k) - \overline{x(k/k)})^T D(k) (x(k) - \overline{x(k/k)}))$$

where $x(k)$ is the true value of the state vector at the kth iteration, and $\overline{x(k/k)}$ is the estimate of the state vector at the kth iteration using invalid information about the uncertain parameters. If $P_a(k)$ reaches a finite steady-state value as k approaches infinity, but is too large to allow the estimate to be of use, "apparent divergence" is said to occur. "True divergence" occurs when $P_a(k)$ becomes unbounded as k approaches infinity [3]. Heffes [4] derived a recursive equation for the actual error covariance matrix considering uncertainty in the covariance of the initial state vector, the plant noise covariance matrix, and the measurement noise covariance matrix. Nishimura [5] extended these results to the continuous case. Neal [6] considered uncertainty in the state transition matrix, and derived a recursive equation for the error covariance matrix. Friedland [7] evaluated the effect of using an erroneous filter gain and derived an expression for the increase in the error covariance matrix of the estimate for a continuous system.

Price [8] considered estimating only a proper subset of the state vector, given noisy measurements of the subset. Uncertainties in the noise covariance matrices were also allowed. He modeled the unmeasured state variables as an additive unknown forcing function and derived conditions for the error covariance matrix stability when the additive forcing function was bounded.

Nishimura [5] proved that if the uncertain parameters

are such that their actual values are less than the values assumed in the Kalman-Bucy filter design, then the actual error covariance matrix will be less than the error covariance matrix predicted by the design.

A more suitable measure of the deleterious effects of uncertainty is the deviation of the Kalman-Bucy filter performance from optimal. Griffin and Sage [9] define the large scale sensitivity of the Kalman filter by the matrix

$$S_i = (P_a - P)/(y_{a_i} - y_i)$$

where P is the optimal value of the error covariance matrix, y_a is the actual value of the uncertain parameter, and y is the assumed value of the uncertain parameter.

For very small variations of the uncertain parameters sensitivity may be measured by the small scale sensitivity matrix, which is the partial derivative of the actual error covariance with respect to the uncertain parameters. The authors derive recursive equations for the small scale sensitivity matrix given uncertainty in the system transition matrix, noise covariance matrices, and initial state covariance matrix.

Any error analysis or sensitivity study has validity limited by the assumed system model and the postulated perturbations of the uncertain parameters; it does not represent a test of the system model and analysis assumptions. In order to verify these sources of error actual system data must be used. Berkovec [10] has derived an algorithm for testing the validity of the system model using the actual data as follows. Consider the kth predicted residual, defined as

$$\tilde{z}_k = z_k - C(k)\,\hat{x}(k/k-1)\,.$$

Assuming the system model is accurate calculate the moments

that the predicted residual should possess. Two important properties are that the predicted residual have zero mean and be "white." Process the actual data to determine the statistical properties of the actual predicted residuals. If the results significantly differ from those calculated using the assumed system model, the system model is invalid. If the statistical properties of the residuals are not significantly different from those predicted by the analysis, no inference can be drawn.

Mehra [11] has also considered the problem of model testing. Assuming the system has reached steady state, he constructs a correlation matrix of the last N samples and using the method of Jenkins and Watts tests the correlation matrix to ascertain whether the residuals possess the predicted properties.

D. The Synthesis of Estimators for Uncertain Systems

Techniques for improving Kalman-Bucy filter performance in an uncertain environment may be classified as bounding techniques, multiple criteria techniques, and adaptive estimation. Bounding techniques are attempts to lower the estimation error below an allowable value, but not necessarily minimize the estimation error. Multiple criteria techniques base the filter design upon a performance index which is a complex combination of mean-square estimation error and filter sensitivity. Adaptive estimation involves simultaneous state and parameter estimation.

1. Adaptive Estimation

A very appealing approach toward the design of a filter that must function in an uncertain environment is to simultaneously estimate the uncertain parameters and the state vector. The estimates of the uncertain parameters are used in the state estimator to obtain the absolute minimum estimation error. There are three basic approaches in adaptive estimation; the Bayesian, maximum likelihood, and error residual approaches.

a. The Bayesian Approach

Consider estimating the state vector of a system with uncertain plant parameter a. Defining the augmented state vector as $y_k = (x_k, a)^T$ one has the nonlinear system equations

$$y_{k+1} = F(y_k) + u_k$$
$$z_k = (C(k),0)y_k + v_k .$$

There exist a variety of non-linear filtering techniques [12]. Some techniques, such as the extended Kalman filter, yield an estimator form no more complex than the usual Kalman filter. Their performance may be as good as more complex formulations, particularly when the plant noise is large [13]. Other techniques are more complex, such as the generalized Kalman filter, but yield better performance when the plant noise is small.

Rather than discuss the advantages, disadvantages, and particular situations in which the various non-linear filtering algorithms are superior, a general approach toward non-linear filtering, the derivation of an expression for the time evolution of the conditional probability density function will be presented.

When estimator performance is measured by the mean-square estimation error it is well known that the optimal estimate is given by

$$\hat{y}_k = \int_Y y_k \, p(y_k/Z^k) \, dy_k$$

where $Z^k = z_0, z_1, \ldots, z_k$. Thus, the optimal estimator design problem is reduced to the derivation of the conditional probability density function (CPDF) of y_k given data Z^k.

If the system is linear with additive Gaussian noise, the CPDF of the state vector y_k, given measurements Z^k, will be Gaussian for all k. A probability density function (PDF) with the property that its form (Gaussian) is invariant of iteration is termed "reproducing." The first two moments of a Gaussian PDF describe it completely. Thus, one need only obtain recursive expressions for these moments to completely describe the evolution of the CPDF $p(y_k/Z^k)$.

Smith [14] assumed Gaussian measurement noise with diagonal covariance matrix $R(k)$, which could be represented as the product of a nominal, $R_{nom}(k)$, and an unknown precision vector g. Each element of the precision vector has an inverted-gamma PDF characterized by location and dispersion parameters. Assuming the dispersion parameters were independent at each iteration, and the precision and state vectors were independent, Smith demonstrated that the PDF of the augmented state vector $(x_k,g)^T$ conditioned on observations Z^k is a reproducing PDF. However, the recursive equations describing the parameters which characterize this CPDF involve the unknown precision vector. In order to obtain real-time, suboptimal solutions Smith approximated the precision vector by its location parameters.

In general, the PDF of the augmented state vector conditioned on the measurements is not a reproducing PDF, and a finite number of moments of the PDF will not completely describe it. Thus, one is usually forced to approximate the CPDF.

For many problems of practical interest the conditional PDF will be approximately Gaussian. Thus, one should consider approximating the CPDF as a series of the Hermite polynomials.

Consider the random variable x with a known density function, and let y be the normalized random variable

$$y = (x - m)/\sigma$$

where m and σ are the mean and standard deviation of x, respectively. Then the PDF of y, denoted $P(y)$, may be expressed by the Gram-Charlièr expansion

$$P(y) = G(y)\ (1 + 1/3!\ C_3 H_3(y) + 1/4!\ C_4 H_4(y) + \cdots$$

where $G(y)$ is a Gaussian PDF with zero mean and unit variance, $H_n(y)$ are the Hermite polynomials, and C_n are the quasi-moments defined by

$$C_n = (-1)^n \int_{-\infty}^{\infty} H_n(y) P(y) dy\ .$$

Sorenson [13] used the Gram-Charlier (G-C) expansion to obtain an approximation for $P(x_k/Z^k)$. Basically, his procedure for approximating $P(x_k/Z^k)$ given a G-C approximation of $P(x_{k-1}/Z^{k-1})$ is as follows.

Use the Bayes rule in the form

$$P(x_k/Z^k) = \frac{P(x_k/Z^{k-1})\ P(z_k/x_k)}{P(z_k/Z^{k-1})}\ .$$

The probability density function $P(z_k/x_k)$ can be found directly from the system equations. Express $P(x_k/Z^{k-1})$ as a G-C expansion. To do this one must calculate the quasi-moments, which are functions of the moments of $P(x_{k-1}/Z^{k-1})$. The latter moments are obtained from the given G-C approximation of $P(x_{k-1}/Z^{k-1})$.

Using Bayes rule combine the results into a form of the G-C expansion. The major result is the derivation of recursive equations for the quasi-moments.

The derived iterative equations for approximating the

time evolution of the CPDF possess an "expanding grid" phenomena in the sense that the approximation of the CPDF at the kth iteration requires knowledge of more moments of the CPDF at the $(k-1)$th iteration than were evaluated. Also, the "skirts" of the approximation CPDF are highly inaccurate and may be negative.

The complexity of the estimator which results from this approach may limit its usefulness. Sorenson [13] has shown that if plant noise is not small less complex estimators are as accurate as those based upon the approximation CPDF.

Klein and Eyman [15] considered estimating the state vector of a continuous linear system with uncertain plant. Assuming the plant uncertainties could be represented by the product of a deterministic scaling matrix and scalar white noise they formulated a non-linear filtering problem by adjoining the noise representing plant uncertainties to the plant noise vector. Differential equations for the mean and covariance matrix of the CPDF were obtained from the Fokker-Planck equation defining the CPDF. The differential equation for the error covariance matrix involved third order moments of the estimation error. The third order term was dropped to yield a suboptimal estimator whose error covariance matrix was a function of the estimate. Thus, the suboptimal filter and its error covariance matrix must be evaluated simultaneously on-line.

A second development of the Bayesian approach is due to Magill [16]. Let the vector a represent the noise covariance, the initial condition, and the plant parameter uncertainties of a linear system. Assume that a can have only r, $0 < r < \infty$, values, and that the probability of each value is known. The conditional mean estimate of the state vector x_k given scalar measurements z_j, $j = 1,\ldots,k$, is

$$\hat{x}_k = \int_X x_k \sum_i P(x_k/a_i;Z^k)\, P_r(a_i/Z^k)dx_k$$

where X is the space of all x_k, and $P_r(a_i/Z^k)$, the probability of a_i given data Z^k, is a non-linear weighting coefficient found by application of Bayes rule. One has

$$P_r(a_i/Z^k) = \frac{P_r(Z^k/a_i)\, P_r(a_i)}{\sum_{j=1}^{r} P_r(Z^k/a_j)\, P_r(a_j)}$$

where $P_r(a_i)$ is given, and $P_r(Z^k/a_i)$ may be evaluated from the system equations.

Assuming that the order of integration and summation can be interchanged, one has

$$\hat{x}_k = \sum_{i=1}^{r} \hat{x}_k(a_i)\, P_r(a_i/Z^k)$$

where

$$\hat{x}_k(a_i) = \int_X x_k\, P(x_k/a_i;Z^k)dx_k\, .$$

Thus, Magill's method for a linear system with uncertainty involves implementing the optimal adaptive filter as the sum of Kalman filters; the Kalman filters operate on the measurements to obtain $\hat{x}_k(a_i)$ for each i. The weighting coefficients of the summation embody the adaptive feature of the filter.

Hilborn and Lainiotis [17] considered a system with a continuous scalar parameter uncertainty, and attempted to reduce the severe complexity and storage requirements of the optimal adaptive filter. The resulting optimal adaptive filter is realizable in delay-feedback form.

b. The Maximum Likelihood Approach

If probability density functions for the system uncertainties are not known, adaptive estimation can be accomplished using the maximum likelihood approach, which is based upon the philosophy that the most likely value of unknown parameters is that value which makes the probability of their occurrence given measurements Z^k greatest.

Shellenbarger [18] used a maximum likelihood approach to obtain suboptimal estimates of unknown plant and measurement noise covariance matrices. Consider only the measurement noise covariance matrix to be uncertain. To estimate the sequence $R = (R_1, \ldots, R_k)$, where R_j is the measurement noise covariance matrix at the jth measurement, Shellenbarger constructed the PDF of the measurements Z^k as a function of the R_n's. Since $P(Z^k/R)$ is an extremely complicated function of the R_n's Shellenbarger used the suboptimal procedure of approximating $P(Z^k/R)$ by

$$P_s(R_k) = P(Z^k/\hat{R}_1, \hat{R}_2, \ldots, \hat{R}_{k-1}; R_k)$$

where $\hat{R}_j$, $j = 1, \ldots, k-1$, is the previously obtained estimate of R_j. The suboptimal estimate of R_k was obtained by maximizing $P_s(R_k)$ with respect to R_k. First-order necessary conditions for a maximum are that the partial derivative of P_s with respect to each element of R_k equal zero. Since

$$P_s(R_k) = P(z_k/Z^{k-1}; \hat{R}_1, \ldots, \hat{R}_{k-1}; R_k)\ P(Z^{k-1}/\hat{R}_1, \ldots, \hat{R}_{k-1})$$

one need only maximize $P(z_k/Z^{k-1}; \hat{R}_1, \ldots, \hat{R}_{k-1}; R_k)$ with respect to R_k. The result is

$$\hat{R}_k = \tilde{z}_k \tilde{z}_k^T - C(k)\ \hat{P}(k/k-1)\ C(k)^T$$

where $\hat{P}(k/k-1)$ is the estimate of the kth value of the error covariance matrix given data up to the $(k-1)$th iteration.

Estimates of Q_k with R known are obtained similarly. However, one needs to take the inverse of $C(k)$, or manipulate it to avoid taking inverses, to completely estimate Q_k. This requires $C(k)$ to have more rows than columns.

Abramson [19] obtains a suboptimal estimator for the state of a system when the statistics of the measurement and plant noise covariance matrices are diagonal and time-invariant, but not precisely know apriori. Simultaneous estimates of the state vector and unknown covariance matrices are found by maximizing the logarithm of the probability density function $P(x_k, Z^k/R, Q)$. The maximizing values for each component of x_k, Q, and R are found by equating the derivative of the logarithm of the PDF with respect to that component equal to zero. As a result of the non-linear nature of the likelihood equations no general closed form solution is available.

To obtain a real-time recursive, suboptimal solution which simultaneously updates estimates of x_k, R, and Q as each new measurement becomes available Abramson uses the following procedure.

1. Obtain estimates of x_k, R, and Q independent of any apriori information about the statistics of R and Q. The apriori information or past estimates of R and Q must be used to solve the estimator equations for the current estimates.

2. Estimates of R are obtained with Q known and of Q with R known. Since the likelihood equations are coupled, biased estimates that fail to distinguish between errors in R and Q can be expected.

If the predicted residual and error covariance matrix are computed as functions of either the apriori estimates of R and Q or some previously obtained estimates, recursive expressions for the estimates of the diagonal elements of R and Q are as follows.

$$\hat{R}_k^{jj} = (k-1)/k\ \hat{R}_{k-1}^{jj} + 1/k(\tilde{z}_k \tilde{z}_k^T + C(k)\ P_k\ C(k)^T)^{jj}$$

$$\hat{Q}_k^{jj} = (k-1)/k\ \hat{Q}_{k-1}^{jj} + 1/k(B(k)^{-1}(\Delta x_k \Delta x_k^T + P_k - U_k)(B(k)^T)^{-1})^{jj}$$

where $\Delta x_k = \hat{x}_k - \hat{x}_{k/k-1}$, and $U_k = \theta_k P_{k-1} \theta_k^T$.

Numerical results for the estimator indicate that if the apriori values of both R and Q are significantly in error, the process will converge to a biased and incorrect estimate of these parameters. However, if only R is uncertain the estimator will be asymptotically unbiased.

Husa [20] considered a more general case than Abramson or Shellenbarger by allowing uncertainties in both the mean and covariance matrices of the plant and measurement noises. In addition, the covariance matrices were not restricted to be diagonal. Estimates of the unknown means, q and r, and the unknown covariance matrices, Q and R, were obtained by maximizing $P(X^k, q, Q, r, R/Z^k)$. Estimator equations for the state vector were obtained by use of the Discrete Maximum Principle. The resulting two-point, boundary-value problem was solved by the method of discrete invariant imbedding to yield the estimator algorithm.

The recursive equations for estimating the statistics of the plant and measurement noises which resulted from the maximization are complicated and somewhat impractical from a computation point-of-view. Practical suboptimal estimator equations were developed by assuming that the covariance matrix for

$$\pi(k/k-1) = 1/(k-1) \sum_{i=1}^{k-1} \tilde{z}_i$$

is zero. The resulting algorithm is similar to that of Abramson.

Simulations showed that the plant noise estimator converged very slowsly, and was very sensitive to the initial conditions for the noise covariance estimators. The measurement noise estimator performed satisfactorily.

In estimation problems where the measurement noise covariance matrix is constant, but uncertain, and the estimation interval is finite the non-linear maximum likelihood estimator may be very sensitive to the initial guesses for R and x_0. For an orbit determination application Tapley and Born [21] obtained the noise covariance estimator by maximizing $P(Z^N, x_k/R)$. The estimator's sensitivity was decreased by modifying $\hat{R}_k$ as follows.

$$\hat{R}_k = (1 - W_k)\, \hat{R}_{k-1} + W_k\, \tilde{z}_k \tilde{z}_k^T$$

where the weighting function W_k is zero for all $k < N$, and as k approaches infinity W_k approaches $1/k$, and N is the number of iterations required before the estimate of the state vector stabilizes. Furthermore, apparent divergence can occur if the initial guess for R is smaller than its actual value. Thus, Tapley and Born suggest choosing the initial guess to be greater than any possible value for R, but less than four times any possible value for R.

c. The Error Residual Approach

The final adaptive estimation approach to be discussed is a statistical approach utilizing error residuals. The approach is more general than the probabilistic approaches discussed previously since neither the apriori PDF of the uncertain parameters nor the PDF's of the plant and measurement noises are required.

Mehra [11] considered a completely observable and

controllable system, which had reached steady-state. The Q and R matrices were assumed bounded, positive-definite matrices. Assuming invalidity of the system model has been verified by the method discussed previously, new estimates for Q and R can be obtained from the error residuals as follows.

The predicted residual has the following statistic.

$$M_k = E(\tilde{z}_i \tilde{z}_{i-k}^T) = \begin{cases} CPC^T + R ; & k = 0 \\ C(\theta(I - KC)^{k-1}\, \theta(PC^T - KM_0)); & k > 0 \end{cases}$$

where K, P are the steady-state Kalman filter gain and error covariance matrix, respectively. Note that the innovation process is not white. Assuming $\{\tilde{z}_k\}$ is a stationary and ergodic sequence consider the estimate of M_k

$$M_k = 1/n \sum_{i=k}^{n} \tilde{z}_i \tilde{z}_{i-k}^T .$$

The estimate of R is

$$\hat{R} = M_0 - C\,\widehat{PC^T}$$

where $\widehat{PC^T}$ can be obtained from M_j as

$$\widehat{PC^T} = K\hat{M}_0 + A\#\,(\hat{M}_1, \ldots, \hat{M}_n)^T$$

and $A\#$ is the pseudo inverse of $(C, \ldots, C(\theta(I - KC)^{n-1}\,\theta)^T$.

The estimate of Q is obtained from manipulations of the steady-state equation for the error covariance matrix. The procedure is quite involved. Since nxr linear relationships between the unknown elements of Q are available only nxr unknown elements of Q can be found. Thus, if Q has more than nxr uncertain elements, Mehra estimates the optimum filter gain matrix directly.

Simulations using Mehra's estimator indicated that the

estimates of the covariance parameters were close to the estimates obtained by the maximum likelihood approach. Also, the filter was insensitive to initial estimates of Q and R.

Jazwinski [22] in investigating the problem of filter divergence due to inaccurate modeling of a deterministic plant has derived an algorithm which may be used for simultaneous state and plant noise covariance matrix estimation of a linear system. Modeling errors are approximated by a white Gaussian plant noise input, whose covariance Q, a diagonal matrix, is determined to produce consistency between residuals and their statistics.

The predicted residual is defined to be

$$r(k+L) = z_{k+L} - E(z_{k+L}/Z^k) \qquad L > 0 .$$

Using the constraint $r^2(k+L) = E(r^2(k+L))$; $L = 1,2,\ldots,n$, which makes the residual values most probable, Jazwinski derives the single residual estimate of Q_k, which is not statistically significant.

To gain statistical significance one may consider the mean of N predicted residuals

$$M_r = 1/N \sum_{L=1}^{N} r(k+L)/(R(k+1))^{\frac{1}{2}} .$$

The estimate of Q_k derived from the residual consistency constraint is

$$\hat{Q}_N(k) = \begin{cases} \dfrac{M_r - E(M_r/Q_k = 0)}{S} ; & \text{if positive} \\ 0 ; & \text{otherwise} \end{cases}$$

where S is a normalizing parameter.

To improve the convergence of the plant noise covariance estimation, Jazwinski utilizes the smoothed estimate

$$\hat{Q}_N^L(k) = \begin{cases} 1/L \sum_{i=k-L+1}^{k} \hat{Q}_N(i) \; ; & \text{if positive} \\ 0 \; ; & \text{otherwise.} \end{cases}$$

Jazwinski demonstrates filter convergence through the use of simulations. However, the procedure for selection of the averaging interval N and the smoothing interval L has not been developed for the general case.

2. Bounding Techniques

Probably the first observed evidence of the sensitivity of the Kalman filter to uncertainties in the system model was in the application of the Kalman filter to orbit estimation [23]. In this application it was observed that modeling errors caused the estimate to diverge from the true state of the system, leading to estimation errors much greater than those predicted in theory.

Divergence of the Kalman filter may be explained as follows. For systems that contain no plant noise the gain and computed error covariance matrix both tend to a zero value as time increases [24]. This means that after a large number of noisy measurements of a deterministic system, the estimator has effectively matched the data with the assumed model and therefore computes each new estimate using only the preceding estimate and the assumed state transition matrix; i.e., independent of any new measurements. Since the assumed model corresponds to the true system model only over a limited time period, either apparent or true divergence of the estimate of the state vector from the true state value will occur.

Apparent divergence of the Kalman filter may be controlled by preventing the Kalman filter from basing its current estimate upon stale data. Two techniques which employ this approach will be discussed. One technique prevents the Kalman filter gain from approaching zero value. This technique is termed "Noise Incrementation." The second technique constrains the Kalman filter to base its current estimate only upon the most recent data. These limited memory techniques are termed "Stale Data Rejection." In both techniques the deleterious effect of the uncertain parameters upon estimation error is not minimized; one desires only to make the estimation error smaller than an allowable value.

a. Noise Incrementation

Schmidt [25] attempted to bound the estimation error of the Kalman filter by constraining the Kalman filter gain to be of the form

$$K_k = (P_{k/k-1} + \alpha)\, C(k)^T\, (C(k)P_{k/k-1}C(k)^T + R(k))^{-1}$$

where α, a constant, is chosen apriori by the designer.

The advantage of Schmidt's method is that it establishes a bound on the acceptable accuracy. Since the limit of the variance in the estimate of $C(k)x_k$ can be shown to equal α if the model is accurate, α forms a lower bound on the estimation accuracy in the case of inaccurate system models.

Schmidt's method has the effect of keeping the Kalman filter gain nonzero regardless of how long measurements are taken. Since this means that new measurements will never be ignored, his method is less sensitive to dynamic errors than the Kalman filter. However, Schmidt has shown that an inaccurate transition matrix used with his method can result in a steady-state bias error.

Tarn and Zaborsky [26] exponentially increased the measurement noise covariance matrix of old observations; e.g., they considered white measurement noise with zero mean and covariance matrix

$$E(v_k v_k^T / t = t_n) = s^{n-k} R_k \; ; \qquad n \geq k$$

where n is the current iteration and k is the iteration at which the noise occurred. $s \geq 1$ is chosen apriori by the designer. This has the effect of escalating exponentially with time the covariance matrix of each past observation; making past observations have less effect upon current estimates.

The Kalman filter equations which result when this form of noise is employed are identical to the usual Kalman filter equations with the exception of the equation for $P_{k/k-1}$, which becomes

$$P_{k/k-1} = s\Theta \, P_{k-1/k-1} \, \Theta^T + Q_k \, .$$

Examples are shown that demonstrate the effectiveness of this approach in bounding the estimation error and its computational attractiveness. However, the performance of this modified Kalman filter may be quite sensitive to the choice of the escalation factor s.

Miller [27] and Sorenson and Sacks [28] studied the asymptotic stability of continuous time estimators with exponential aging. Miller considered a deterministic plant and aged the observations by using

$$E(v(t)v(t_1)^T) = e^{at} R(t) \, \delta(t - t_1)$$

where $a > 0$ is the aging parameter. He showed that the inverse of the product of e^{at} and the error covariance matrix was stable and converged to a steady-state solution if

$$a > -2R_e(\lambda_i) \; ; \qquad i = 1, \ldots, n$$

where λ_i are the eigenvalues of the plant matrix, and $R_e(.)$ is the real part of the argument. This condition provides a lower bound for the admissible exponential aging value.

Sorenson and Sacks exponentially aged the covariance matrices for the state vector initial condition, the plant noise, and the measurement noise by weighting these matrices by the factor

$$\exp \int_t^T a(t_1)dt_1$$

where $a(t_1)$ is a positive aging function chosen by the designer, T is the current time, and $t \in [t_0, T]$. The estimator employing this type of aging has form identical to the usual Kalman-Bucy filter except that the plant matrix $A(t)$ is modified to be

$$A_m(t) = A(t) + a(t)/2\ I .$$

Thus, the stability properties of the Kalman-Bucy filter may be utilized to establish the stability properties of Sorenson and Sacks' exponentially aged filter.

b. Stale Data Rejection

Limited memory filtering (LMF) is based upon the observation that if the assumed model corresponds to the true system only over a limited period of time, the processing of data "older" than the correspondence period by the Kalman filter to form an estimate of the current state of the system will lead to unacceptable estimation errors.

Jazwinski [29] considered a deterministic plant, and approached the LMF problem from a probabilistic point-of-view by considering $P(x_k/Z_k^N)$, where $Z_k^N = z_k, z_{k-1}, \ldots, z_{k-N}$, and N is indicative of the memory length of the filter.

In order that the LMF will be independent of old observations one must constrain $P(x_k/Z_k^N)$ to be independent of $P(x_0)$. This is accomplished by evaluating the limit of $P(x_k/Z_k^N)$ as the standard deviation in x_0 approaches infinity.

A maximum likelihood approach is used on the resulting PDF to obtain an estimator that requires two Kalman filters and a predictor for its implementation. N observations must be stored in memory. Obviously, this scheme is no solution since it involves the difference between two diverging Kalman filters.

To circumvent the divergence problem Jazwinski proposes using the following procedure. The conditioning of the estimate on old data is discarded in batches of N. One filters N observations and also predicts over the same time period. This makes $\hat{x}_{k/k}$ and $\hat{x}_{m/k}$, $m > k$, both finite memory filters. The last estimate of one batch of data is used as the initial condition for the next batch of data.

The IMF described above is not a "batch" processor since it derives estimates with each new observation. Its memory varies between N and 2N.

Simulations have shown that the IMF is computationally stable and does result in less estimation error than the extended Kalman filter when the latter diverges. However, no general procedure for selection of the memory length, or the sensitivity of the IMF performance to memory length is available.

Crump [30] bases his approach to IMF upon the observation that the Kalman filter computes an estimate which represents a weighted least-squares curve fit of the assumed model to all past data. Considering scalar observations of the state of a deterministic system, he represents the model of the system as a truncated power series in time. His IMF is constrained to be a linear weighted sum of the current observation and the last $N-1$ estimates, termed "the augmented memory estimator (AME)."

To evaluate the weighting coefficients optimum from the

point-of-view of minimizing mean-square error over the memory period the problem is transformed into a scalar optimization problem by adjoining all the state vectors over the memory period into an augmented state vector. The weighting coefficients which result from the minimization are complicated functions of a matrix of noise and previous estimation error covariance matrices.

Crump demonstrates that the AME is as good as the Jazwinski IMF, and in situations where the system model is accurate the AME will perform better than Jazwinski's IMF. In cases where the system model is not accurate, the AME exhibits a bias in estimation error, which may be controlled to some extent by a proper choice for the memory length N.

3. Multiple Criteria Techniques

In several applications of Kalman filtering the estimator design is constrained by the facts that nothing is known about the characteristics of the parameter uncertainties except bounds on their values, and that other storage requirements limit the amount of computer storage available for the estimation algorithm. In such cases, adaptive estimation algorithms cannot be utilized.

One approach toward the design of an estimator subject to these constraints is to include the filter's performance sensitivity to the uncertain parameters in the design criteria. The object of the estimator design is to gain the optimum trade off between filter sensitivity and minimum mean-square error. This is accomplished by replacing the original performance index by a modified performance index which reflects the sensitivity of the primary performance index (mean-square error) to the uncertain parameters. The estimator that results from minimization of the modified performance index will have a form identical to the usual Kalman filter, and incorporates open-loop compensations for the uncertain parameters.

Two modified performance indices that have been used to measure performance sensitivity are the absolute and relative performance measures defined by [31]

$$J_1(t) = \max_{b \in B} \left(J(a,b;t) - J(a_b,b;t) \right)$$

$$J_2(t) = \max_{b \in B} \frac{J(a,b;t) - J(a_b,b;t)}{J(a_b,b;t)}$$

where $J(a,b;t)$ is the primary performance index evaluated at t, A is the set of admissible design choices, B is the parameter uncertainty set, and a_b is the design that is optimal when the uncertain parameter value is b.

Assuming only the noise covariance matrices to be uncertain D'Appolito [31] obtained a procedure for the determination of the steady-state estimators, termed "minimax sensitivity filters," which yield the minimax values of J_1 and J_2. A closed form solution for the steady-state, minimax sensitivity filter is not available since the algebraic performance sensitivities $J_1(\infty)$ and $J_2(\infty)$ do not have saddle points [31]. However, D'Appolito has shown that the optimizing uncertain parameter value is an extreme point of the uncertain parameter set B, and the optimizing estimator design is optimum with respect to the primary performance index for some element of B. The optimizing steady-state design is found by a search over the set of extreme points of B, B_E, and the set of designs optimum for parameter values belonging to B_E. Unfortunately, the set B_E is usually prohibitively large.

The multiple criteria approach toward the design of estimators for uncertain systems will be the major topic of this chapter. Less restrictive modified performance criteria will be employed so that the design evaluation will be simplified and the estimator design of time-varying systems will be possible. The resulting filter will not necessarily be optimum for each value of the uncertain parameter as a filter using D'Appolito's approach is. However, the filter will be designed such that the overall estimation error is minimum.

III. DESIGN OF SYSTEMS WITH UNCERTAINTY

A. Introduction

The problem of optimal design of a system with uncertainty may be posed as follows. Given a system S with uncertain parameters b belonging to a known set of possible values B choose an element a from the set A of possible design choices that results in the "minimizing performance" as measured by $K(a,b)$.

If the uncertainty set B is not void, minimization of the performance index without regard to the system sensitivity to the uncertain parameters should not be attempted. Mesarovic and Takahara [32] have shown that the optimization process may increase the system sensitivity to such an extent that the performance becomes unsatisfactory for some element of the uncertainty set. Likewise, the sensitivity of the system may be such that the performance of the system is much better than normal for some values of the uncertain parameters. Thus, minimization of the sensitivity of the system may lead to the design of a system whose overall performance is much worse than that of a system designed without sensitivity considerations. Design of a system with uncertainty is a trade off between minimum performance and performance sensitivity.

Two traditional means of solving optimization problems with uncertainties are (1) assuming a probability density function for the uncertain parameters and minimizing the expected value of the performance index, and (2) correcting the control optimal for the nominal value of the uncertain parameters by an amount equal to the control necessary to minimize the variation of the system's performance from optimal due to a variation in the uncertain parameter. The first method is limited in that the probability density function of the uncertain parameters is usually not known, while the second approach is valid only for small variations.

In this chapter an alternate approach will be presented. The presence of uncertainty will be acknowledged and the sensitivity of the system's performance to the uncertainty will be incorporated into the formulation of the process by which the design is chosen.

B. General Theory

A cursory development of the general theory for design of systems with uncertainty will now be given. Similar discussions are contained in Mesarovic and Takahara [32], and Witsenhausen [33].

The problem of design of a system with uncertainty may be mathematically formulated as follows. Given a system represented by the continuous mapping

$$S : AxB \rightarrow X$$

where A is the set of admissible design choices, B is the compact, convex set of uncertain parameter values, and X is the n-dimensional Euclidean space of the state vector. Find the design choice $a_0 \in A$ which results in the minimum value of the performance index defined by the mapping

$$K : AxX \rightarrow [0, \infty) .$$

It is important to note that the set A does not include adaptive design choices since they may result in very complex designs which are difficult to implement in a system with limited computer storage. The design approach considered herein is an attempt to make compensations for parameter uncertainties through consideration of performance value and performance sensitivity.

The design objective is to choose the particular design that results in the optimum trade off between minimum performance and performance sensitivity. This selection requires comparisons of the overall performances of the systems that result when each of the possible design choices is utilized in the system.

Suppose a design choice $a_1 \in A$ is chosen. The system output is

$$S_{a_1} = S/\{a_1\} : \{a_1\} \times B \to X$$

and the overall performance of the system is

$$K_{a_1}(B) = K/\{a_1\} : B \to [0,\infty) .$$

The set $K_{a_1}(B)$, termed "the performance set corresponding to a_1," is representative of both the relative optimality of the design, and the performance sensitivity of the design choice.

A comparison of performance sets to obtain the set whose overall performance is minimum may be accomplished by partial ordering.

DEFINITION 1. Equivalence classes of designs are termed "partially ordered," written $a_1 \leq a_2$, whenever $K_{a_1}(b) \leq K_{a_2}(b)$ for all $b \in B$.

If all of the performance sets corresponding to elements of A can be partially ordered, the choice of the design parameters $a*$ that has lowest order will result in the optimum performance. If $a*$ obeys the partial ordering $a* \leq a$ for all $a \in A$, $a*$ is termed "the dominant design."

Since the performance sets are measures of the sensitivity of the system, this design process includes both optimality, e.g., minimization of the original performance index, and sensitivity considerations.

Generally, the performance sets cannot all be partially ordered. Under such circumstances transformations called "supercriterion," denoted by $J : E \to [0,\infty)$, are sought which enable partial ordering of the elements E of A that cannot be partially ordered with respect to the minimum of all performance sets that can be partially ordered. The design element that is dominant with respect to partial ordering of the supercriterion is chosen as the optimizing design.

Consider supercriterion functions of the form

$$J(a) = V(K_a(B)) .$$

The functional V relating the performance sets to the supercriterion is termed "the evaluator." A major task in the design of systems with uncertainties is the selection of evaluators that are meaningful and mathematically convenient. Properties that an evaluator must possess are given below.

DEFINITION 2. An evaluator for the system $S : AxB \to X$ with performance measure $K : AxX \to [0,\infty)$ is the functional $V : C_E \to [0,\infty)$, where C_E is the class of performance sets corresponding to admissible design choices that cannot be partially ordered. An evaluator satisfies the following conditions.

1. If two performance sets p_1 and p_2 are partially ordered $p_1 \leq p_2$, then $V(p_1) \leq V(p_2)$.

2. If for a design $a \in A$, $K_a(B) = e \geq 0$ (a point on the positive half of the real line), then $V(K_a(B)) = e$.

3. Consider two noncomparable design choices a_1 and

$a_2 \in A$ such that $\text{lub } K_{a_1}(B) > \text{lub } K_{a_2}(B)$. Let $y \in K_{a_1}(B) \cap K_{a_2}(B)$, and

$$U(x) = (x = b/b \in (K_{a_1}(B)/K_{a_2}(B) > y))$$

$$L(x) = (x = b/b \in \text{ either } K_{a_1}(B)/K_{a_2}(B) < y, \text{ or } K_{a_2}(B)/K_{a_1}(B) < y).$$

Translate $U(x)$ and $L(x)$ to the origin by considering

$$U'(z) = (z = x - y_2/x \in U(x) \text{ and } y_2 = \text{glb } U(x))$$

$$L'(x) = (z = x - y_3/x \in L(x) \text{ and } y_3 = \text{glb } L(x)).$$

If $U'(z) \supset L'(z)$, then $V(K_{a_1}(B)) \geq V(K_{a_2}(B))$.

The first two conditions guarantee that the multiple criterion approach will yield results identical to the usual minimization problem results when the uncertain set B degenerates to a point.

The third condition requires the evaluator to partially order performance sets such that designs with lowest order are such that their increases or decreases in performance sensitivity relative to that of other performance sets are experienced primarily as a decrease in performance value as compared to that of other performance sets. This will result in optimization of the trade off between minimum performance and performance sensitivity.

C. Vector-Valued Performance Indices

Use of the definition of partial ordering given above to obtain the dominant design is equivalent to minimization of a vector-valued performance index whose dimension is equal to the number of elements that belong to B. Unfortunately, B

is usually composed of such a large number of elements that the dimension of the equivalent vector-valued performance index is prohibitively large.

If $K(a,b)$ is linear in the elements of B and if B is a convex, compact set, then a reduction in the dimension of the equivalent vector-valued performance index is possible. Consider the extreme points of the set B [34].

DEFINITION 3. A point x in a convex set X is called an extreme point of X if there exist no points $x_1, x_2 \in X$ such that $x = \lambda x_1 + (1-\lambda)x_2$ for some λ, $0 \leq \lambda \leq 1$, where $x_1 \neq x_2$.

The following theorem from Karlin [34] indicates the importance of extreme points.

THEOREM 1. A compact, convex set B is spanned by its extreme points. That is, every x in X can be represented in the form

$$x = \sum_{i=1}^{r} \lambda_i x^i; \quad \lambda_i \geq 0, \qquad i = 1,\dots,r; \quad \sum_{i=1}^{r} \lambda_i = 1$$

where $x^1,\dots,x^r$ are the extreme points of X.

Thus, if $K(a,b)$ is linear in the elements of B and if B is a compact, convex set, one can write the performance index in the form

$$K_s(a,b) = \sum_{i=1}^{r} \lambda_i K(a,b^i); \quad 0 \leq \lambda_i \leq 1, \quad \sum_{i=1}^{r} \lambda_i = 1 .$$

Now consider an optimization problem with vector-valued performance index. Given a system represented by the continuous mapping

$$S : AxB \rightarrow X .$$

Find the design choice $a^* \in A$ which results in the minimum value of the performance index defined by the continuous mapping

$$K_1' : AxX \rightarrow Y$$

where Y is the r-dimensional Euclidean space of the performance index, r is the dimension of the uncertain parameters, and the jth element of the vector K_1' is $(K_1')_j = K(a,S(a,b^j))$.

Polak and daCunha [35] have derived conditions under which the solutions to the vector-valued optimization problem and the uncertain system problem with primary performance index given by $K_s(a,b)$ are equivalent. Their results are contained in the following theorem and corollary.

THEOREM 2. Suppose that $K_1'(a,b)$ is a componentwise convex functional of the elements of the convex set A of design choices. Then the set N of all solutions of the uncertain system problem contains the set L of all solutions of the vector-valued optimization problem.

This theorem states that if one solves the vector-valued optimization problem he will obtain at least one solution of the uncertain system problem if the conditions of the theorem are met All solutions of the uncertain system problem may be obtained via the vector-valued optimization problem under the following conditions.

COROLLARY 1. If the set A is convex and K_1' is a componentwise strictly convex function of A, the sets N and L are equivalent.

D'Appolito [31] has shown that in the case of noise covariance matrix uncertainties the performance index is linear in the elements of the parameter uncertainty set B and strictly convex in the elements of the set of design choices A. Thus, solutions of the noise covariance matrix uncertainty problem may be obtained via solutions of a vector-valued optimization problem of reduced dimension. However, using D'Appolito's approach it may also be shown that if the plant matrix is uncertain, the performance index will be convex in the elements of the parameter uncertainty set. Thus, if the approach of transforming the original design problem into a vector-valued optimization problem of reduced dimension is used, one must be satisfied with obtaining solutions which minimize an upper bound of the original performance index.

A theory for the solution of optimization problems with vector-valued performance indices has been developed by Zadeh [36]. Consider comparisons of the vector-valued performance index evaluated for a design choice a_1 with the vector-valued performance index evaluated for all other design choices belonging to the set A. The result is a division of A into the disjoint subsets; $A_>(a_1)$ of all designs that are better than (superior to) a_1, $A_{\leq}(a_1)$ of all designs that are inferior to or equal to a_1, and $A_\sim(a_1)$ of all designs that are not comparable to a_1 in the sense that some of the elements of their vector performance measures are greater than those of a_1, and other elements of their vector performance measures are less than the corresponding ones of a_1.

The optimality of a design choice for a system whose measure of performance is vector-valued may not be determined directly from a partial ordering of performance measures since it is possible to have design choices whose performance measures are not comparable with the minimum of all performance measures which can be partially ordered. However, it is possible to determine a subset E of A in which the optimal design will reside.

DEFINITION 4. The subset E of A, termed "the set of noninferior design choices," is composed of elements $a_j \in A$ such that $A_>(a_j)$ is empty.

The primary task in the solution of an optimization problem with vector-valued performance index is the definition of the set of noninferior designs. Then, supercriterion functionals may be used to partially order the noninferior designs to arrive at the dominant design. However, definition of the set of noninferior designs is not a simple task. If the vector-valued performance index is strictly convex in A, then some elements of the set of noninferior designs can be identified [37].

THEOREM 3. If $K_1'(a,b)$ is strictly convex in elements of A, the set A_0 of all designs that are optimal with respect to $K(a,b^j)$ for some b^j belonging to the set of extreme points B_E of B is contained in E.

Beeson [38] has derived an algorithm for solving static optimization problems with vector-valued performance criteria. His approach is basically to solve the optimization problem as a sequence of optimization problems with vector-valued performance indices whose dimensions increase from 2 to r. The components of the lower dimensional performance criteria form a subset of the set of all elements of the r-dimensional performance criteria. Evaluation of the noninferior set for an m-dimensional performance criteria, $2 \leq m \leq r$, provides candidates for the noninferior set for the problem with $(m+1)$-dimensional performance criteria. In addition, the noninferior set of the m-dimensional performance criteria may be used to calculate upper and lower bounds for the $(M+1)$th component of the $(m+1)$-dimensional performance criteria [38]. This increases the efficiency of the computer search program.

D. A Two-Dimensional Vector-Valued Performance Index

For problems with matrix uncertainties the number of extreme points belonging to B may be quite large. Suppose only the plant noise covariance matrix Q is uncertain. If Q is a qxq-diagonal matrix, there are 2^q extreme points of B. This will make the dimension of even $K_1'(a,b)$ extremely large. In such cases, a more mathematically covenient, albeit suboptimal, vector/valued performance index must be considered.

Consider the following 2-dimensional vector performance index K_1'' that measures the performance of a subset of the extreme points of B.

$$K_1''(a) = (K(a,b'), K(a,b^*)) = (\underset{b \in B}{\text{Max}}\, K(a,b), \underset{b \in B}{\text{Min}}\, K(a,b)) .$$

If the performance index $K(a,b)$ is linear in elements of the set B, then b^* and $b' \in B_E$. If $K(a,b)$ is convex in elements of B, $b' \in B_E$.

The primary effect of the utilization of K_1'' in place of K_1' is that a greater number of performance sets will be comparable using K_1'' than when using K_1'. Thus, performance sets noninferior to the dominant design of all performance sets which can be partially ordered with respect to K_1', may be inferior to the dominant design of all performance sets which can be partially ordered with respect to K_1''. Therefore, it would appear that utilization of K_1'' will result in a suboptimal selection. However, one should note that a meaningful evaluator which operates on the set of noninferior designs relative to partial orderings with respect to K_1' may partially order these designs such that those comparable with respect to K_1'' will be listed as inferior anyway.

The class of performance sets which constitute noninferior designs relative to $K_1''(a)$ can be determined by considering two important performance sets, termed "the limiting performance sets."

Consider $a_p \in A$ and $b_p \in B$ which satisfy

$$K(a_p, b_p) = \underset{a \in A}{\text{Min}}\ \underset{b \in B}{\text{Max}}\ K(a,b) .$$

The performance set associated with a_p is called "the performance set corresponding to pessimistic design." Its upper and lower bounds are

$$P_{up} = K(a_p, b_p), \quad \text{and} \quad P_{10} = \underset{b \in B}{\text{Min}}\ K(a_p, b) .$$

Consider $a_0 \in A$ and $b_0 \in B$ which satisfy

$$K(a_0, b_0) = \underset{a \in A}{\text{Min}}\ \underset{b \in B}{\text{Min}}\ K(a,b) .$$

The performance set associated with a_0 is called "the performance set corresponding to optimistic design." Its upper and lower bounds are

$$O_{up} = \underset{b \in B}{\text{Max}}\ K(a_0, b), \quad \text{and} \quad O_{10} = K(a_0, b_0) .$$

A subset of A which contains the set E of noninferior designs is defined as follows [37].

THEOREM 4. The set of noninferior designs E is the set of design choices a_j whose performance sets belong to D_1 defined by

$$D_1 = ((K_{up}(a_j),\ K_{10}(a_j))/O_{10} \le K_{10}(a_j) \le P_{10}\ , \quad \text{and}$$

$$P_{up} \le K_{up}(a_j) \le O_{up}) .$$

A drawing of the set D_1 of designs noninferior with respect to K_1'' is given in Figure 1. The dimensions of the rectangle representing D_1 indicate which limiting performance set is superior. In the case of the figure, the dimension of the rectangle along the ordinate is smaller than its dimension along the abscissa, indicating that the performance set corresponding to pessimistic design is superior.

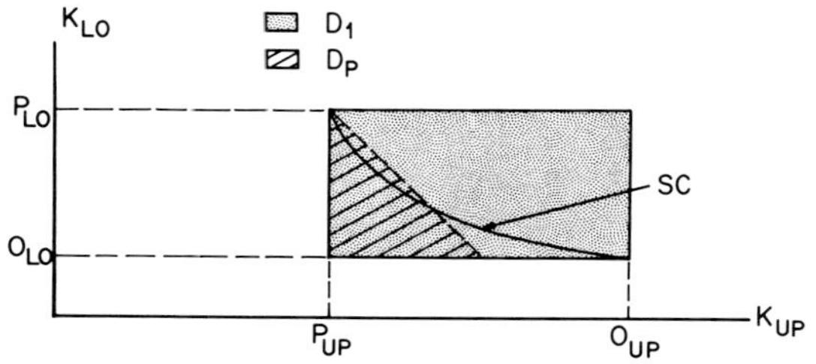

Fig. 1. An Illustration of D_p, D_1, and SC when the Pessimistic Design is Superior.

Sensitivity considerations allow one to find a subset D of D_1 which will contain performance sets with superior performance. Pick a particular value of the upper bound K_{up}, $P_{up} \leq K_{up} \leq O_{up}$, and consider all performance sets which have K_{up} as their upper bound. Since the sensitivity of the system to the uncertain parameters will vary with the choice of design parameters $a \in A$, the lower bounds of these performance sets will all be different Consider the performance set with the smallest lower bound. Since this set has the smallest lower bound, there exists a subset of the uncertain parameter set B such that if the uncertain parameter is restricted to lie in this subset, the performance set with smallest lower bound will be a dominant set with respect to all the other performance sets with upper bound K_{up}. On the other hand, if the uncertain parameter lies in the other half of B, the performance of the set with smallest lower bound can be no worse than that of any other performance set with upper bound K_{up}. Thus, the overall performance of the set with smallest lower bound is most favorable, and the other performance sets with upper bound K_{up} can be deleted from D_1. If this procedure of ordering sets with respect to their lower bounds is repeated for each $K_{up} \in [P_{up}, O_{up}]$, the result can be represented by a curve in the set D_1 with endpoints (P_{up}, P_{10}) and (O_{up}, O_{10}). This curve is called the Sensitivity Curve (SC) [39]. Figure 1 illustrates the sensitivity curve.

Medanic [39] has shown that the sensitivity curve formed by ordering performance sets with identical upper bounds K_{up} with respect to their lower bounds results in a SC which is identical to the SC formed by ordering performance sets with identical lower bounds K_{10} with respect to their upper bounds.

The exact form of the SC for a particular problem cannot be determined without extensive computations. However, an important property which must hold for any SC is that it be monotone nonincreasing [39].

The second step in the determination of D involves deletions of performance sets belonging to SC which are inferior to other sets belonging to the SC. This will be done by considering Definition 2 for two particular situations.

Consider a situation in which the pessimistic design is superior to the optimistic design. In this case the set D_1 is of the form shown in Figure 1. Approach the design problem from the point-of-view of seeking a design whose performance is superior to the pessimistic design's performance. Take a step δ along the SC from the endpoint designating the pessimistic design. The new point designates another design a_1. Since the performance set corresponding to a_1 contains the pessimistic design's performance set, the design a_1 is more sensitive to the uncertain parameters. However, the design a_1 will yield superior performance if the increased sensitivity is mainly experienced as a decrease in the original performance index. Mathematically, a_1 will yield superior performance if its upper and lower bounds belong to the set

$$D_p = ((K_{up}(a_j),\ K_{10}(a_j))/P_{10} - K_{10}(a_j) \geq K_{up}(a_j) - P_{up}) .$$

The portion of the SC which does not belong to D_p is such that the increase in sensitivity is experienced mainly as an increase in the original performance index. Thus, the design which yields superior performance when the pessimistic design is favored over optimistic design lies on the portion of the SC which is located below a line of minus one slope, drawn through the point designating the pessimistic performance set. Figure 1 illustrates the relationship between D_p and D_1.

If the optimistic design is favored over pessimistic design, approach the design problem from the point of view of seeking a design whose performance is superior to the optimistic design's performance. Take a step δ along the SC from the endpoint designating the optimistic design. The new point designates another design a_1 that is less sensitive to parameter uncertainties. In this case, the design a_1 will result in superior performance if the upper and lower bounds of its performance set belong to the set

$$D_0 = ((K_{up}(a_j), K_{lo}(a_j))/O_{up} - K_{up}(a_j) \geq K_{lo}(a_j) - O_{lo}) .$$

The relationship between D_0 and D_1 is illustrated in Figure 2.

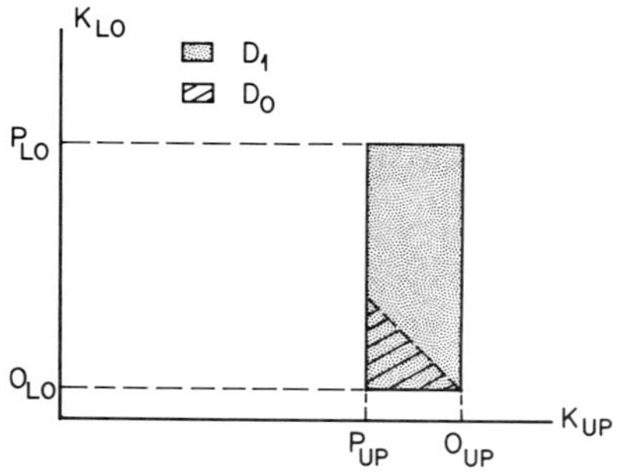

Fig. 2. An Illustration of D_0 and D_1 when the Optimistic Design is Superior.

If neither of the limiting performance sets is more

favorable, the set D_1 will be a square, and D_0 will be equal to D_p.

Thus, the subset D of D_1 which contains sets belonging to D_1 that possess performance superior to other sets belonging to D_1 may be defined as

$$D = \begin{cases} SC \cap D_p; & \text{if the pessimistic design is more favorable} \\ SC \cap D_0; & \text{if the optimistic design is more favorable .} \end{cases}$$

An immediate consequence of this development is that if the SC is a concave function of K_{up}, the design which yields superior performance is either the pessimistic or the optimistic design.

E. Evaluator Functionals

Relative to selection of possible functionals for the evaluator, the discussion of Section III, D restricts the choice of functionals V to functionals with the following characteristics. If SC is concave in K_{up}, V must be such that a choice of either pessimistic or optimistic designs will occur. Also, if SC is convex in K_{up}, V must be such that a choice of performance set extreme points belonging to D_p if a pessimistic design is more favorable, or D_0 if an optimistic design is more favorable will occur. These characteristics require the partial derivative of K_{10} with respect to K_{up} for constant V to satisfy

$$\left.\frac{\partial K_{10}}{\partial K_{up}}\right|_{V=c} = \begin{cases} < -1 & \text{in a neighborhood of } (P_{10}, P_{up}) \\ > -1, \text{ but } < 0 & \text{in a neighborhood of } (O_{10}, O_{up}). \end{cases}$$

Evaluators order the elements of the set of noninferior designs in such a manner that a dominant design can be chosen.

The dominant design should have a performance set which has minimum value in some sense. Thus, a second characteristic of an evaluator functional is that it penalize excursions of the performance set extreme points $K_{10}(a)$ and $K_{up}(a)$ from the origin of the $K_{up} \times K_{10}$ plane.

Consider the following candidates for evaluator functionals.

$$V_1(a) = w_1 K_{10}(a) + w_2 K_{up}(a)$$

where w_i are non-negative constants, and

$$V_2(a) = (K_{10}(a)^2 + K_{up}(a)^2)^{\frac{1}{2}} .$$

Both candidates penalize excusions from the origin of the $K_{up} \times K_{10}$ plane. However, $V_2(a)$ satisfies the inequality defining the first characteristic of an evaluator only if the pessimistic performance set is superior to the optimistic performance set, while $V_1(a)$ satisfies the inequality as long as the ratio of w_2 to w_1 is greater than zero, but less than or equal to one. Thus, $V_1(a)$ may always be used as an evaluator, but $V_2(a)$ should be used only if the pessimistic performance set is superior.

It is important to note that the various evaluators each view the trade off between sensitivity and minimum performance value differently. Evaluators with very large negative slopes near (P_{10}, P_{up}), or very small negative slopes near (O_{10}, O_{up}) penalize sensitivity more than performance value, while evaluators with slopes near -1 penalize performance value more than sensitivity.

IV. NOISE COVARIANCE UNCERTAINTY

A. Introduction

The design of an estimator that is to estimate the state of a system subject to uncertainties in the plant and measurement noise covariance matrices will now be considered. The estimation problem will be reformulated as an equivalent deterministic optimization problem. The first step in the design will be the evaluation of the limiting performance sets. Then, evaluator functionals will be utilized to partially order the set of noninferior designs to obtain the dominant design.

B. Problem Formulation

Assume $Q(t)$ and $R(t)$ are uncertain, but constrained to belong to the following sets.

$$Q(t) \in \{Q(t) = \|q_{ij}\| / Q(t) \geq 0,\ (q_{ij})_{min} \leq q_{ij} \leq (q_{ij})_{max}\} \tag{7}$$

$$R(t) \in \{R(t) = \|r_{ij}\| / R(t) > 0,\ (r_{ij})_{min} \leq r_{ij} \leq (r_{ij})_{max}\} . \tag{8}$$

The bounds $(q_{ij})_{min}$, $(q_{ij})_{max}$, for $i,j = 1,\ldots,q$, and $(r_{ij})_{min}$, $(r_{ij})_{max}$, for $i,j = 1,\ldots,r$, are based upon the limited knowledge of the system gained from the few data available. However, the bounds must be chosen such that every matrix which satisfies the inequalities of Eq. (7) is positive semi-definite, and such that every matrix which satisfies the inequalities of Eq. (8) is positive definite.

It is proposed to find an estimate $\hat{x}(t)$ of $x(t)$ that is of the form

$$d\hat{x}/dt = F(t)\,\hat{x}(t) + G(t)\,z(t) . \tag{5}$$

The estimator matrices $F(t)$ and $G(t)$ are to be chosen such that the error criterion J, defined by Eq. (9), is minimized.

$$J(F,G) = \int_{t_0}^{T} E((x - \hat{x})^T D(x - \hat{x}))dt = \int_{t_0}^{T} \mathrm{Trace}(D\gamma)dt . \tag{9}$$

It should be noted that the mean-square estimation error averaged over the estimation interval $[t_0,T]$ is being used as the primary performance measure instead of the mean-square error criteria employed by Kalman and Bucy so that the solution of the supercriterion problem will be meaningful.

To transform the stochastic estimation problem into a mathematically convenient form consider the first two moments of the estimation error. In general, these moments are not sufficient statistics to describe the estimation error of a Kalman filter. However, they may be used as "goodness" properties, and serve as measures of the relative performance of estimators.

Nahi [40] defines an estimate $\hat{x}(t)$ of a stochastic process $x(t)$ to be unconditionally unbiased if

$$E(x(t) - \hat{x}(t)) = E(e(t)) = 0 .$$

To determine the constraints unbiasedness places on the form of the estimator consider the following. Using Eqs. (1) and (5) the differential equation for the estimation error can be written as

$$de/dt = (A - F - GC)\, x(t) + Fe(t) + Bu(t) - Gv(t) . \tag{10}$$

Taking the expectation of the above equation, one has

$$d\bar{e}/dt = F\, \bar{e}(t) + (A - F - GC)\, \bar{x}(t) .$$

The solution of this differential equation is

$$\overline{e(t)} = \Theta_{11}(t,t_0)\, \overline{e(t_0)} + \Theta_{12}(t,t_0)\, \overline{x(t_0)} \tag{11}$$

where

$$d\Theta_{11}/dt = F\, \Theta_{11}(t,t_0) \quad ; \quad \Theta_{11}(t_0,t_0) = I$$

$$d\Theta_{12}/dt = F\, \Theta_{12}(t,t_0) - (A - F - GC)\, \Theta_{22}(t,t_0); \quad \Theta_{12}(t_0,t_0) = 0$$

$$d\Theta_{22}/dt = A\, \Theta_{22}(t,t_0) \quad ; \quad \Theta_{22}(t_0,t_0) = I.$$

Using Eq. (11) one can list two restrictions upon estimator forms which will guarantee unconditional unbiasedness of the estimate. First, one must have $\overline{e(t_0)} = 0$. The initial filter estimate, $\hat{x}(t_0)$, must be equal to the mean of the initial plant state, $\overline{x(t_0)}$. Second, one must have $F = A - GC$. The mean of the initial plant state $x(t_0)$ is generally not zero. Therefore, one must constrain $\theta_{12}(t,t_0)$ to be zero. This is accomplished by requiring $A - F - GC = 0$.

The first constraint determines the initial condition of the estimator, while the second constraint defines $F(t)$. By imposing the unconditionally unbiased constraint one has reduced the original problem to that of selecting the filter gain $G(t)$ such that the filter $\hat{x}(t)$ from the set of all unconditionally unbiased estimators minimizes Eq. (9).

If the estimate is unconditionally unbiased, it is possible to write a single matrix differential equation which describes the time behavior of the error covariance matrix $\gamma(t)$. Taking the derivative of $\gamma(t)$ one has

$$d\gamma/dt = E(de/dt\ e^T) + E(e\ de^T/dt)\ .$$

Substituting Eq. (10) into the above equation one can derive the following equation for $\gamma(t)$ [1].

$$\begin{aligned} d\gamma/dt &= (A - GC)\gamma + \gamma(A - GC)^T + BQB^T + GRG^T \\ \gamma(t_0) &= \gamma_0\ . \end{aligned} \tag{12}$$

A deterministic optimization problem whose solution yields an estimator that is optimal with respect to the second order statistics of the estimation error may now be stated.

<u>Noise Covariance Uncertainty Problem</u>

Find a matrix $G(t)$ for $t \in [t_0, T]$ which minimizes

$$J(G,\gamma) = \text{Trace} \int_{t_0}^{T} D(t)\gamma(t)dt\ .$$

Subject to the constraints given by Eqs. (7), (8), and (12).

The noise covariance uncertainty problem stated above will be solved by defining the set of noninferior designs and obtaining the dominant design of that set.

C. Limiting Performance Sets

The limiting performance sets are the keys to the design of systems with uncertainty. They provide a measure of the possible improvement in performance, indicate the evaluator functionals which are applicable, and serve as "initial conditions" for iterative optimization procedures used in suboptimal design.

Consider first the performance set corresponding to pessimistic design. To find the upper bound on this performance set one must find matrices $G^*(t)$, $Q^*(t)$, and $R^*(t)$ defined on $t \in [t_0, T]$ such that

$$J(G^*, Q^*, R^*) = \underset{G}{\text{Min}}\ \underset{Q,R}{\text{Max}}\ \text{Trace} \int_{t_0}^{T} D(t)\gamma(t)dt\ .$$

Optimization is subject to the constraints given by Eqs. (7), (8), and (12).

For this problem the elements of the error covariance matrix, $\gamma_{ij}(t)$, are considered state variables, the elements of the noise covariance matrices, $q_{ij}(t)$ and $r_{ij}(t)$, are considered maximizing variables, and the elements of the Kalman-Bucy gain matrix, $g_{ij}(t)$, are considered as minimizing variables.

The general procedure for solving differential games consists essentially of two steps [41]. First, use stationary conditions to establish a two-point, boundary-value problem (TPBVP) whose solution yields the optimizing controls Q^*, R^*, and G^* Second, verify the saddle-point condition

$$J(G^*, Q, R) \leq J(G^*, Q^*, R^*) \leq J(G, Q^*, R^*)$$

by solving two one-sided optimization problems, using G^*, Q^*, and R^* as derived from the stationary conditions.

Stationary conditions will be established through the use of a Hamiltonian, which is defined as follows.

$$H(\gamma, G, Q, R, \beta) = \sum_{i=1}^{n} \sum_{j=1}^{n} ((d\gamma_{ij}/dt)\beta_{ij} + d_{ij}\gamma_{ij})$$

$$= \mathrm{Trace}(d\gamma/dt\beta^T + D\gamma) .$$

The adjoint matrix $\beta(t)$ with elements β_{ij}; $i,j = 1,\ldots,n$, satisfies

$$d\beta_{ij}/dt = -\partial H/\partial\gamma_{ij}$$

when evaluated along the minimax path.

A necessary condition for a minimax solution is [42]

$$H^0(\gamma, G^*, Q^*, R^*, \beta) = \underset{G}{\mathrm{Min}}\ \underset{Q,R}{\mathrm{Max}}\ H(\gamma, G, Q, R, \beta) .$$

Taking partial derivatives of the Hamiltonian with respect to the elements of G, Q, and R and setting the results equal to zero, one obtains the stationary conditions

$$(\beta + \beta^T)(-\gamma C^T + G^*R^*) = 0 \tag{13}$$

$$G^{*T}\beta G^* = 0 \tag{14}$$

$$B^T\beta B = 0 . \tag{15}$$

Let $\theta(t,t_1)$ be the transition matrix corresponding to $-F(t)^T$. Then, the solution to the differential equation for the adjoint matrix is [43]

$$\beta(t) = \int_t^T \theta(t,t_1)\ D(t_1)\ \theta(t,t_1)^T\ dt_1 . \tag{16}$$

Obviously, $\beta(t)$ is symmetric and positive definite. Therefore, Eq. (13) reduces to

$$G^* = \gamma C^T (R^*)^{-1} . \tag{17}$$

Since the terms of the Hamiltonian involving Q and R are linear in Q and R, stationary conditions yield no information about the optimal control. One must consider bang-bang and singular controls. For bang-bang control one utilizes the following rules for selection of Q^* and R^*.

$$q^*_{ij} = S_w((q_{ij})_{max}, (q_{ij})_{min}; (B^T \beta B)_{ij}) \tag{18}$$

$$r^*_{ij} = S_w((r_{ij})_{max}, (r_{ij})_{min}; (G^{*T} \beta G^*)_{ij}) \tag{19}$$

where $S_w(p,y;e)$ is p if e is greater than zero, y if e is less than zero, and indeterminate if e equals zero; and $(Y)_{ij}$ is the ijth element of the matrix Y.

The selection rule for Q^* can be simplified if one partitions B into columns. Consider $B = (b_1 \vdots b_2 \vdots \cdots \vdots b_n)$. Then, the ijth element of the switching matrix is $b_i^T \beta b_j$, a quadratic form of a positive definite matrix. Therefore, if B has rank q, the diagonal elements of the switching matrix will be greater than zero, and

$$q^*_{ii} = (q_{ii})_{max} \qquad i = 1, \ldots, q . \tag{20}$$

Likewise, it follows that if G^* has rank r

$$r^*_{ii} = (r_{ii})_{max} \qquad i = 1, \ldots, r . \tag{21}$$

Evaluation of the off-diagonal elements of Q^* and R^* requires knowledge of $\beta(t)$. Thus, one must simultaneously evaluate β, γ, and the off-diagonal elements of Q^* and R^* by solving the TPBVP given by Eqs. (12) and (16)-(19).

Verification of the saddle-point condition can easily be demonstrated [37].

The lower bound on the performance set corresponding to

pessimistic design is obtained by finding matrices $Q^{**}(t)$ and $R^{**}(t)$ defined on $t \in [t_0,T]$ such that

$$J(G^*,Q^{**},R^{**}) = \underset{Q,R}{\text{Min}} \int_{t_0}^{T} \text{Trace}(D\gamma)dt \; .$$

Minimization is performed subject to the constraints given by Eqs. (7), (8), (12), (17), and (19).

The solution of the above minimization problem is obtained by utilization of the Matrix Minimum Principle [44]. The results are

$$q_{ij}^{**} = S_w((q_{ij})_{min}, (q_{ij})_{max}; (B^T\beta B)_{ij}) \tag{22}$$

$$r_{ij}^{**} = S_w((r_{ij})_{min}, (r_{ij})_{max}; (G^{*T}\beta G^*)_{ij}) \; . \tag{23}$$

Consider now the performance set corresponding to optimistic design. To find the lower bound on this performance set one must find matrices $G'(t)$, $Q'(t)$, and $R'(t)$ defined on $t \in [t_0,T]$ such that

$$J(G',Q',R') = \underset{G}{\text{Min}}\;\underset{Q,R}{\text{Min}} \left(\int_{t_0}^{T} \text{Trace}(D\gamma)dt \right) .$$

Optimization is subject to the constraints given by Eqs. (7), (8), and (12).

Games of this type are called "cooperative games" by Sarma and Ragade [45]. The solution can be obtained by adjoining the uncertainty matrices Q and R to the filter gain matrix G, and solving the resulting one-sided optimization problem with respect to the new control matrix.

Define the Hamiltonian as in the pessimistic design problem. The first order necessary conditions for a minimum are the following.

$$d\alpha/dt = -(A - G'C)^T\alpha - \alpha(A - B'C) - D; \quad \alpha(T) = 0 \tag{24}$$

$$q'_{ij} = S_w((q_{ij})_{nin}, (q_{ij})_{max}; \quad (B^T\alpha B)_{ij}) \tag{25}$$

$$r'_{ij} = S_w((r_{ij})_{nin}, (r_{ij})_{max}; \quad (G'^T\alpha G')_{ij}) \tag{26}$$

$$G' = \gamma C^T(R')^{-1} . \tag{27}$$

As in the case of the pessimistic design, the diagonal elements of the switching matrices are positive if B has rank q and G' has rank r. Thus, one has

$$q'_{ii} = (q_{ii})_{min} \qquad i = 1,\ldots,q \tag{28}$$

$$r'_{ii} = (r_{ii})_{min} \qquad i = 1,\ldots,r . \tag{29}$$

The off-diagonal elements of Q' and R' must be evaluated simultaneously with the error covariance matrix by solving the TPBVP indicated by Eqs. (12), (24), (25), (26), and (27).

The upper bound on the performance set corresponding to optimistic design is evaluated by finding matrices $Q''(t)$ and $R''(t)$ defined on $t \in [t_0,T]$ such that

$$J(G',Q'',R'') = \underset{Q,R}{\text{Max}} \int_{t_0}^{T} \text{Trace}(D\gamma)dt .$$

Maximization is performed subject to the constraints (7), (8), (12), (25), (26), and (27).

The solution of the above maximization problem is obtained using the Matrix Minimum Principle. The results are

$$q''_{ij} = S_w((q_{ij})_{max}, (q_{ij})_{min}; (B^T\alpha B)_{ij}) \tag{30}$$

$$r''_{ij} = S_w((r_{ij})_{max}, (r_{ij})_{min}; (G'^T\alpha G')_{ij}) . \tag{31}$$

D. The Dominant Design

Evaluation of the limiting performance sets defines the

set of noninferior designs. The next step toward the selection of the dominant design involves the use of evaluator functionals to partially order the performance sets corresponding to the noninferior designs. A design which is dominant relative to the evaluator functional employed may then be chosen.

The set D defines which evaluator functionals are admissible. From these one selects the admissible evaluator with the desired tradeoff between filter sensitivity and minimum mean-square error. Consider use of the admissible evaluator V_i. The dominant filter design is obtained from the set of noninferior designs by finding the filter gain $G^*(t)$ such that

$$J(G^*) = \underset{G}{\mathrm{Min}}\; V_i(K_{up}(G),\; K_{lo}(G))$$

where K_{up} and K_{lo} are the upper and lower bounds of the performance set corresponding to the filter gain $G(t)$. Optimization is performed subject to constraints given by Eqs. (7), (8), and (12).

The supercriterion problem is a cascade of two optimization problems; one with respect to Q and R assuming an arbitrary $G(t)$, and a second with respect to G given the solutions of the first problem. The first optimization problem may be readily solved. If the results are incorporated into the second optimization problem, the supercriterion problem can be restated as follows.

Equivalent Supercriterion Problem

Find a matrix $G^*(t)$ defined on $t \in [t_0, T]$ such that

$$J(G^*) = \underset{G}{\mathrm{Min}}\; V_i(K_{lo}(G), K_{up}(G))\;.$$

Subject to

1. $$K_{10}(G) = \int_{t_0}^{T} \mathrm{Trace}(D\gamma^*)dt$$

$$K_{up}(G) = \int_{t_0}^{T} \mathrm{Trace}(D\gamma')dt .$$

2. $$d\gamma^*/dt = (A - GC)\gamma^* + \gamma^*(A - GC)^T + BQ^*B^T + GR^*G^T; \quad \gamma^*(t_0) = \gamma_0$$

$$d\gamma'/dt = (A - GC)\gamma' + \gamma'(A - GC)^T + BQ'B^T + GR'G^T; \quad \gamma'(t_0) = \gamma_0$$

3. $$q^*_{ij} = S_w((q_{ij})_{min}, (q_{ij})_{max}; (B^T\beta B)_{ij})$$

$$r^*_{ij} = S_w((r_{ij})_{min}, (r_{ij})_{max}; (G^T\beta G)_{ij})$$

$$q'_{ij} = S_w((q_{ij})_{max}, (q_{ij})_{min}; (B^T\beta B)_{ij})$$

$$r'_{ij} = S_w((r_{ij})_{max}, (r_{ij})_{min}; (G^T\beta G)_{ij})$$

4. $$d\beta/dt = -(A - GC)^T\beta - \beta(A - GC) - D; \quad \beta(T) = 0 .$$

The optimization problem posed above is unusual in the sense that the differential constraints are discontinuous with respect to the filter gain since the switching matrices defining Q^*, Q', R^*, and R' are functions of G. If B and G are of full rank, the diagonal elements of the switching matrices are positive for all $t \in [t_0, T]$. Since the switching matrices are continuous with respect to elements of G and β, small perturbations in the elements of G or β from their assumed optimum values will have no effect upon the sign of the diagonal elements of the switching matrices. However, the sign of the off-diagonal elements of the switching matrices may vary for $t \in [t_C, T]$. If the sign of an off-diagonal element does vary, perturbations in the elements of G or β may shift the point at which the sign changes. Changing the points where the elements of the switching matrices change sign can increase or decrease the value of the performance index J. Thus, the derivation of first order necessary conditions for optimality of the filter gain $G(t)$ must include the influence of elements of G and β upon the off-diagonal elements of the switching matrices.

Basically, the elements of the optimum filter gain influence the off-diagonal elements of the switching matrices in such a manner that the signs of the off-diagonal elements switch at the optimum times $t_i^* \in (t_0, T)$, $i = 1, \ldots, m$. A derivation of first order necessary conditions will include an evaluation of the t_i^*'s if one includes constraints of the form

$$(N_1(t_k))_{ij} = (B^T \beta B(t_k))_{ij} = 0 \qquad t_k \in (t_0, T) \tag{32}$$

$$(N_2(t_L))_{ij} = (G^T \beta G(t_L))_{ij} = 0 \qquad t_L \in (t_0, T) \tag{33}$$

in the formulation of the Equivalent Supercriterion Problem.

This formulation of the supercriterion problem forces switchings of the signs of the off-diagonal elements of the switching matrices to occur. However, the optimum solution may not have switchings. Thus, one should first solve the problem assuming that no switchings occur. Then the no switching assumptions should be verified. If switchings do occur, the following solution should then be utilized.

To simplify the derivation a second order system will be considered. It is assumed that one switching of the off-diagonal elements of $G^T \beta G$ occurs at $t = t_2$, and that one switching of the off-diagonal elements of $B^T \beta B$ occurs at $t = t_1$. Extensions to higher order systems with more than one switching are straightforward.

Disregard Eqs. (32)-(33) for the moment and evaluate the first variation of the performance index. To simplify the notation let $X_1 = \gamma'$, $X_2 = \gamma^*$, and $X_3 = \beta$. Define the Hamiltonian as

$$H_1(X_i, \lambda_i, G;\ i = 1,2,3) = \text{Trace}\left(\sum_{i=1}^{3} Y(X_i)\lambda_i^T \right)$$

where $Y(X_1) = (A - GC)X_1 + X_1(A - GC)^T + BQ'B^T + GR'G^T$, etc. Also define

$$D_1(t) = \text{Trace}\left(\sum_{i=1}^{3} (dX_i/dt)\lambda_i^T \right).$$

Then, the first variation of the performance index is

$$\delta J = \delta V_i + \delta \int_{t_0}^{T} (H_1 - D_1)dt .$$

But, the first variation of the evaluator functional is

$$\delta V_i = \alpha_1(T)\ \delta K_{10} + \alpha_2(T)\ \delta K_{up}$$

where $\alpha_1(T) = \partial V_i/\partial K_{10}$ and $\alpha_2(T) = \partial V_i/\partial K_{up}$ are evaluated at $G = G^*$. Thus,

$$\delta J = \delta \int_{t_0}^{T} (H_1' - D_1)dt \tag{34}$$

where $H_1' = H_1 + \text{Trace}(D(\alpha_1(T)\gamma^* + \alpha_2(T)\gamma'))$.

To evaluate the first variation of the integral term given above one must separate the integral into three parts to allow for discontinuities in the integrand at $t = t_1$ and $t = t_2$. Leibnitz's rule is used on each part separately. The result is

$$\delta J = \sum_{i=1}^{2} (H_1' - D_1)dt_i \Big|_{t_i^+}^{t_i^-} + \int_{t_0}^{T} \delta(H_1' - D_1)dt .$$

Integrate the above equation by parts and use the identity [41]

$$d(X_i(t_k))_{nm} = \begin{cases} \delta(X_i(t_k^-))_{nm} + (d(X_i(t_k^-))_{nm}/dt)dt_k \\ \delta(X_i(t_k^+))_{nm} + (d(X_i(t_k^+))_{nm}/dt)dt_k \end{cases} \tag{35}$$

to simplify the results. Collecting terms one has

$$\delta J = \sum_{i=1}^{2} \left(\left(H_1' dt_i - \text{Trace}\left(\sum_{j=1}^{3} \lambda_j dX_j^T \right) \right) \Bigg|_{t_i^+}^{t_i^-} \right)$$
$$+ \text{Trace}(\delta X_3 \lambda_3 (t_0)^T - \delta X_1 \lambda_1 (T)^T - \delta X_2 \lambda_2 (T)^T)$$
$$+ \int_{t_0}^{T} \text{Trace}\left(\sum_{j=1}^{3} (((H_1')_{X_j} + d\lambda_j/dt)\delta X_j^T) + (H_1')_G \delta G^T \right) dt$$

where $(H_1')_{X_j}$ is a matrix with nmth element $\partial H_1'/\partial(X_j)_{nm}$, and δX_j is the matrix with nmth element $\delta(X_j)_{nm}$.

To avoid having to calculate variations in the elements of X_i, $i = 1,2,3$, due to variations in the elements of G via sensitivity differential equations, define the adjoint matrices as follows.

$$\begin{aligned} d(\lambda_i)_{nm}/dt &= -\partial H_1'/\partial(X_i)_{nm}; \qquad && i = 1,2,3 \\ \lambda_i(t_j^+) &= \lambda_i(t_j^-) \quad ; && i = 1,2,3 \quad j = 1,2 \\ \lambda_i(T) &= 0 \quad ; && i = 1,2 \\ \lambda_3(t_0) &= 0 . \end{aligned} \tag{36}$$

Then, the first variation of the performance index becomes

$$\delta J = \sum_{i=1}^{2} (H_1') dt_i \Bigg|_{t_i^+}^{t_i^-} + \int_{t_0}^{T} \text{Trace}((H_1')_G \, \delta G^T) dt . \tag{37}$$

Now, consider the constraints given by Eqs. (32) and

(33). Eq. (33) is an explicit function of the elements of G. Since the class of admissible variations in the elements of G is not restricted, the constraint may be trivially satisfied using the condition $N_2(t_2) = 0$ to define $G(t_2)$. This will introduce a point discontinuity in the elements of G. However, since the time interval over which G is discontinuous is of zero measure, the effect on the system performance is negligible. Thus, the optimal value for t_2 is arbitrary. A convenient choice for t_2^* is the time instant when the design based upon the assumption of no switchings of the signs of the elements of $G^T\beta G$ experienced a switching in the off-diagonal elements of $G^T\beta G$.

Consider Eq. (32) as a performance index. Adjoining the differential constraints to the "performance index" one has

$$N_1(t_1) = N_1(t_1) + \int_{t_0}^{T} (H_2 - D_2)dt$$

where

$$H_2 = H_1(X_1, X_2, X_3, \lambda_4, \lambda_5, \lambda_6, G, t)$$

$$D_2 = D_1(X_1, X_2, X_3, \lambda_4, \lambda_5, \lambda_6) .$$

Using the approach that was employed to evaluate the first variation of the performance index, J, the differential of $N_1(t_1)$ can be shown to be

$$dN_1(t_1) = H_2 dt_1 \Big|_{t_1^+}^{t_1^-} + \mathrm{Trace}(M(\lambda_6, 1, t_1)\delta X_3^T) + \int_{t_0}^{T} \mathrm{Trace}((H_2)_G \delta G^T)dt \tag{38}$$

where $M(y, a, t_i) = y(t_i^+) - y(t_i^-) + a(N_1)_{X_3}$, and

$$\begin{aligned}
&d(\lambda_{j+3})/dt = -\partial H_2/\partial X_j && ; \quad j = 1,2,3 \\
&\lambda_{j+3}(t_i^+) = \lambda_{j+3}(t_i^-) && ; \quad j = 1,2 \\
& && \quad\;\; i = 1,2 \\
&\lambda_{j+3}(T) = 0 && ; \quad j = 1,2 \\
&\lambda_6(t_0) = 0 \,.
\end{aligned} \tag{39}$$

The minimizing solution will now be found by constructing time histories for the elements of δG, and selecting values for dt_1 and $\delta X_3(t_1)$ that produce $\delta J < 0$ and satisfy $dN_1(t_1) = 0$. Adjoin the differential of the constraint $N_1(t_1)$ to the first variation of the performance index.

$$\delta J + \upsilon dN_1(t_1) = H''dt_1\Big|_{t_1^+}^{t_1^-} + \mathrm{Trace}(M(\lambda_8,\upsilon,t_1)\delta X_3^T)$$
$$+ \int_{t_0}^{T} \mathrm{Trace}\;((H'')_G\,\delta G^T)dt$$

where $H'' = H_1' + \upsilon H_2 : \lambda_8 = \lambda_3 + \upsilon\lambda_6$; and υ is a constant-valued Lagrange multiplier. Choose

$$\begin{aligned}
\delta X_3(t_1) &= -k_1\; M(\lambda_8,\upsilon,t_1) && ; \quad k_1 > 0 \\
dt_1 &= -k_2(H''(t_1^-) - H''(t_1^+)) && ; \quad k_2 > 0 \\
\delta G &= -k_3 H''_G && ; \quad k_3 > 0 \,.
\end{aligned} \tag{40}$$

Then, the first variation of the performance index augmented with the first differential of the constraint becomes

$$\delta J + \upsilon\; dN_1(t_1) = -k_2\|H''(t_1^-) - H''(t_1^+)\|^2 - k_1\|M(\lambda_8,\upsilon,t_1)\|^2$$
$$- k_3 \int_{t_0}^{T} \|H''_G\|^2\; dt \,.$$

Thus, the performance index is always decreasing unless the stationary conditions defined by Eqs. (41) are satisfied.

$$H''(t_1^-) = H''(t_1^+)$$
$$\lambda_8(t_1^-) = \lambda_8(t_1^+) + \upsilon N_1(t_1)_{X_3} \tag{41}$$
$$(H'')_G = 0\ .$$

The value of υ that guarantees $dN_1(t_1) = 0$ when the stationary conditions are satisfied can be found by substituting Eq. (40) into Eq. (38). The result is

$$\upsilon = -\left(\int_{t_0}^{T} \mathrm{Trace}((H_2)_G(H_2)_G^T)dt\right)^{-1}\left(\int_{t_0}^{T} \mathrm{Trace}((H_2)_G(H_1')_G^T)dt\right)\ . \tag{42}$$

The existence of the inverse in Eq. (42) is the controllability condition required for solution of the constrained optimal control problem.

The evaluation of Eq. (41) results in the following choice for the optimal filter gain, G^*.

$$G^* = (\alpha_1(T)\gamma^* + \alpha_2(T)\gamma' - (\lambda_8 + \lambda_8^T))C^T(\alpha_1(T)R^* + \alpha_2(T)R')^{-1}\ . \tag{43}$$

It can be shown that $\lambda_8(t) = 0$ for $t \in [t_0, t_1)$ so that the constraint given by Eq. (32) has no effect on the filter gain over this time interval. However, for time greater than t_1 the constraint does affect the filter gain by changing the effective covariance matrix. For an effective filter, A-GC is an asymptotically stable matrix. Thus, the effect of the discontinuity in λ_8 should diminish as time increases past t_1.

The solution of multiple-point, boundary-value problems

such as the one listed above is extremely difficult. Bryson and Ho [41] suggest using steepest descent algorithms to solve such problems. As the dimension of the problem increases, the number of possible time instants at which switching may occur increases, making this approach impractical for large dimensional systems.

If it is known that the elements of the noise covariance matrices change slowly with respect to the estimation interval's duration, or are constants, suboptimal techniques may be used. The switching matrix becomes [37]

$$\int_{t_0}^{T} B^T \beta B \, dt .$$

Thus, $\beta(t)$ need not be considered as a differential constraint, and the optimal filter gain for a system with constant, but uncertain noise covariance matrices will be

$$G^* = (\alpha_1(T)\gamma^* + \alpha_2(T)\gamma')C^T(\alpha_1(T)R^* + \alpha_2(T)R')^{-1} . \tag{44}$$

E. A Second Order Design Example

The design technique developed above will now be illustrated by an example which involves a second-order system with uncertain plant noise covariance matrix. Consider the linear system

$$dx/dt = \begin{bmatrix} -1 & 0 \\ 0 & -2 \end{bmatrix} x(t) + u(t); \qquad x(t_0) = 0$$

$$z(t) = x(t) + v(t) .$$

The measurement noise covariance matrix is the identity matrix, while the plant noise covariance matrix is

$$Q = \begin{bmatrix} 10 & q \\ q & 10 \end{bmatrix} \qquad 2 \leq q \leq 9 .$$

One wishes to find the estimator design optimum with respect to

$$V_2(G) = ((K_{lo}(G))^2 + (K_{up}(G))^2)^{\frac{1}{2}}$$

over the time period of 0 to 1 seconds.

The first step in the design is the determination of which limiting performance set is more favorable. It can be shown that there are no switchings for the minimin and minimax noise covariance values, q' and q^*, respectively, and that β_{12} is less than zero [37]. Thus, $q' = 9$ and $q^* = 2$. The performance sets that result from these optimizing design choices are shown in Figure 3. Obviously, the minimax design yields the more favorable performance set. Thus, use of the evaluator V_2 is valid.

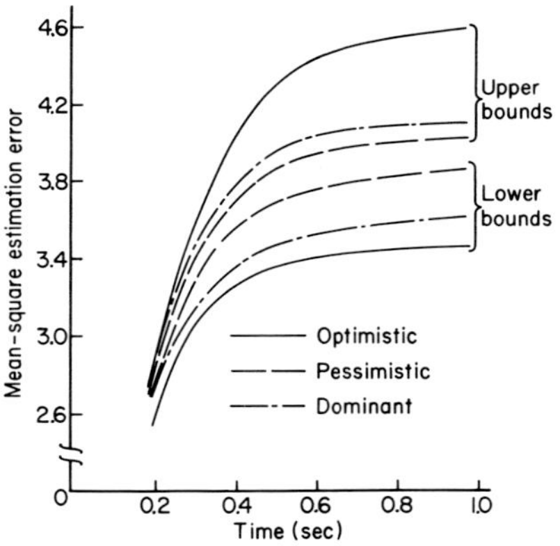

Fig. 3. Time Histories of Performance Set Bounds for System with Uncertain Plant Noise Covariance Matrix.

The filter gain matrix optimum with respect to the evaluator V_2, as determined by Eq. (44), is

$$G^* = (\gamma^* + \delta(T)\gamma')/(1 + \delta(T))$$

where $\delta(T) = \alpha_2(T)/\alpha_1(T)$.

To complete the design one must evaluate $\delta(T)$. This requires the solution of the fixed-point problem

$$0 = f(\delta) = -\delta(T) + \left(\mathrm{Trace} \int_{t_0}^{T} \gamma'(\delta,t)dt\right) / \left(\mathrm{Trace} \int_{t_0}^{T} \gamma^*(\delta,t)dt\right)$$

First-order gradient algorithms may be used to evaluate $\delta(T)$; i.e., the $(n+1)$th guess for $\delta(T)$ can be found from

$$\delta_{n+1} = \delta_n - \left((\partial f/\partial\delta)\Big|_{f_n}\right)^{-1} f_n \, .$$

Note that $\delta(T)$ must satisfy

$$P_{up}/P_{10} \leq \delta(T) \leq O_{up}/O_{10} \, .$$

Thus, one initial condition for the gradient algorithm is

$$\delta(T)_0 = 0, \; f_0 = O_{up}/O_{10} \, .$$

The filter gain matrix that results from the solution of the fixed-point problem is

$$G^* = 0.4728\gamma^* + 0.5276\gamma' \, .$$

The transient response of the dominant performance set is shown in Figure 3. Obviously, a significant improvement in performance has been obtained. The dominant performance set's lower bound is 60% of the total possible lower bound reduction, while it's upper bound is only 14% of the total possible increase.

V. PLANT MATRIX UNCERTAINTY

A. Introduction

The general estimator design approach for uncertain systems developed in Section III will now be applied to systems with plant matrix uncertainties. The estimation problem will be reformulated as a deterministic optimization problem.

Equations for the dominant estimator design of the set of non-inferior designs will be derived. The special case of constant, uncertain plant matrices will be shown to result in a greatly simplified estimator algorithm.

B. Problem Formulation

Consider the linear continuous system described in Section I, A. Assume that the plant matrix $A(t)$ is uncertain, but constrained to belong to the following set.

$$A(t) \in \{A(t) = \|a_{ij}\| / (a_{ij})_{min} \leq a_{ij} \leq (a_{ij})_{max}\} . \tag{45}$$

The bounds $(a_{ij})_{min}$ and $(a_{ij})_{max}$, $i,j = 1,\ldots,n$, are based upon the limited knowledge of the system gained from the few data available. It is not necessary to assume that $A(t)$ is a stable matrix in order to derive the dominant estimator. However, the design of Kalman-Bucy filters for systems that are known to have uncontrollable instability is not practical.

It is proposed to find an estimate $\hat{x}(t)$ of $x(t)$ that is of the form

$$d\hat{x}/dt = F(t)\,\hat{x}(t) + G(t)\,z(t) . \tag{5}$$

The estimator matrices $F(t)$ and $G(t)$ are chosen such that the error criterion J, defined by Eq. (9), is minimized.

To determine if one can reformulate this stochastic optimization problem as an equivalent deterministic optimization problem one must consider the conditions derived in Section IV, B that guarantee that the estimate is unconditionally unbiased. First, the estimator's initial condition must be equal to the mean value of the state vector's initial condition. Secondly, the estimator matrix $F(t)$ must be set equal to $A(t) - G(t)C(t)$. The first condition may be met. However, the second condition can be satisfied only if $A(t)$ is known exactly. Thus, one should expect the estimate to be biased when the plant matrix $A(t)$ is uncertain.

Since the estimate may be biased, it is not possible to write a single matrix differential equation which describes the time behavior of the mean-square estimation error matrix $\gamma(t)$. However, making the definitions

$$P(t) = E(x(t)x(t)^T)$$
$$\hat{P}(t) = E(\hat{x}(t)\hat{x}(t)^T)$$
$$S(t) = E(x(t)\hat{x}(t)^T) .$$

Eq. (9) can be written as

$$J(F(t),G(t)) = \text{Trace} \int_{t_0}^{T} D(t)(P(t) + \hat{P}(t) - 2S(t))dt . \tag{46}$$

Differential equations that describe the time behavior of the matrices $P(t), \hat{P}(t)$, and $S(t)$ can be derived using the same approach as utilized in Section IV, B in the derivation of a differential equation for $\gamma(t)$. The resulting equations are [37]

$$\begin{aligned} dP/dt &= AP + PA^T + BQB^T &&; \quad P(t_0) = \gamma_0 + S(t_0) \\ d\hat{P}/dt &= F\hat{P} + \hat{P}F^T + GCS + (GCS)^T + GRG^T; &&\quad \hat{P}(t_0) = S(t_0) \\ dS/dt &= SF^T + AS + PC^TG^T &&; \quad S(t_0) = E(x(t_0))E(x(t_0))^T . \end{aligned} \tag{47}$$

A deterministic optimization problem whose solution yields an estimator that is optimal with respect to the mean-square estimation error may now be stated.

Plant Matrix Uncertainty Problem

Find matrices $F(t)$ and $G(t)$ for $t \in [t_0, T]$ such that

$$J(F(t), G(t)) = \text{Trace} \int_{t_0}^{T} D(t)(P(t) + \hat{P}(t) - 2S(t))dt$$

is minimized subject to

1. The differential constraint Eq. (47).
2. The admissible plant matrix constraint Eq. (45).

A solution for the plant matrix uncertainty problem stated above will be obtained by defining the set of non-inferior designs and evaluating the dominant design of that set.

C. Limiting Performance Sets

Consider first the performance set corresponding to pessimistic design. To find the upper bound on this performance set one must find matrices $F^*(t)$, $G^*(t)$, and $A^*(t)$ defined on $t \in [t_0, T]$ such that

$$J(F^*, G^*, A^*) = \underset{F,G}{\text{Min}}\ \underset{A}{\text{Max}} \int_{t_0}^{T} \text{Trace}(D(P + \hat{P} - 2S))dt .$$

Optimization is subject to the constraint Eqs. (45) and (47).

For this problem the elements of the matrices $P(t)$, $\hat{P}(t)$, and $S(t)$ are considered state variables, the elements of the plant matrix $A(t)$ are considered maximizing control variables, and the elements of the Kalman-Bucy filter matrices $F(t)$ and $G(t)$ are considered minimizing control variables.

The general procedure for solving differential games consists essentially of two steps [41]. First, use stationary conditions to establish a two-point, boundary-value problem (TPBVP) whose solution yields the optimizing controls F^*, G^*, and A^*. Second, verify the saddle-point condition

$$J(F^*, G^*, A) \leq J(F^*, G^*, A^*) \leq J(F, G, A^*)$$

by solving the two one-sided optimization problems associated with the saddle-point condition, using F^*, G^*, and A^* derived from the stationary conditions.

Stationary conditions will be established through the use of a Hamiltonian, which is defined as follows.

$$H(P,\hat{P},S,F,G,A,\lambda_P,\lambda_{\hat{P}},\lambda_S) = \text{Trace}(dP/dt\lambda_P^T + d\hat{P}/dt\lambda_{\hat{P}}^T + dS/dt\lambda_S^T + D(P + \hat{P} - 2S))$$

where λ_P, $\lambda_{\hat{P}}$, and λ_S are the adjoint matrices associated with P, $\hat{P}$, and S, respectively.

When evaluated along the minimax path, the elements of the adjoint matrices satisfy

$$\begin{aligned}
d\lambda_P/dt &= -A^T\lambda_P - \lambda_P A - \lambda_S GC - D \quad ; &\lambda_P(T) = 0\\
d\lambda_{\hat{P}}/dt &= -F^T\lambda_{\hat{P}} - \lambda_{\hat{P}}F - D \quad ; &\lambda_{\hat{P}}(T) = 0\\
d\lambda_S/dt &= -\lambda_S F - A^T\lambda_S - 2(\lambda_{\hat{P}}GC)^T + 2D; &\lambda_S(T) = 0\ .
\end{aligned} \tag{48}$$

It can be easily shown that $\lambda_{\hat{P}}$ satisfies [43]

$$\lambda_{\hat{P}}(t) = \int_t^T \Theta(t,t_1)D\Theta(t,t_1)^T\, dt_1 \tag{49}$$

where $\Theta(t,t_1)$ is the transition matrix corresponding to $-F(t)^T$. Thus, $\lambda_{\hat{P}}$ is symmetric and positive definite for $t \neq T$.

A necessary condition for a minimax solution is [42]

$$H^0(P,\hat{P},S,\lambda_P,\lambda_{\hat{P}},\lambda_S) = \underset{F,G}{\text{Min}}\ \underset{A}{\text{Max}}\ H(P,\hat{P},S,F,G,A,\lambda_P,\lambda_{\hat{P}},\lambda_S)\ .$$

Since F, G, and A appear separately in Eq. (47), it is clear that the order of minimization and maximization is immaterial.

The selection of the optimizing $F(t)$ and $G(t)$ proceeds as follows [46]. Since $F(t)$ appears linearly in H and is not constrained a singular optimal control problem must be solved. The general procedure for solving singular optimal control problems is to take successive time derivatives of the partial derivative of the Hamiltonian with respect to the elements of the singular control matrix until an expression that is an explicit function of the singular control matrix occurs. The final expression is solved for the optimum singular control. In the case of $F(t)$, two time derivatives yield [47]

$$F^*(t) = A^*(t) - G^*(t)C \ . \tag{50}$$

This singular control satisfies Tait's necessary condition for singular control minimality [37].

Since the Hamiltonian is convex in elements of $G(t)$ the optimizing filter gain matrix may be obtained by taking partial derivatives of the Hamiltonian with respect to elements of $G(t)$ and setting the results equal to zero. One obtains

$$G^*(t) = \gamma C^T R^{-1} \ . \tag{51}$$

The uncertain plant matrix $A(t)$ appears linearly in the Hamiltonian and is constrained by Eq. (45). Thus, bang-bang and singular controls must be considered. The maximizing bang-bang control is given by [47]

$$a^*_{ij} = S_w((a_{ij})_{max}, (a_{ij})_{min}; \ (\lambda_{\hat{P}}\gamma)_{ij}) \ . \tag{52}$$

In general, one must evaluate $\lambda_{\hat{P}}$, γ, F^*, G^*, and A^* simultaneously by solving the TPBVP given by Eqs. (47)-(52). However, if the plant dynamics and measurements are uncoupled, and R, Q, and D are diagonal, a simplification is possible. $\lambda_{\hat{P}}$ will be diagonal, and the maximizing strategy will be

$$a^*_{ii} = (a_{ii})_{max} \qquad i = 1,\ldots,n \ .$$

This result agrees with D'Appolito and Hutchinson's result for a scalar system in steady-state [48].

If both A and Q are uncertain, but satisfy Eqs. (45)

and (7), respectively, a minimax filter can still be derived. Q, F, G, and A appear separately in the Hamiltonian. Thus, the order of minimization and maximization is still immaterial. F^*, G^*, and A^* will be given by Eqs. (50), (51), and (52). Q^* will be given by

$$q_{ij}^* = S_w((q_{ij})_{max}, (q_{ij})_{min}; (B^T \lambda_P B)_{ij}) . \tag{53}$$

Since λ_P is positive definite [47] the diagonal elements of the switching matrix are positive. Thus,

$$q_{ii}^* = (q_{ii})_{max} \qquad i = 1, \ldots, q .$$

R(t) and G(t) do not appear separately in the Hamiltonian. Thus, an extension of the above results to include uncertainty in the measurement noise covariance matrix is not direct. One must use convexity arguments to interchange the orders of minimization and maximization.

The lower bound on the performance set corresponding to pessimistic design is obtained by finding the matrix $A^{**}(t)$ defined on $t \in [t_0, T]$ such that

$$J(F^*, G^*, A^{**}) = \underset{A}{\text{Min}} \int_{t_0}^{T} \text{Trace}(D(P + \hat{P} - 2S))dt .$$

Minimization is performed subject to the constraint Eqs. (45), (47), (50)-(52).

The solution of the above minimization problem is obtained by use of the Matrix Minimum Principle [44]. The optimizing control is given by

$$a_{ij}^{**} = S_w((a_{ij})_{min}, (a_{ij})_{max}; ((\lambda_P P + \lambda_P^T P + \lambda_S S)_{ij})\Big|_{A=A^{**}}) . \tag{54}$$

In this case, no further simplification is possible, one must evaluate A^{**} and γ^{**} simultaneously by solving the TPBVP given by Eqs. (47), (48), (50)-(54).

If both Q and A are uncertain, the minimizing value of Q, Q^{**}, is given by

$$q_{ij}^{**} = S_w\left((q_{ij})_{min},\ (q_{ij})_{max};\ ((B^T \lambda_P B)_{ij})\Big|_{A=A^{**}}\right).$$

However, λ_P may no longer be positive definite. Thus, the diagonal elements of Q^{**} are not obvious.

Consider now the performance set corresponding to optimistic design. To find the lower bound on this performance set one must find matrices $F'(t)$, $G'(t)$, and $A'(t)$ defined on $t \in [t_0, T]$ such that

$$J(F', G', A') = \underset{F,G}{\text{Min}}\ \underset{A}{\text{Min}} \int_{t_0}^{T} \text{Trace}(D(P + P - 2S))dt\ .$$

Minimization is performed subject to the constraint Eqs. (45) and (47).

This "cooperative game" [45] can be solved by adjoining the uncertainty matrix A to the filter matrices F and G, and solving the resulting one-sided optimization problem with respect to the new control matrix.

Define the Hamiltonian as in the pessimistic design problem. The first order necessary conditions for the minimum are the following.

$$F' = A' - G'C \tag{55}$$

$$G' = \gamma' C^T R^{-1} \tag{56}$$

$$a'_{ij} = S_w((a_{ij})_{min},\ (a_{ij})_{max};\ (\lambda_{\hat{P}} \gamma')_{ij}) \tag{57}$$

where $\lambda_{\hat{P}}$ is evaluated using Eq. (49) with F replaced by F'.

The upper bound on the performance set corresponding to optimistic design is evaluated by finding a matrix $A''(t)$

defined on $t \in [t_0, T]$ such that

$$J(F', G', A'') = \max_{A} \int_{t_0}^{T} \mathrm{Trace}(D(P + \hat{P} - 2S))dt .$$

Maximization is performed subject to the constraint Eqs. (45), (47), (55)-(57).

First order necessary conditions for a maximum are found using the Matrix Minimum Principle. The result is

$$a''_{ij} = S_W\left((a_{ij})_{max}, (a_{ij})_{min}; \quad (\lambda_P P + \lambda_P^T P + \lambda_S S)_{ij}\Big|_{A=A''}\right) \quad (58)$$

where λ_P and λ_S are evaluated using Eq. (48) with F, G, and A replaced by F', G', and A'', respectively.

D. <u>The Dominant Design</u>

Consider use of the admissible evaluator V_i. The dominant filter design is obtained from the set of noninferior designs by finding the filter matrices $F^*(t)$ and $G^*(t)$ such that

$$J(F , G) = \min_{F,G} V_i(K_{up}(F,G), K_{lo}(F,G))$$

where K_{up} and K_{lo} are the upper and lower bounds of the performance set corresponding to the filter matrices $F(t)$ and $G(t)$. Optimization is performed subject to constraints given by Eqs. (45) and (47).

The supercriterion problem is a cascade of two optimization problems; one with respect to $A(t)$ assuming arbitrary $G(t)$ and $F(t)$, and a second with respect to $F(t)$ and $G(t)$ given the solution of the first problem. The first optimization problem may be readily solved using the Matrix Minimum Principle [44]. If the results are incorporated into the second optimization problem, the supercriterion problem can be restated as follows.

Equivalent Supercriterion Problem

Find matrices $F^*(t)$ and $G^*(t)$ defined on $t \in [t_0,T]$ such that

$$J(F^*,G^*) = \underset{F,G}{\text{Min}}\ V_i(K_{10}(F,G),K_{up}(F,G))$$

subject to

1. $$K_{up} = \int_{t_0}^{T} \text{Trace}(D(P' + \hat{P}' - 2S'))dt$$

$$K_{10} = \int_{t_0}^{T} \text{Trace}(D(P^* + \hat{P}^* - 2S^*))dt$$

2. $$dP'/dt = A'P' + P'A'^T + BQB^T;\ P'(t_0) = \gamma_0 + S_0$$

$$d\hat{P}'/dt = F\hat{P}' + \hat{P}'F^T + GCS' + (GCS')^T + GRG^T;\ \hat{P}'(t_0) = S_0$$

$$dS'/dt = S'F^T + A'S' + P'C^TG^T;$$

$$S'(t_0) = S_0 = E(x(t_0))\ E(x(t_0))^T$$

3. $$dP^*/dt = A^*P^* + P^*A^{*T} + BQB^T;\ P^*(t_0) = \gamma_0 + S_0$$

$$d\hat{P}^*/dt = F\hat{P}^* + \hat{P}^*F^T + GCS^* + (GCS^*)^T + GRG^T;\ \hat{P}^*(t_0) = S_0$$

$$dS^*/dt = S^*F^T + A^*S^* + P^*C^TG^T;\ S^*(t_0) = S_0$$

4. $$d\lambda_{P^*}/dt = -A^{*T}\lambda_{P^*} - \lambda_{P^*}A^* - \lambda_{S^*}GC - D;\ \lambda_{P^*}(T) = 0$$

$$d\lambda_{S^*}/dt = -\lambda_{S^*}F - A^{*T}\lambda_{S^*} - 2C^TG^T\lambda_{\hat{P}} + 2D;\ \lambda_{S^*}(T) = 0$$

$$d\lambda_{\hat{P}}/dt = -F^T\lambda_{\hat{P}} - \lambda_{\hat{P}}F - D;\ \lambda_{\hat{P}}(T) = 0$$

$$d\lambda_{P'}/dt = -A'^T\lambda_{P'} - \lambda_{P'}A' - \lambda_{S'}GC - D;\ \lambda_{P'}(T) = 0$$

$$d\lambda_{S'}/dt = -\lambda_{S'}F - A'^T\lambda_{S'} - 2C^TG^T\lambda_{\hat{P}} + 2D;\ \lambda_{S'}(T) = 0$$

5. $$(A')_{ij} = S_w((a_{ij})_{max}, (a_{ij})_{min};$$

$$\left(((\lambda_P + \lambda_P^T)P + \lambda_S S^T)\Big|_{A=A'}\right)_{ij}$$

$$(A^*)_{ij} = S_w\left((a_{ij})_{min}, (a_{ij})_{max}; \left(((\lambda_P + \lambda_P^T)P + \lambda_S S^T)\Big|_{A=A^*}\right)_{ij}\right).$$

In this formulation of the supercriterion problem the elements of the matrices defined by differential constraints 2, 3, and 4 are considered as state variables. The elements of $F(t)$ and $G(t)$ are considered as minimizing variables.

One cannot guarantee that the elements of the switching matrices defining A^* and A' do not change sign over the estimation interval. Since the switching matrices are functions of the "state" matrices, variations in the elements of F and G can change the point at which the elements of the switching matrices change sign. Changing the points where the elements of the switching matrices change sign can increase or decrease the value of the performance index, J. Basically, the elements of the optimum filter matrices, $F^*(t)$ and $G^*(t)$, influence the elements of the switching matrices in such a manner that the signs of the elements of the switching matrices switch at the optimum times $t_i^* \in (t_0, T)$, $i = 1,\ldots,m$. A derivation of first order necessary conditions must include a derivation of the t_i^*'s. This may be accomplished by including constraints of the form

$$(N_1(t_L))_{ij} = (((\lambda_{P^*} + \lambda_{P^*}^T)P^* + \lambda_{S^*}S^{*T})(t_L))_{ij} = 0 \quad i,j = 1,\ldots,n \quad L = 1,\ldots,m$$

$$(N_2(t_k))_{ij} = (((\lambda_{P'} + \lambda_{P'}^T)P' + \lambda_{S'}S'^T)(t_k))_{ij} = 0 \quad i,j = 1,\ldots,n \quad k = 1,\ldots,p$$

in the formulation of the supercriterion problem. The t_k's and t_L's must belong to (t_0, T).

This formulation of the supercriterion problem forces

switchings of the signs of the elements of the switching matrices defining A^* and A' to occur. However, the optimum solution may not have switchings. Thus, one should first solve the problem assuming that no switchings occur. Then, the derived forms for $F^*(t)$ and $G^*(t)$ should be utilized to evaluate $N_1(t)$ and $N_2(t)$ and thereby verify the assumption of no switchings. If switchings occur, the solution of the supercriterion problem to be developed below must be utilized.

To simplify the derivation only one switching of only one element, say the ijth, of each switching matrix will be considered. Extension to higher order systems with more than one switching is straightforward. Define

$$N_1(t_1) = (N_1(t_1))_{ij} = 0 \tag{59}$$

$$N_2(t_2) = (N_2(t_2))_{ij} = 0 \tag{60}$$

where t_1 and t_2 are the times at which N_1 and N_2 change sign, respectively.

To simplify the notation let $X_1 = P'$, $X_2 = \hat{P}', \ldots, X_{11} = \lambda_{S'}$. Define the Hamiltonian by

$$H_1(X_i, \beta_i, F, G, t;\ i = 1, \ldots, 11) = \mathrm{Trace}\left(\sum_{i=1}^{11} Y(X_i)\beta_i^T \right)$$

where $Y(X_1) = A'X_1 + X_1A'^T + BQB^T$, etc.

Also, define

$$D_1(t) = \mathrm{Trace}\left(\sum_{i=1}^{11} (dX_i/dt)\beta_i^T \right).$$

Let υ_1 and υ_2 be constant valued Lagrange multipliers. Then, the performance index augmented with the constraint equations can be written as

$$J(F,G) = V_i(K_{10},K_{up}) + \upsilon_1 N_1(t_1) + \upsilon_2 N_2(t_2) + \int_{t_0}^{T} (H_1 - D_1)dt \; .$$

The first variation of the augmented performance index is

$$\delta J = \delta V_i + \upsilon_1 dN_1(t_1) + \upsilon_2 dN_2(t_2) + \delta \int_{t_0}^{T} (H_1 - D_1)dt \; .$$

To simplify the first variation of the augmented performance index note that

$$\delta V_i = \alpha_1^{\cdot}(T)\; \delta K_{10} + \alpha_2(T)\; \delta K_{up}$$

where α_1 and α_2 are defined in Section IV, D. Thus,

$$\delta J = \upsilon_1 dN_1(t_1) + \upsilon_2 dN_2(t_2) + \delta \int_{t_0}^{T} (H_1' - D_1)dt$$

where H_1' is defined in Section IV, D.

The differentials of the constraint Eqs. (59)-(60) are

$$dN_1 = \mathrm{Trace}((N_1)_{X_7} dX_7^T + (N_1)_{X_4} dX_4^T + (N_1)_{X_8} dX_8^T + (N_1)_{X_6} dX_6^T) \quad (61)$$

$$dN_2 = \mathrm{Trace}((N_2)_{X_{10}} dX_{10}^T + (N_2)_{X_1} dX_1^T + (N_2)_{X_{11}} dX_{11}^T + (N_2)_{X_3} dX_3^T) \quad (62)$$

where for example, $(N_1)_{X_4}$ is the matrix with ijth element $\partial N_1/\partial (X_4)_{ij}$ and dX_4 is a matrix with ijth element $d(X_4)_{ij}$.

To evaluate the first variation of the integral term given above one must separate the integral into three parts to allow for discontinuities in the integrand at $t = t_1$ and $t = t_2$. Leibnitz's rule is used on each part separately. The result is

$$\delta \int_{t_0}^{T} (H_1' - D_1)dt = \sum_{i=1}^{2} (H_1' - D_1)dt_i \Big|_{t_i^+}^{t_i^-} + \int_{t_0}^{T} \delta(H_1' - D_1)dt .$$

Integrate the above equation by parts and use Eq. (35) to simplify the results. Collecting terms one has

$$\delta \int_{t_0}^{T} (H_1' - D_1)dt = \sum_{i=1}^{2} (H_1' dt_i - \sum_{n=1}^{11} \mathrm{Trace}(\beta_n dX_n^T)) \Big|_{t_i^+}^{t_i^-}$$

$$- \sum_{i=1}^{11} \mathrm{Trace}(\delta X_i \beta_i^T) \Big|_{t_0^+}^{T} + \int_{t_0}^{T} \mathrm{Trace}\Bigg(\Bigg(\sum_{i=1}^{11} ((H_1')_{X_i}$$

$$+ d\beta_i/dt)\ \delta X_i^T \Bigg) + (H_1')_G\ \delta G^T + (H_1')_F\ \delta F^T \Bigg) dt .$$

To avoid having to calculate variations in the elements of X_n, $n = 1,\ldots,11$, due to variations in the elements of G and F via sensitivity differential equations, define the adjoint matrices as follows.

$$d(\beta_n)_{ij}/dt = -\partial H_i'/\partial(X_n)_{ij} \tag{63}$$

with boundary-values given by $\beta(T) = 0$ if $X_n(T)$ is undefined, or $\beta_n(t_0) = 0$ if $X_n(t_0)$ is undefined.

Then, the first variation of the augmented performance index becomes

$$\delta J = \upsilon_1 dN_1(t_1) + \upsilon_2 dN_2(t_2) + \sum_{i=1}^{2} \left(H_1' dt_i - \sum_{n=1}^{11} \mathrm{Trace}(\beta_n dX_n^T) \right) \Bigg|_{t_i^+}^{t_i^-}$$

$$+ \int_{t_0}^{T} \mathrm{Trace}((H_1')_G\ \delta G^T + (H_1')_F\ \delta F^T) dt .$$

Now choose $H_1'(t_i^-)$ and $\beta_n(t_i^-)$ for $i = 1,2$ and $n = 1,\ldots,11$, to cause the coefficients of $dX_n(t_i)$ and dt_i for $i = 1,2$, to vanish, e.g., choose

$$H_1'(t_i^-) = H_1(t_i^+) \qquad i = 1,2 \qquad (64)$$

$$\beta_i(t_1^-) = \beta_i(t_1^+) + \upsilon_1(N_1)_{X_i} \qquad i = 4,6,7,8 \qquad (65)$$

$$\beta_i(t_2^-) = \beta_i(t_2^+) + \upsilon_2(N_2)_{X_i} \qquad i = 1,3,10,11 \qquad (66)$$

$$\beta_i(t_1^-) = \beta_i(t_1^+) \qquad i = 1,2,3,5,10,11 \qquad (67)$$

$$\beta_i(t_2^-) = \beta_i(t_2^+) \qquad i = 2,4,5,6,7,8,9\,. \qquad (68)$$

Therefore, the first variation of the augmented performance index is reduced to

$$\delta J = \int_{t_0}^{T} \mathrm{Trace}((H_1')_G\ \delta G^T + (H_1')_F\ \delta F^T)dt\,.$$

The first order necessary conditions for optimality of F^* and G^* are found by setting δJ equal to zero. However, one cannot automatically say that stationarity of J implies that $(H_1')_G$ and $(H_1')_F$ equal zero since δG and δF are not arbitrary, but must produce $dX_n(t_i)$ for $i = 1,2$ consistent with Eqs. (61)-(62). A controllability argument similar to the one employed in Section IV, D regarding admissible variations, must be used to justify placing $(H_1')_G$ and $(H_1')_F$ equal to zero. The argument is straightforward, but lengthy. Thus, it will not be documented here.

The filter matrices that result in the stationary value for the performance index can be evaluated as follows. For the filter gain, G, one has from the stationary condition, $(H_1')_G = 0$, the following.

$$G^* = -(2(\alpha_1+\alpha_2)\lambda_{\hat{P}})^{-1}(P^{(2)} + P^{(1)} - 2\lambda_{\hat{P}}S'')C^TR^{-1} \qquad (69)$$

where

$$P^{(2)} = \beta_3^T P' - \lambda_{S'}^T \beta_{10}$$
$$P^{(1)} = \beta_6^T P^* - \lambda_{S^*}^T \beta_7$$
$$S'' = \alpha_2(T)S'^T + \alpha_1(T)S^{*T} - \beta_{11}^T - \beta_8^T .$$

Since $F(t)$ appears linearly in H_1' and is not constrained a singular optimal problem must be solved. Following the general procedure for solving singular optimal control problems one takes the first time derivative of the partial derivative of H_1' with respect to each element of the matrix $F(t)$. Using the gradient matrix notation [49] one has

$$(H_1')_F = 2\lambda_{\hat{P}} \hat{P}'' + \beta_3^T S' - \lambda_{S'}^T \beta_{11} + \beta_6^T S^* - \lambda_{S^*}^T \beta_8 = 0 \tag{70}$$

where $\hat{P}'' = \alpha_2(T)\hat{P}' + \alpha_1(T)\hat{P}^* - \frac{1}{2}(\beta_9 + \beta_9^T)$, and

$$d((H_1')_F)/dt = 2D \sum_{i=1}^{2} (\alpha_i(T)(S^{(i)} - \hat{P}^{(i)})) = 0 \tag{71}$$

where

$$S^{(2)} = S' - \beta_{11}/\alpha_2; \qquad \hat{P}^{(2)} = \hat{P}' - (\beta_9 + \beta_9^T)/(4\alpha_2)$$
$$S^{(1)} = S^* - \beta_8/\alpha_1 ; \qquad \hat{P}^{(1)} = \hat{P}^* - (\beta_9 + \beta_9^T)/(4\alpha_1) .$$

Since $F(t)$ did not appear explicitly in Eq. (71), a second time derivative must be taken. The result is

$$d^2((H_1')_F)/dt^2 = (A'' - F - GC)\hat{P}'' + (P'' - \hat{P}'')C^TG^T - (\alpha_1 + \alpha_2)GRG^T = 0$$

where

$$A'' = (\alpha_2 A'S^{(2)} + \alpha_1 A^* S^{(1)})(\hat{P}'')^{-1}$$
$$P'' = \alpha_2 P' - \beta_{10} + \alpha_1 P^* - \beta_7 .$$

Assuming that the inverse of $\hat{P}''$ exists, one has for $F^*(t)$ the following.

$$F^*(t) = A'' - GC + ((P'' - \hat{P}'')C^T - (\alpha_1 + \alpha_2)GR)G^T(\hat{P}'')^{-1} . \qquad (72)$$

The adjoint variables β_7 through β_{11} satisfy

$$d\beta_7/dt = A^*\beta_7 + \beta_7 A^{*T} \quad ; \qquad \beta_7(t_0) = 0$$

$$\beta_7(t_1^-) = \beta_7(t_1^+) + \upsilon_1(N_1)X_7$$

$$d\beta_8/dt = A^*\beta_8 + \beta_8 F^T + \beta_7 C^T G^T \quad ; \qquad \beta_8(t_0) = 0$$

$$\beta_8(t_1^-) = \beta_8(t_1^+) + \upsilon_1(N_1)X_8$$

$$d\beta_9/dt = F\beta_9 + \beta_9 F^T + 2GC(\beta_8 + \beta_{11}); \qquad \beta_9(t_0) = 0$$

$$d\beta_{10}/dt = A'\beta_{10} + \beta_{10}A'^T \quad ; \qquad \beta_{10}(t_0) = 0$$

$$\beta_{10}(t_2^-) = \beta_{10}(t_2^+) + \upsilon_2(N_2)X_{10}$$

$$d\beta_{11}/dt = A'\beta_{11} + \beta_{11}F^T + \beta_{10}C^T G^T \quad ; \qquad \beta_{11}(t_0) = 0$$

$$\beta_{11}(t_2^-) = \beta_{11}(t_2^+) + \upsilon_2(N_2)X_{11} .$$

Note that $\beta_n(t) = 0$, $n = 7,8,9,10,11$, for $t \in [t_0, \mathrm{Min}(t_1,t_2))$ so that the constraint Eqs. (59)-(60) do not affect the filter matrices over this time interval. However, for time greater than $\mathrm{Min}(t_1,t_2)$ the constraints do affect the filter matrices by changing the effective error covariance matrix. However, if A^*, A', and F^* are asymptotically stable matrices the effect will diminish as time increases.

For large dimensional systems the solution of the required multiple-point, boundary-value problem given by Eqs. (47), (63), (69), and (72) may not be practical. If it is known that the elements of the plant matrix change slowly with respect to the estimation interval's duration, or are constants, sub-optimal techniques may be used. The switching matrix defining

A^* and A' becomes [37]

$$\int_{t_0}^{T} ((\lambda_P + \lambda_P^T)P + \lambda_S S^T)dt .$$

Thus, λ_S, λ_P, and $\lambda_{\hat{P}}$ need not be considered as differential constraints, and the optimal filter matrices for a system with constant, but uncertain plant matrix are

$$G^* = -(2(\alpha_1 + \alpha_2)\lambda_{\hat{P}})^{-1}(\beta_3^T(P' - S') + \beta_6^T(P^* - S^*))C^T R^{-1} \tag{73}$$

$$F^* = A_v - GC + ((\alpha_1 \gamma^* + \alpha_2 \gamma')C^T - (\alpha_1 + \alpha_2)GR)G^T(\hat{P}'')^{-1} \tag{74}$$

where

$$A_v = (\alpha_1 A^* S^* + \alpha_2 A' S')(\hat{P}'')^{-1}$$

$$\hat{P}'' = \alpha_1 \hat{P}^* + \alpha_2 \hat{P}'$$

$$d\beta_3/dt = -A'^T \beta_3 - \beta_3 F - 2C^T G^T \lambda_{\hat{P}} + 2\alpha_2 D ; \qquad \beta_3(T) = 0$$

$$d\beta_6/dt = -A^{*T} \beta_6 - \beta_6 F - 2C^T G^T \lambda_{\hat{P}} + 2\alpha_1 D ; \qquad \beta_6(T) = 0 .$$

E. A Design Example

To illustrate the design techniques developed above consider the following scalar system with constant-valued, uncertain time constant.

$$dx/dt = ax + u$$

$$z(t) = x + v .$$

It is known that a is greater than or equal to -3, but less than or equal to -1.

The measurement noise covariance matrix has value 4, while the plant noise covariance matrix has value 1. The initial condition for the plant is zero mean, and has covariance matrix value 0.001.

One desires to find the filter design optimal with respect to the evaluator functional

$$V_1(f,g) = \tfrac{1}{2}(K_{10}(f,g) + K_{up}(f,g))$$

over the time period of 0 to 1.0 seconds.

The first step in the design is the evaluation of the limiting performance sets. For a scalar system, one can immediately determine the minimax and minimin designs. Since $\lambda_{\hat{p}}$ is positive definite $a^* = -1$ and $a' = -3$. Numerical integration of Eq. (47) yields the results given in Figure 4.

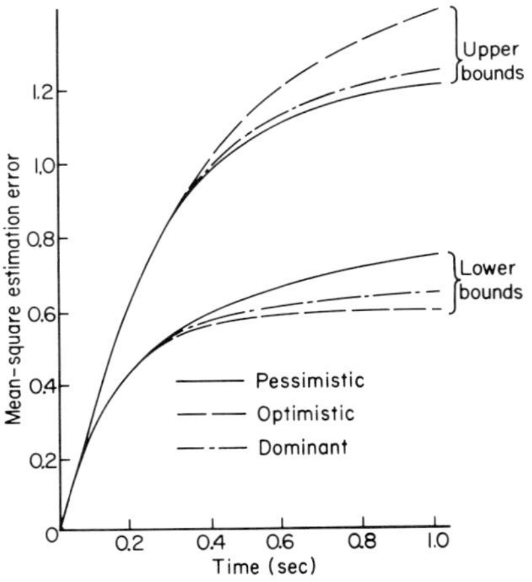

Fig. 4. Time Histories of Performance Set Bounds for System with Uncertain Time Constant

The dominant estimator design is defined by Eqs. (73)-(74). One must solve a TPBVP involving $\lambda_{\hat{p}}$, β_3, β_6, s', s^*, p', and p^* in order that the transient response of the dominant performance set may be evaluated. Note that the dimension of the TPBVP is four while the dimension of the plant is one. The TPBVP can be solved by iterating on the initial conditions for the adjoint variables until convergence to the desired terminal values is realized. Define $\beta^T = (\lambda_{\hat{P}}, \beta_3, \beta_6)$. Then, the (n + 1)th guess for the initial conditions is

$$\beta(t_0)_{n+1} = \beta(t_0)_n - \varepsilon W^{-1}\beta(T)_n$$

where W is the Jacobian matrix with elements $\partial\beta_i(T)/\partial\beta_j(t_0)$, and ε, $0 \le \varepsilon \le 1$, is chosen to make the algorithm computationally stable.

The optimal filter gain, g^*, is proportional to the ratios $\beta_3/\lambda_{\hat{P}}$ and $3_6/\lambda_{\hat{P}}$. Thus, g^* may be extremely sensitive to the values chosen as the initial conditions for the adjoint variables. This sensitivity aspect of the computation may be avoided by evaluating $\lambda_{\hat{P}}$ in terms of the other two adjoint variables using the first order necessary condition $H_f = 0$, e.g., use

$$\lambda_{\hat{P}} = -.5(s'\beta_3 + s^*\beta_6)/\hat{P}'' .$$

The solution of the TPBVP is shown in Figure 4. A comparison of the dominant performance set with the minimax performance set indicates that a significant improvement in performance has been achieved. The upper bound of the dominant performance set is only 3% larger than the upper bound of the minimax performance set; while its lower bound is 13% below the minimax performance set lower bound.

VI. CONCLUSIONS AND COMMENTS

The problem of estimating the state of a system when the time-varying plant and measurement noise covariance matrices as well as the plant dynamics are uncertain, but bounded with known bounds has been considered. The approach taken toward a solution was motivated by the following desired properties, which restricted the estimator form to one that makes open loop compensations for the uncertain system parameters. First, the estimator storage requirements must be comparable to those of the Kalman-Bucy filter. Second, one must be able to evaluate the performance expected before the estimator is actually employed without resorting to complex computer simulations.

A general design theory for the performance optimization of uncertain systems was presented. It was shown that since the performance expected from an uncertain system is a set of values, one can view the design problem as seeking the optimum of a vector-valued performance index. Zadeh's theory for optimizing non-scalar-valued performance indices was discussed. A two-dimensional, suboptimal performance index, whose components are the upper and lower bounds of the performance set, was introduced. The set of designs noninferior with respect to the suboptimal performance index was derived. A restriction on the noninferior set was obtained utilizing the tradeoff between system sensitivity and performance value.

The problem of designing a filter to estimate the state of a system with uncertain noise covariance matrices was formulated, the estimator was constrained to be unbiased, and the filter designs that yield the limiting performance sets were developed. Then, the estimator design dominant with respect to the two-dimensional, suboptimal performance index was derived. Eq. (43) defines the dominant estimator gain matrix when the uncertain noise covariance matrices are time-varying. A multiple-point, boundary-value problem, whose dimension is at least as great as the number of uncertain parameters, must be solved to completely specify the dominant design. This computation aspect negates the usefulness of the dominant design defined by Eq. (43). For many applications, the uncertain noise covariance matrices are slowly varying compared to the estimation time interval or are unknown constants. In this case, the dominant estimator is defined by the greatly simplified filter gain matrix of Eq. (44). Note that the dominant filter design for constant-valued uncertainties has form identical to the Kalman-Bucy filter, However, the noise covariance matrices employed to evaluate the filter gain matrix are weighted sums of the two extremes of their uncertainty sets. The weighting, which may be evaluated by the solution of a fixed-point problem, embodies the tradeoff between filter sensitivity and performance value. The design defined by Eq. (44) is relatively easy to evaluate, and should be of great utility.

Estimators for systems with uncertainty in the plant dynamics cannot be constrained to be unbiased. Therefore, the formulation of the supercriterion problem is more complicated; three matrix differential equations are required to define the mean-square error matrix. The limiting performance designs are defined by Eqs. (50)-(52) and (55)-(57). One must solve a TPBVP to evaluate the performance of these designs. However, if the plant dynamics and measurement equations are both sets of uncoupled equations one can immediately select the optimizing plant and estimator matrices without solving the TPBVP. The dominant design for a system with time-varying plant matrix uncertainties is defined by Eqs. (69) and (72). Again, a multiple-point, boundary-value problem must be solved to evaluate the dominant design's performance. If the uncertain plant matrix is slowly varying with respect to the duration of the estimation interval the greatly simplified design given by Eqs. (73)-(74) is dominant. Note that the estimator matrix $F^*(t)$ is identical to the corresponding Kalman-Bucy matrix with the exceptions that $A(t)$ is replaced by a weighted average of the two extremes of the plant matrix and there is an added term representing the deviation of $G^{*}(t)$ from a weighted sum of the extremes of the mean-square error matrix. This estimator design is relatively easy to evaluate, and should have many applications in the physical world.

VII. REFERENCES

1. M. ATHANS and E. TSE, "A Direct Derivation of the Optimal Linear Filter Using the Maximum Principle," IEEE, Trans. Auto. Control, AC-12: 690-698 (1967).
2. R. B. MERRICK, "Simplified Kalman Estimator for an Aircraft Landing Display," J. Aircraft, 8: 44-49 (1971).
3. R. J. FITZGERALD, "Divergence of the Kalman Filter," IEEE, Trans. Auto. Control, AC-16: 736-747. (1971).
4. H. HEFFES, "The Effect of Erroneous Models on the Kalman Filter Response," IEEE, Trans. Auto, Control, AC-11: 541-543 (1966).
5. T. NISHIMURA, "Error Bounds of Continuous Kalman Filters and the Application to Orbit Determination Problems," IEEE, Trans. Auto, Control, AC-12: 268-275 (1967).
6. S. R. NEAL, "Linear Estimation in the Presence of Errors in Assumed Plant Dynamics," IEEE, Trans. Auto Control, AC-12: 592-594 (1967).
7. B. FRIEDLAND, "On the Effect of Incorrect Gain in Kalman Filter," IEEE, Trans. Auto. Control, AC-12: 610-611 (1967).
8. C. F. PRICE, "An Analysis of the Divergence Problem in the Kalman Filter," IEEE, Trans. Auto. Control, AC-13: 699-702 (1968).
9. R. E. GRIFFIN and A. P. SAGE, "Sensitivity Analysis of Discrete Filtering and Smoothing Algorithms," AIAA J. 7: 1890-1897 (1969).

10. J. W. BERKOVEC, "A Method for the Validation of Kalman Filter Models," Joint Auto. Control Conf., August 5-7, 1969, pp. 488-493. Univ. of Colorado, Boulder, Colorado, 1969.

11. R. K. MEHRA, "On the Identification of Variances and Adaptive Kalman Filtering," IEEE, Trans. Auto. Control, AC-15: 175-184 (1970).

12. R. P. WISHNER, J. A. TABACZNSKI, and M. ATHANS, "On the Estimation of the State of Noisy Nonlinear Multivariable Systems," IFAC Symp. Multivariable Control Syst. Dusseldorf, Germany, Oct. 7-8, 1968.

13. H. W. SORENSON, "A Nonlinear Perturbation Theory for Estimation and Control of Time-Discrete Stochastic Systems," Univ. of Calif., Los Angeles, Dept. of Eng., Technical Report No. 68-2, Jan. 1968.

14. G. L. SMITH, "Sequential Estimation of Observation Error Variances in a Trajectory Estimation Problem," AIAA J. 5: 1964-1970 (1967).

15. R. L. KLEIN and E. D. EYMAN, "A Suboptimal Filter and Filter Sensitivity for Stochastic Parameter Systems," Seventh Annual Allerton Conf. on Circuit and Syst. Theory, Univ. of Illinois, Oct. 8-10, 1969, pp. 440-449.

16. D. T. MAGILL, "Optimal Adaptive Estimation of Sampled Stochastic Processes," IEEE, Trans. Auto. Control, AC-10: 434-439 (1965).

17. C. G. HILBORN and D. G. LAINIOTIS, "Optimal Adaptive Filter Realizations for Sample Stochastic Processes with and Unknown Parameter," 1967 SWIEEECO Record, pp. 1-4-1 to 1-4-8, April 19-21, 1967.

18. J. C. SHELLENBARGER, "Estimation of Covariance Parameters for an Adaptive Filter," National Electronics Conference, McCormick Place, Chicago, Ill., Oct. 3-5, 1966, pp. 698-702

19. P. D. ABRAMSON, Jr., "Simultaneous Estimation of the State and Noise Statistics in Linear Dynamical Systems," National Aeronautics and Space Administration, Tech. Report No. R-332, March, 1970.

20. G. W. HUSA, "Adaptive Bayes Filtering with Unknown Prior Statistics," Ph. D., Elect. Eng., SMU, 1969.

21. B. D. TAPLEY and G. H. BORN, "Sequential Estimation of the State and Observation Error Covariance Matrix," AIAA J. 9: 212-217 (1971).

22. A. H. Jazwinski, "Adaptive Filtering," IFAC Symp. Multivariable Control Syst., Dusseldorf, Germany, Oct. 7-8, 1968.

23. F. H. SCHLEE, C. J. STANDISH, and N. F. TODA, "Divergence in the Kalman Filter," AIAA J. 5: 1114-1120 (1967).

24. H. W. SORENSON, "On the Error Behavior in Linear Minimum Variance Estimation Problems," IEEE, Trans. Auto. Control, AC-12: 557-562 (1967).

25. S. F. SCHMIDT, "Estimation of State with Acceptable Accuracy Constraints," Analytical Mechanics Associates, Inc., Report No. 67-4, 1967.

26. T. J. TARN and J. ZABORSKY, "A Practical Nondiverging Filter," AIAA J. 8: 1127-1133 (1970).

27. R. W. MILLER, "Asymptotic Behavior of the Kalman Filter with Exponential Aging," AIAA J. 9: 537-539 (1971).

28. H. W. SORENSON and J. E. SACKS, "Recursive Fading Memory Filtering," Information Sciences, 3: 101-119 (1971).

29. A. H. JAZWINSKI, "Stochastic Processes and Filtering Theory," Academic Press, New York, New York, 1970.

30. N. D. CRUMP, "Estimation of Discrete Signals Containing a Non-random Component," Ph.D., Elect. Eng., Georgia Institute of Technology, 1969.

31. J. A. D'APPOLITO, "Minimax Design of Low Sensitivity

Filters for State Estimation," Ph.D., Eng. Dept., Univ. of Massachusetts, Amherst, Mass., 1969.

32. M. D. MESAROVIC and Y. TAKAHARA, "On Global Sensitivity," Fourth Annual Allerton Conference on Circuit and Syst. Theory, Univ. of Illinois, Oct. 5-7, 1966, pp. 1-7.

33. H. S. WITSENHAUSEN, "On Performance Bounds for Uncertain Systems," SIAM J. Control, 8: 55-89 (1970).

34. S. KARLIN, "Mathematical Methods and Theory in Games, Programming, Economics," Vol. II, Addison-Wesley, Reading, Mass., 1959.

35. N. O. daCUNHA and E. POLAK, in "Mathematical Theory of Control," (A. V. Balakrishnan and L. W. Neustadt, eds.), p. 96, Academic Press, New York, 1967.

36. L. A. ZADEH, "Optimality and Non-Scalar-Valued Performance Criteria," IEEE, Trans. Auto. Control, AC-8: 59-60 (1963).

37. J. O. PEARSON, "Suboptimal Estimation in the Presence of Uncertainties Using Sensitivity Relationships," Ph.D., Engr., Univ. of California, Los Angeles, 1971.

38. R. M. BEESON, "Optimization with Respect to Multiple Criteria," Ph.D., Engr., University of Southern California, 1971.

39. J. MEDANIC, "Three Segment Method in the Sensitivity Design of Control Systems," Fifth Annual Allerton Conference on Circuit and System Theory, Univ. of Illinois, Oct. 4-6, 1967, pp. 439-450.

40. N. E. NAHI, "Estimation Theory and Applications," Wiley, New York, New York, 1969.

41. A. A. BRYSON, Jr. and Yu-Chi HO, "Applied Optimal Control," Blaisdell Publishing Co., Waltham, Mass., 1969.

42. Yu-Chi HO, "Optimal Terminal Maneuver and Evasion Strategy," J. SIAM Control, 4: 421-428 (1966).

43. W. A. PORTER, "On the Matrix Riccati Equation," IEEE,

Trans. Auto. Control, AC-12: 746-749 (1967).

44. M. ATHANS, "The Matrix Minimum Principle," Information and Control, 11: 592-606 (1967).

45. I. G. SARMA and R. K. RAGADE, "Some Considerations in Formulating Optimal Control Problems as Differential Games," Int. J. Control, 4: 265-279 (1966).

46. J. S. LEE, "Optimal Linear Filter and Its Relation to Singular Control," Joint Auto. Control Conf. June 22-26, 1970, pp. 313-319, Georgia Institute of Technology, Atlanta, Georgia, 1970.

47. C. T. LEONDES and J. O. PEARSON, "A Minimax Filter for Systems with Large Plant Uncertainties," IEEE, Trans. Auto. Control, AC-17: 266-268 (1972).

48. J. A. D'APPOLITO and C. E. HUTCHINSON, "Low Sensitivity Filters for State Estimation in the Presence of Large Parameter Uncertainties," IEEE, Trans. Auto. Control, AC-14: 310-312 (1969).

49. M. ATHANS and F. C. SCHWEPPE, "Gradient Matrices and Matrix Calculations," MIT Lincoln Lab., Tech. Note No. 1965-53 (1965).

Application of Modern Control and Optimization Techniques to Transportation Systems

DANIEL TABAK[1]
Department of Electrical Engineering,
University of the Negev,
Beer-Sheva, Israel

---

[1]On leave of absence from RPI of Connecticut, Hartford, Connecticut, 06120.

I. INTRODUCTION

During the past few years we have witnessed a considerable growth of interest and research work dedicated to the solution of various societal problems. The problem of Transportation is certainly one of them. Until about ten years ago, the discipline of Transportation Systems Analysis and Planning, was being dealt with mainly by Civil Engineers, by scientists employed by automobile producers (Ford, General Motors), and by State and Federal Transportation agencies (with a few exceptions, of course). However, recently, more and more scientists and engineers in various other fields, became interested in many aspects of transportation systems. Both theoreticians and practitioners of Control Systems Engineering are no exception; research and development work in transportation problems has been initiated in many universities and industrial companies by professionals whose primary field of interest is control theory and its applications. This growing interest of control systems professionals in transportation problems has been expressed in numerous publications as well as by formal actions of the professional societies. For instance, the IEEE Control Systems Society (formerly the Automatic Control Group) has established a Transportation Systems Committee in 1970, under the chairmanship of D. F. Wilkie (Ford Motor Company).

The literature reporting research work in transportation systems, performed by control scientists, is growing and a considerable amount of it has already been generated.

At the moment, this literature is scattered in many different journals, conferences proceedings, theses and internal reports. A partial goal of this survey article is to bring a considerable part of this literature under a single cover. This will permit the control scientist who is interested, but not yet experienced, in transportation systems to become acquainted with work previously accomplished in this area, in a relatively short time.

A second goal of this article is to present in a concise manner and under unified notation, some mathematical models of transportation systems which have been used in conjunction with application of modern control and optimization techniques to these systems. This information should also be useful to the control specialist starting to work in the transportation systems area.

The word "transportation" carries a wide variety of possibilities and concepts. There is the air and ground transportation. There is private and public transportation of both kinds. On the ground — we have automobile and railroad transportation, and so on. This article concentrates on the ground — automobile transportation problems, for the following reasons:

1. Most of the work involving application of modern control and optimization techniques, has been performed in this particular branch of transportation.

2. Traffic congestion on the roads and in urban areas has become one of the toughest and urgent problems facing our society. It is true that the Air Traffic Control problem is equally important and urgent, however, so far, very little results in this area have been published. Perhaps, in a year or two, it may become a good subject for a separate survey article.

3. It is practically impossible to cover more than this subject within the space allotted to a single article in the "Advances". Even this particular branch of transportation is wide enough to fill an article.

It will be assumed throughout the article that the readers

are familiar with the theory and practices of modern control and optimization. In any case, references have been provided where appropriate.

II. MODELS OF TRAFFIC FLOW

Mathematical models of highway traffic flow have been under active development since the early 1950's [1,2]. There exists a vast amount of literature dedicated to this subject. Among many results, reported to date, we can distinguish two main and basic approaches to traffic flow modeling:

1. The macroscopic, or continuum model, which describes the flow of highway traffic in a manner analogous to fluid flow. The population of vehicles on the highway is treated as a continuous flow and its dynamics are described by partial differential equations.

2. The microscopic, or car-following model, which observes each vehicle individually and takes into account the average response of the driver who follows the string of vehicles ahead of him.

The detailed formulation of these models will be given in the following subsections:

A. The Continuum Model

The continuum model was originally proposed by Lighthill and Whitham [3]. It is therefore often referred to as: Lighthill-Whitham theory. Similar results were also reported independently by Richards [4]. The theory is also presented in several texts [2,5,6].

The continuum model is valid primarily for roads carrying a large population of automobiles. Only in this case we can regard the flow of automobile traffic on the road as a continuous flow. In order to develop the basic equations of the model, we will consider a small element of the road, as shown in Figure 1. The segment of the road, shown in Figure 1, includes traffic going in one direction. It will be assumed that the traffic flows in the positive direction of the x-axis. The x-coordinate values will therefore represent locations along the road, and differences between them will represent distances along it.

Consider now a segment of the road of length dx, with the coordinate x centered in its midpoint. Suppose now that the total number of vehicles crossing the element dx during the time interval $\left(t - \frac{dt}{2}\right) \leq t \leq \left(t + \frac{dt}{2}\right)$ is n. That is, the time interval during which the count of vehicles was performed is of length dt. It will be assumed that dt was of sufficient length so that the number of vehicles n is of significant value.

We will now define the <u>flow</u>, q, as the number of vehicles per unit time crossing the road interval dx at x. That is,

$$q = n/dt\,[\text{vehicles/hour}] . \tag{1}$$

We will define the <u>concentration</u>, c, as the number of vehicles on the road per unit length of the road [vehicles/mile]. A third important entity is the <u>space-mean speed</u>, v [miles/hour], defined as:

$$v = q/c . \tag{2}$$

The space-mean speed is an average of vehicle speeds weighted according to the time they remain on the road interval dx [3]. Eq. (2) is usually expressed in the form

$$q = vc . \tag{3}$$

We will assume that there are no on- or off-ramps along the segment dx. If the flow entering the segment is q, the flow at its exit is $q + \frac{\partial q}{\partial x}dx$ (see Figure 1). Assuming that the concentration at the beginning of the time interval dt was c, the concentration at its end is $c - \frac{\partial c}{\partial t} dt$. (No vehicles are generated along the segment, of course). The total number of vehicles on the segment at the beginning of interval dt is cdx, and at its end, $\left(c - \frac{\partial c}{\partial t} dt\right)dx$. On the other hand, the total number of vehicles entering the segment within dt is qdt, while the number of exiting vehicles is $\left(q + \frac{\partial q}{\partial x} dx\right)dt$.

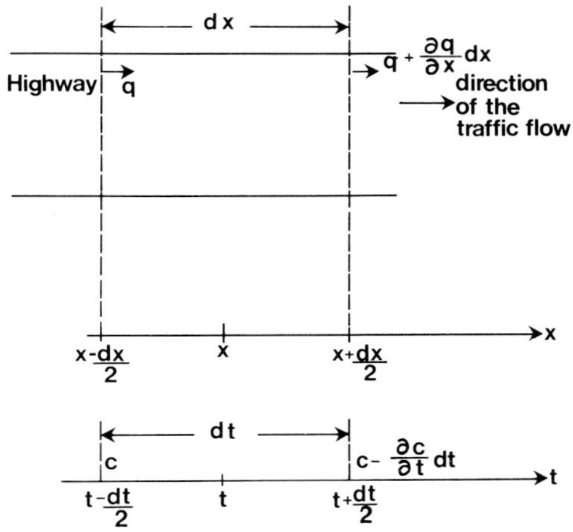

Fig. 1. An Element of the Highway and the Space and Time Axes.

Hence, the balance of vehicles can be expressed as:

$$cdx - \left(c - \frac{\partial c}{\partial t} dt\right)dx = qdt - \left(q + \frac{\partial q}{\partial x} dx\right)dt$$

or

$$\frac{\partial c}{\partial t} + \frac{\partial q}{\partial x} = 0 . \tag{4}$$

Equation (4) is called the <u>continuity equation</u>. Its analogy to the continuity equation of fluid dynamics is obvious. Equations (3) and (4) are the basic equations of the continuum model.

Equation (4) was written under the assumption that there are no external sources which change the total number of vehicles on the highway interval considered. In general, we do have sources of this kind. Usually they take many different forms of on- or off-ramps. We will therefore introduce the term s, which will represent the external source (or input, or forcing function) in the continuity equation:

$$\frac{\partial c}{\partial t} + \frac{\partial q}{\partial x} = s \ . \tag{5}$$

A more detailed formulation of the source term will be given in the next section. The units of s are in [vehicles/hour/mile]. It represents the rate of change in vehicle density contributed by the external source.

Substituting Eq. (3) into Eq. (4), we obtain:

$$\frac{\partial c}{\partial t} + v\,\frac{\partial c}{\partial x} = 0 \tag{6}$$

since

$$\frac{\partial q}{\partial x} = \frac{\partial q}{\partial c}\,\frac{\partial c}{\partial x} = v\,\frac{\partial c}{\partial x} \ .$$

The solution to Eq. (6) has the general form of:

$$c(x,t) = C(x - vt) \ . \tag{7}$$

As we can see from Eq. (7), the space-mean speed v can be interpreted as the speed of propagation of perturbations along the highway [1-3]. It is also referred to as the kinematic wave velocity [2].

A somewhat different description of the continuum model of vehicular traffic was given by Prigogine, Herman and their associates [1,2]. A distribution function $f(x,u,t)$ is introduced, where u represents the speed (velocity) of individual vehicles. Consider a segment of the highway between x and $x + dx$. Assume that the speed of vehicles on this segment varies from u to $u + du$. The number of vehicles dN, that at a given time t are on this segment and whose speeds are within the specified limits, is given by the following expression:

$$dN = f(x,u,t)\, dxdu\,. \tag{8}$$

Based on this fundamental expression, other traffic functions can be formulated. For instance, the concentration can be expressed as:

$$c(x,t) = \int_0^\infty f(x,u,t)du \tag{9}$$

and the flow:

$$q(x,t) = \int_0^\infty uf(x,u,t)du = c(x,t)\, \bar{u}(x,t) \tag{10}$$

where $\bar{u}(x,t)$ is the average velocity of the vehicles.

The distribution function $f(x,u,t)$ can in many cases be obtained experimentally [2]. It represents the space-velocity distribution of traffic actually achieved. In addition to this distribution, Prigogine [7] introduced the notion of the desired speed distribution, $f_0(x,u,t)$. In general, f_0 differs from f. However, integration over all velocities yields the same result:

$$\int_0^\infty uf_0(x,u,t)du = \int_0^\infty uf(x,u,t)du = c(x,t)\bar{u}(x,t)\,. \tag{11}$$

The desired speed distribution represents the general driving pattern the driver's wish to follow along the road. Due to unpredicted obstacles and interactions between the cars this pattern cannot be achieved (not immediately anyway) precisely as desired. Instead, the vehicles follow the f-distribution.

Under ideal conditions, assuming the absence of any obstacles or interactions between vehicles, the distribution function f will satisfy the continuity equation as the concentration c in Eq. (6):

$$\frac{\partial f}{\partial t} + u \frac{\partial f}{\partial x} = 0 . \tag{12}$$

When obstacles exist, f differs from f_0, as argued before. However, if interactions between cars are neglected, f will approach f_0 as time goes on. In a simpliefied model, adopted in [2], this approach is characterized by a single time constant; the so-called <u>relaxation time</u> T. Equation (12) takes the form of:

$$\frac{\partial f}{\partial t} + u \frac{\partial f}{\partial x} = - \frac{f - f_0}{T} . \tag{13}$$

In the more general case, when interaction between cars is taken into account, it can be shown [2] that Eq. (13) becomes:

$$\frac{\partial f}{\partial t} + u \frac{\partial f}{\partial x} = - \frac{f - f_0}{T} + (1 - p)c(x,t)[\bar{u}(x,t) - u(x,t)]f \tag{14}$$

where p is the probability of a car passing another car. The first term on the right side of Eq. (14) is called the <u>relaxation term</u>, while the second one is called the <u>interaction term</u>.

It is easy to see that Eq. (14) can be transformed into the continuity equation, Eq. (4). Integrating Eq. (14) over all velocities, on both sides and taking into account Eqs. (9)-(11), we obtain [2]:

$$\frac{\partial c}{\partial t} + \frac{\partial q}{\partial x} = - \frac{1}{T} \int_0^{\infty} (f - f_0)du + (1 - p)\, c\left[\bar{u} \int_0^{\infty} f du - \int_0^{\infty} u f du\right] = 0$$

since both terms on the right-hand side vanish separately.

In the works described in this article, the continuum model characterized by Eqs. (3), (4) was used in most cases. Some further analysis, involving this model, for a signalized one-way artery, was recently performed by Preparata [8]. A somewhat more complicated model, involving second order derivatives, was proposed by Franklin [9]. To the best of the author's knowledge, this model has hardly been utilized.

It should be noted that highway traffic is basically a discrete process involving a large number of individual vehicles. On the other hand, the model described by Eq. (4) is definitely continuous. While in many physical processes the continuous representation is closer to reality and the discrete — an approximation, in the case of highway traffic — the continuum model is a continuous approximation of a basically discrete process. As will be seen in Section III, the discretized form of Eqs. (4) or (5) was actually utilized in control and optimization applications.

B. The Car-Following Model

The car-following model considers each vehicle and its driver on an individual basis. The model consists of a string of vehicles following each other in a single lane. It takes explicitly into account the reaction time (or rather an average reaction time) of the drivers to outside stimula. An example of a stimulus could be a sudden change in distance between the vehicle and the vehicle preceding it. For instance, if the preceding vehicle slows down, the driver would have to hit the brake and slow down accordingly. If the preceding vehicle speeds up — the driver may hit the accelerator in order to keep up with the flow of vehicles.

In the car-following model it is assumed that the connection between the outside stimulus and the driver's reaction, at any time t, is of the following general form [1,2]:

$$(\text{driver reaction})_{t+T} = \lambda(\text{outside stimulus})_t \qquad (15)$$

where

T = delay, reaction time lag

λ = sensitivity coefficient

$(\ )_t$ = entity in the brackets at time t .

As in the continuum model, it is assumed that the highway runs in parallel to the x-axis and the vehicles travel in the positive direction. The coordinates of the vehicles traveling along the road will be designated as $x_1(t), x_2(t), \ldots, x_n(t), \ldots$ with the understanding that index 1 represents the first vehicle, 2 - the second, and so on. It has been shown both theoretically and verified experimentally [1,2,10-15] that the more specific form of Eq. (15) is:

$$\frac{d^2 x_n(t+T)}{dt^2} = \lambda \left[\frac{dx_{n-1}(t)}{dt} - \frac{dx_n(t)}{dt} \right] . \qquad (16)$$

In the <u>linear car-following</u> model, the sensitivity coefficient λ is assumed to be a constant. In the more general <u>nonlinear</u> model, λ has the form:

$$\lambda = \lambda_0 \frac{[dx_n(t)/dt]^m}{[x_{n-1}(t) - x_n(t)]^l} \qquad (17)$$

where λ_0 is a constant and l, m are integer exponents. Various nonlinear car-following models are characterized by pairs of specific values (l,m). For instance, the linear model is characterized by $(0,0)$.

Assuming a linear model, integrate Eq. (16). We obtain [2]:

$$\frac{dx_n}{dt} = \lambda_0(x_{n-1} - x_n) + b \tag{18}$$

where b is a constant of integration.

Introduce the notion of <u>spacing</u> between vehicles, s

$$s = x_{n-1} - x_n = 1/c . \tag{19}$$

Equation (18) now becomes:

$$v = \lambda_0 s + b \tag{20}$$

where v is the average speed of the vehicles. We can establish the constant b from the jammed state boundary condition. In this case the vehicles come to a standstill, i.e., $v = 0$. The spacing in this case will be denoted by s_j and the concentration by c_j. Naturally,

$$s_j = 1/c_j . \tag{21}$$

Using the jammed state boundary condition we obtain:

$$0 = \lambda_0 s_j + b$$

or

$$b = - \lambda_0 s_j = - \lambda_0/c_j .$$

Equation (20) now becomes:

$$v = \lambda_0(s - s_j) = \frac{\lambda_0}{c}\left(1 - \frac{c}{c_j}\right). \tag{22}$$

Multiplying Eq. (22) by c we obtain:

$$q = vc = \lambda_0(1 - c/c_j) . \tag{23}$$

In a different boundary case, when the concentration c is very small, we have $c/c_j \ll 1$ and this term can be neglected. The speed in this case is denoted as the <u>free</u>

speed, $v_f = \lambda_0/c$. In view of this, Eqs. (22) and (23) can be rewritten as:

$$\left.\begin{aligned} v &= v_f\left(1 - \frac{c}{c_j}\right) \\ q &= v_f c\left(1 - \frac{c}{c_j}\right) \end{aligned}\right\} \tag{24}$$

The functional relationship of v and q with respect to c is illustrated in Figure 2.

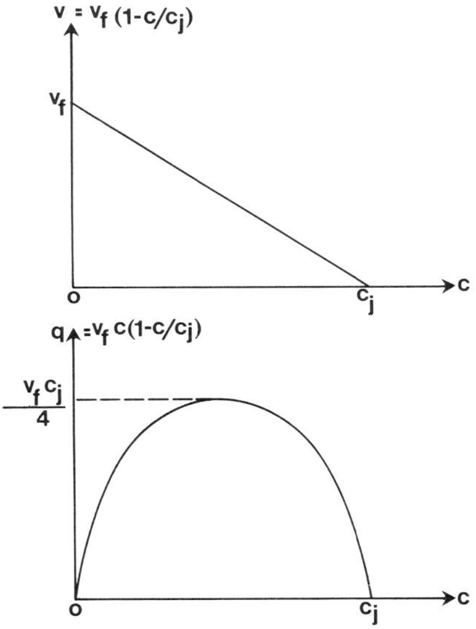

Fig. 2. Velocity and Flow vs. Concentration Diagrams in a Linear Car-Following Model.

In the nonlinear case (1,0), integration of Eq. (16) yields [2]:

$$v = \lambda_0 \ln s + b \,. \tag{25}$$

Using the jammed boundary condition $v = 0$; $s = s_j$, we obtain:

$$v = \lambda_0 \ln(s/s_j) = \lambda_0 \ln(c_j/c) \tag{26}$$

or

$$q = vc = \lambda_0 c\ln(c_j/c) . \tag{27}$$

The quantities of speed and flow, used above, should be interpreted as ensemble averages taken over the population of vehicles on the road.

III. OPTIMAL CONTROL OF HIGHWAY TRAFFIC

A. Ramp Control Problems

One of the most efficient ways of controlling the flow of vehicles on a highway is by regulating the number of vehicles permitted to enter the highway. Usually, vehicles enter the highway through on-ramps. By controlling the flow of vehicles passing through the ramps onto the highway, we can exert direct control on the flow and concentration of vehicles on it. Considerable number of projects, implementing control and optimization techniques, involved ramp control problems.

One of the earliest works, involving a systematic approach and direct application of optimization techniques to ramp control problems, was the work of Wattleworth and his associates [16-18, and Ref. 6, p. 326]. In this project, linear programming (LP) was used. The approach of Wattleworth will be discussed next.

1. Control of On-Ramp Volumes by Linear Programming

A sketch of a part of a highway considered is shown in Figure 3. The highway has been segmented into a finite number of sections. The coordinates of each partition point are: $x_0, x_1, \ldots, x_k, x_{k+1}, \ldots, x_K$, where K is the total number of sections considered. Each partition point will also be designated by the appropriate index $0, 1, \ldots, k-1, k, k+1, \ldots, K$ (see Figure 3). The partition is performed in such a manner,

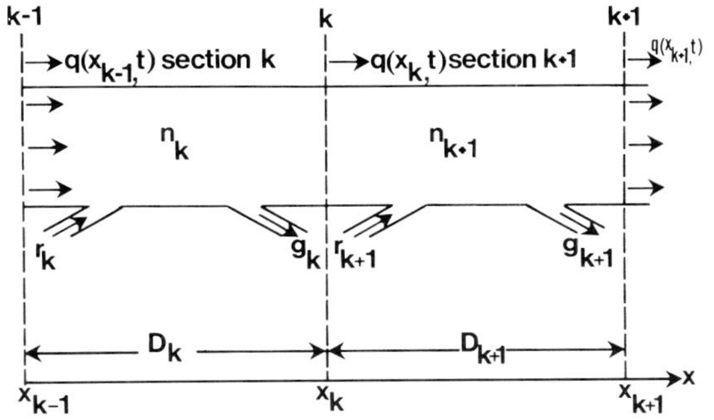

Fig. 3. Segmented Model of a Highway.

that each section contains one on- and one off-ramp. The flow rate of vehicles through the on-ramp of section k is denoted as r_k [vehicles/hour] and that of the off-ramp: g_k [vehicles/hour]. This notation will be used throughout the article in the discussion of other models as well. In the original model of Wattleworth [17,18], a section could contain more than one on- and off-ramps. This distinction does not create any conceptual difference in the formulation of the LP problem.

If we denote the rate of storage or accumulation of vehicles in section k, at the time t, as $n_k(t)$ [vehicles/hour], we can write the following conservation of vehicles equation:

$$n_k(t) = [r_k(t) + q(x_{k-1},t)] - [g_k(t) + q(x_k,t)] \qquad (28)$$

where $q(x_k,t)$ is the flow rate at $x = x_k$ at the time t.

Let us consider a time interval from t_0 to t_1. The cumulative number of vehicles which entered and exited section k, respectively, is:

$$R(t_1) = \int_{t_0}^{t_1} [r_k(t) + q(x_{k-1},t)]dt$$

$$G(t_1) = \int_{t_0}^{t_1} [g_k(t) + q(x_k, t)]dt$$

and the total number of vehicles in section k is:

$$N(t_1) = N_0 + R(t_1) - G(t_1) \tag{29}$$

where N_0 is the total number of vehicles at $t = t_0$. The integral

$$I_t = \int_{t_0}^{t_1} N(t)dt \tag{30}$$

can be viewed [18] as the number of vehicle-hours accumulated in the section during the time $(t_1 - t_0)$. For the time interval considered, the integral in Eq. (30) represents the total time of occupancy of all vehicles while they are in section k. If there is no parking in the section (and this is indeed the case on a highway), the time of occupancy would also be the total travel time through the section. It should of course be realized that the actual units of I_t are [(vehicles) × (hours)]. The same reasoning could similarly be repeated for a larger highway system containing several sections.

A very reasonable and realistic choice of a performance criterion, or an objective function, would be the minimization of the total travel time through the system. Taking into account Eqs. (29), (30), and assuming N_0, t_0, t_1 and $\int_{t_0}^{t_1} R(t_1)dt$ as constants, this would be equivalent to maximizing the cumulative system output [18]

$$\int_{t_0}^{t_1} G(t)dt .$$

The number of vehicles entering the system equals the number of vehicles exiting plus the storage. Since the goal of the project is the elimination of storage or congestion along the system, it is assumed that it approaches zero in the steady state. Following this line of reasoning, Wattleworth [17,18] actually chose as his objective function the sum of the on-ramp flow rates:

$$\text{maximize} \quad z = \sum_{k=1}^{K} r_k . \tag{31}$$

The individual on-ramp flow rates r_k [vehicles/hour] are the unknown variables in this LP model. They are in fact the control variables in this and many other ramp control models.

There are two types of basic constraints which were formulated for this model. The first type was to assure the prevention of congestion at bottlenecks. It assured that the total demand should not exceed the capacity at each section partition. The general form of this constraint was [18]:

$$\sum_{k=1}^{K} a_{kj}\, r_k \leq b_j ; \qquad j = 1,\ldots,M \tag{32}$$

where

a_{kj} = decimal fraction of vehicles entering at the on-ramp k, which pass through partition j.

b_j = the capacity of the overall section terminating at partition j. The partitions j are those for which the constraint in Eq. (32) has been formulated. Those are the places where a bottleneck congestion is likely to develop. In other words, constraints of the type in Eq. (32) are in general not formulated for all of the partitions (although, sometimes that could be the case as well).

M = total number of bottleneck partitions considered.

The second type of constraint assures that the input flow rates should not exceed the demand:

$$r_k \leq d_k ; \qquad k = 1,\ldots,K \quad (33)$$

where d_k = hourly demand at the on-ramp k. Naturally, all the variables are nonnegative.

$$r_k \geq 0 ; \qquad k = 1,\ldots,K . \quad (34)$$

As we can easily see, Eqs. (31)-(34) constitute a linear programming problem with K variables and (M + K) inequality constraints (not counting the trivial nonnegativity constraints). the LP problem is repeated again for the reader's convenience:

$$\max z = \sum_{k=1}^{K} r_k ;$$

subject to

$$\sum_{k=1}^{K} a_{kj} r_k \leq b_j ; \; j = 1,\ldots,M; \; 0 \leq r_k \leq d_k ; \; k = 1,\ldots,K . \qquad (35)$$

The numerical solution of LP problems, by various versions of the simplex algorithm, is well known and presented in numerous textbooks. Only a few chosen references are given here [19,20]. A numerical example applied to a three-sectioned portion of a highway in the city of Chicago, was solved by Wattleworth [17,18]. Six entrance ramps were considered. The a_{kj}, b_j and d_k parameters were obtained experimentally by direct observation of the highway. A similar LP model, where the objective function is the maximization of the total flow, was recently proposed by Chen et. al. [21]. As an illustration of the approach, a numerical example, reported in [17], will be presented next.

Example III-1

A part of the Chicago Freeway was considered. It contained six on-ramps $(K = 6)$ and the constraint in Eq. (32) was formulated for three partitions $(M = 3)$. The origin-destination fractions, a_{kj}, are given in the following table:

Origin	Destination: Laramie	Central	Austin	Harlem	Through
1. Cicero mainline	0.137	0.086	0.158	0.100	0.519
2. Cicero on-ramp	0.009	0.022	0.047	0.098	0.824
3. Central on-ramp				0.067	0.933
4. Austin on-ramp				0.051	0.949
5. Harlem on-ramp					1.000
6. DesPlaines on-ramp					1.000

The maximum hourly demand at each origin is as follows:

Origin	Program Variable r_k	Maximum Hourly Demand d_k [veh./hour]
1. DesPlaines on-ramp	r_1	600
2. Harlem on-ramp	r_2	475
3. Austin on-ramp	r_3	450
4. Central on-ramp	r_4	500
5. Cicero on-ramp	r_5	825
6. Cicero mainline	r_6	6800

It should be noted that in this example the mainstream of vehicles at the starting section of the portion of the freeway, under consideration, is treated as one of the on-ramps. The partitions where bottlenecks may occur and their corresponding capacities are:

	Partition Location	Capacity b_j [vehicles/hour]
1.	Mainline output	5900
2.	Austin Ave. (past on-ramp)	6000
3.	Austin Ave. (approaching off-ramp)	6450

Based on these data and the model in Eq. (35), the following LP problem is formulated:

$$\max z = \sum_{k=1}^{6} r_k$$

subject to:

$$\begin{aligned}
r_1 + r_2 + 0.949r_3 + 0.933r_4 + 0.824r_5 + 0.519r_6 &\leq 5900 \\
r_3 + r_4 + 0.922r_5 + 0.619r_6 &\leq 6000 \\
r_4 + 0.969r_5 + 0.777r_6 &\leq 6450 \\
r_1 &\leq 600 \\
r_2 &\leq 475 \\
r_3 &\leq 450 \\
r_4 &\leq 500 \\
r_5 &\leq 825 \\
r_6 &\leq 6800
\end{aligned}$$

$$r_k \geq 0; \quad k = 1,\ldots,6 .$$

There is a total of 9 nontrivial inequality constraints and 9 slack variables, $S_1,\ldots,S_9$ are added. They constitute the initial basis. Any slack variable appearing in the basis in the final solution, has a definite physical interpretation. If any of the slack variables S_1, S_2, S_3, associated with the first three constraints, are positive in the final solution, it means that the freeway capacity at the corresponding location is not fully utilized. If a slack variable, associated with the demand limitation constraints $(S_4, \ldots, S_9)$, is positive in the final solution, it means that the actual volume at that particular on-ramp cannot satisfy the demand.

The optimal solution obtained in this case (in the 6-th tableau) was:

$r_1 = 447 < d_1$; $\quad S_4 = 153$
$r_2 = 475 = d_2$
$r_3 = 450 = d_3$
$r_4 = 367 < d_4$; $\quad S_7 = 133$
$r_5 = 825 = d_5$
$r_6 = 6800 = d_6$
$S_2 = 213$, number of additional vehicles which could be handled at the second partition.
$z_0 = 9364$, total flow of vehicles/hour served by the system under an optimal ramp flow rate configuration.

As we can see, ramps 1 and 4 cannot handle a total of $S_4 + S_7 = 288$ vehicles/hour. That is, particular control should be applied at the DesPlaines and Central on-ramps.

2. Discretized Continuity Models

As we could see, the model adopted by Wattleworth resulted in a relatively simple linear programming problem. A different approach, resulting in a more complicated model, is the discretization of the continuity equation, Eq. (4), and the use of it as a part of the optimization problem. One of the earliest works, utilizing this approach was performed by Yuan and Kreer [22,23]. Similar results were subsequently reported by Kaya [24,25]. It should be noted that some of the earliest ideas concerning ramp metering and control, were expressed by May [26].

The continuity equation, Eq. (4) is discretized in space, in the following manner (see Figure 3):

$$\frac{dc_k}{dt} = \frac{1}{D_k} \left(q_{k-1} - q_k + r_k - g_k \right) \qquad k = 1,\ldots,K \qquad (36)$$

where

c_k = vehicle density in section k (averaged value over the section)

D_k = length of section k

q_k = flow rate exiting section k along the mainstream

r_k = on-ramp flow rate in section k

g_k = off-ramp flow rate in section k

Equation (36) is a first order, ordinary differential equation.

In addition to the continuum model, Eq. (24), derived from the linear car-following model is used. It should be noted that Eq. (24) is valid for vehicles traveling in a single lane. Multiplicity of lanes is accounted by multiplying the right-hand side by the number of lanes, l [22,23]

$$q = l v_f c \left(1 - c/c_j\right) . \qquad (37)$$

The off-ramp flow, g_k, is related to the flow q_{k-1} coming into the section by mainstream through a linear relationship [22].

$$g_k = f_{k-1}\, q_{k-1} \qquad (38)$$

where f_{k-1} is the fraction of q_{k-1} leaving through the off-ramp. Substituting Eqs. (37), (38) into Eq. (36) we obtain [22,23]:

$$\left.\begin{aligned}\frac{dc_1}{dt} &= \frac{1}{D_1}\,[q_0 - l_1 v_{f1}\,(c_1 - c_1^2/c_{j1}) + r_1] \\ \frac{dc_k}{dt} &= \frac{1}{D_k}\,[l_{k-1} v_{f,k-1}(c_{k-1} - c_{k-1}^2/c_{j,k-1})(1 - f_{k-1}) \\ &\quad - l_k v_{fk}(c_k - c_k^2/c_{jk}) + r_k] \qquad k = 2,3,\ldots,K\end{aligned}\right\} \qquad (39)$$

where

l_k = number of lanes in section k

v_{fk} = free speed in section k

c_{jk} = concentration in the jammed state in section k.

A <u>steady state</u> solution of Eq. (39) is defined as the solution in which the concentration in each section remains constant, that is $dc_k/dt = 0$ for all k. This steady state solution will be denoted by an additional subscript s. It is easily obtained from Eq. (39):

$$\left.\begin{aligned}q_{1s} &= r_{1s} \\ q_{ks} &= q_{k-1,s}\,(1 - f_{k-1}) + r_{ks}; \qquad k = 2,3,\ldots,K\end{aligned}\right\} \qquad (40)$$

where

q_{ks} = steady state flow from section k

r_{ks} = on-ramp steady state flow rate in section k.

From Eq. (40) we can derive the following recursive relation [22]:

$$q_{ks} = \prod_{i=1}^{k-1}(1 - f_i)\,r_{1s} + \prod_{i=2}^{k-1}(1 - f_i)\,r_{2s} + \cdots + (1 - f_{k-1})\,r_{k-1,s} + r_{ks}$$

or

$$q_{ks} = \sum_{j=1}^{k-1}\prod_{i=j}^{k-1}(1 - f_i)\,r_{js} + r_{ks}; \qquad k = 2,3,\ldots,K. \qquad (41)$$

Equation (41) is used in the steady state optimization problem. The following linear objective function was used originally by Yuan and Kreer [22]:

$$\text{maximize} \quad z = \sum_{k=1}^{K} (w_{qk}\, \eta_k + w_{dk}\, d_k) r_{ks} \tag{42}$$

where

η_k = queue length at the on-ramp in section k
d_k = demand rate at the on-ramp section k
w_{qk} = weighting factor for the queue term at section k
w_{dk} = weighting factor for the demand term at section k.

It should be noted that this objective function assigns higher priorities to on-ramps with longer queues and higher demands. Conceptually, it is similar to the objective function used by Wattleworth [17,18]. The constraints used in [22] were as follows:

$$q_{ks} \le Q_k; \qquad k = 1,\ldots,K \tag{43}$$

$$0 \le r_{ks} \le \begin{cases} R_k & \text{for all} \quad \eta_k > 0 \\ \min(R_k, d_k) & \text{for all} \quad \eta_k = 0 \end{cases} \tag{44}$$

where

Q_k = maximum allowable flow rate in section k
R_k = maximum allowable on-ramp flow rate in section k.

The q_{ks} in Eq. (43) were expressed in terms of the r_{ks} (the LP problem variables) using Eqs. (40) and (41). It is easy to see that Eqs. (42)-(44) constitute a linear programming problem, the solution of which includes the steady state on-ramp flows $r_{ks}(k = 1,\ldots,K)$, from which the steady state flows q_{ks} are directly calculated. The steady state concentrations, c_{ks} can be calculated by solving the quadratic equation, Eq. (37), and choosing the lowest of the two solutions obtainable:

$$c_{ks} = \frac{1}{2}\left[c_{jk} - \sqrt{c_{jk}^2 - \frac{4c_{jk}q_{ks}}{l_k v_{fk}}}\right]. \tag{45}$$

Once the steady state solution is obtained, a dynamic linear regulator problem is solved. This will be described later on in the section. In a subsequent publication [23], Yuan and Kreer have used a quadratic objective function in the steady state problem:

$$\text{minimize} \quad z = \sum_{k=1}^{K} w_k (r_{ks} - e_{dk})^2 \tag{46}$$

$$e_{dk} = d_k + \eta_k / T$$

T = time interval between computations of the reference ramp metering rate.

The constraints are identical to the previous linear constraints in Eq. (43). The problem now is a quadratic programming (QP) one, and can be solved by several known techniques, for instance, by that of Frank and Wolfe [27,28], which indeed was used in [23].

Example III-2

A portion of the highway, consisting of three sections is considered [23]. The first two sections have on-ramps. Only the second section has an off-ramp. The numerical data were as follows:

Section No.:	1	2	3
No. of lanes, l_k	1	1	1
Section length D_k [miles]	0.2	0.3	0.15
Jam concentration c_{jk} [vehicles/mile]	130	132	133
Free speed v_{fk} [miles/hour]	65	63	61
Initial concentration $c_k(0)$ [vehicles/mile]	54	57	55
Initial queue length $\eta_k(0)$ [vehicles]	3	4	0

The demand rate in Section 1, d_1, was 1500 vehicles/hour during the first 6 minutes and 500 after that. In Section 2, d_2 was 1000 vehicles/hour during the first 6 minutes, and 200 after that. The exit fraction, f_2, in Section 2, was 0.3 during the first 6 minutes, and 0.1 after that.

Using these data, Eq. (41) can be written for this example:

$$q_{1s} = r_{1s}$$
$$q_{2s} = r_{1s} + r_{2s}$$
$$q_{3s} = (1 - f_2)\,(r_{1s} + r_{2s})\,.$$

This example was simulated by Yuan and Kreer [23] both under local and optimal control conditions. The results for the queue lengths developing at the on-ramps, under local control, is shown in Figure 4(a), and under optimal control in Figure 4(b). It is obvious that applying the optimal control procedure shortens the queue at the on-ramp 2 at the expense of the on-ramp 1.

The steady state optimal control results were obtained by solving a quadratic programming problem, defined by Eqs. (43) and (46). The queues functions were calculated using the differential equation [23]:

$$\dot{\eta}_k(t) = d_k(t) - q_k(t)\,.$$

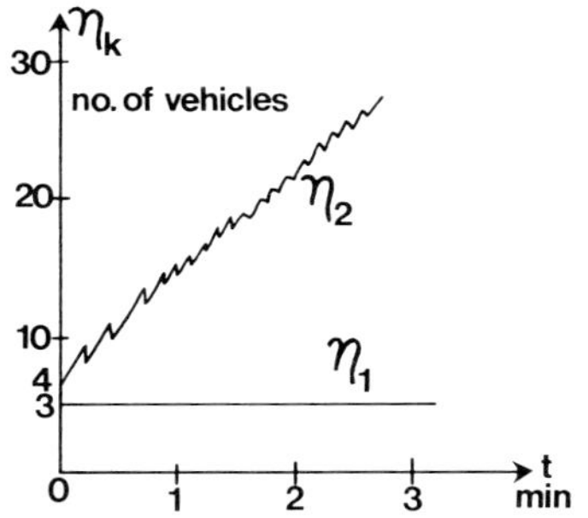

Fig. 4(a). Queue Lengths Under Local Control [23].

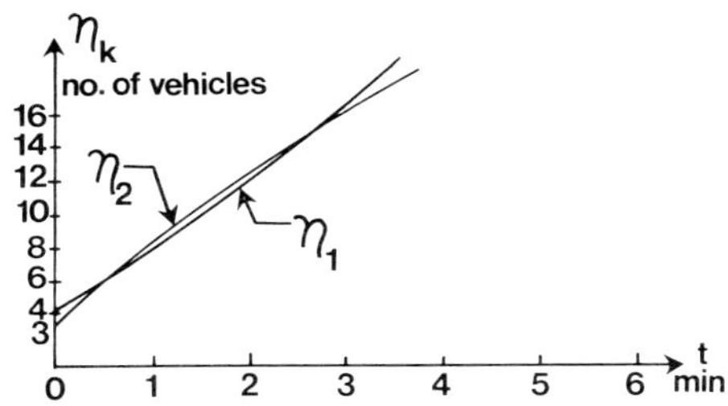

Fig. 4 (b). Queue Lengths Under Optimal Control [23].

The overall throughput of the freeway, within the first 3 minutes was 98 vehicles under local control and 101 under optimal control. The main significant achievement of the optimal control procedure, in this example, is the alleviation of congestion at the on-ramp in Section 2.

A discretization of the continuum equation, identical to the one in Eq. (36), was also used by Kaya [24,25]. His objective function is different however. Conceptually, as in the work of Wattleworth [17,18], the goal is to minimize the total travel time for all vehicles involved. The exact formulation of Kaya's performance criterion is as follows:

$$\text{minimize} \quad J = \int_{t_0}^{t_1} \sum_{k=1}^{K} [(w_{tk} t_{wk} + w_{gk} \eta_k + w_{ek} e_k) r_k + w_{ck}(c_k - c_{dk})] dt \tag{47}$$

where:

$t_{wk}(t)$ = time elapsed for a vehicle on the on-ramp k between arriving and starting the merging process.

$\eta_k(t)$ = length of queue at the kth on-ramp.

$e_k(t)$ = sum of the waiting times of vehicles at on-ramp k.

$c_k(t)$ = vehicle concentration in section k.

c_{dk} = concentration corresponding to maximum flow in section k.

$w_{tk}, w_{gk}, w_{ek}, w_{ck}$ = weighting factors.

The discretized form of the continuity equation is used in the appropriate expressions. The constraints used with this model include:

(1) Flow restrictions

(2) Maximum limit on r_k

as in the Yuan and Kreer's model, described above. The performance criterion has been reformulated as a nonlinear, separable objective function. A solution was obtained using a polygonal approximation of the objective function [28].

3. The Aggregate Variables Model

A somewhat different discrete model, denoted as the aggregate variables model, was developed by Payne [29,30]. This model was implemented in simulation and optimization problems of highway traffic control by Isaksen, Payne and their associates [31-34]. A linear programming highway traffic optimization problem, based on Payne's model, was formulated by Tabak [35,36].

In Section II, we have discussed the following continuum variables associated with the highway traffic:

concentration	$c(x,t)$
flow	$q(x,t)$
speed	$u(x,t)$.

The corresponding aggregate variables were defined by Payne [29] over a finite interval of the highway and a finite interval of time. Consider a highway section with the interval of

$$x_{k-1} \leq x \leq x_k$$

and during the time interval

$$t_{i-1} \leq t \leq t_i \, .$$

The aggregate variables are defined as follows:

aggregate concentration

$$c(x_{k-1}, x_k; t) = \frac{1}{D_k} \int_{x_{k-1}}^{x_k} c(x,t)dx$$

$$D_k = x_k - x_{k-1} \tag{48}$$

aggregate flow

$$q(x_k; t_{i-1}, t_i) = \frac{1}{T_i} \int_{t_{i-1}}^{t_i} q(x_k, t)dt$$

$$T_i = t_i - t_{i-1} \tag{49}$$

aggregate speed

$$u(x_{k-1},x_k;t) = \frac{1}{D_k} \int_{x_{k-1}}^{x_k} u(x,t)dx \,. \tag{50}$$

In addition, we define the following variables:

the total number of vehicles at time t within the interval

$$x_{k-1} \leq x \leq x_k :$$

$$n_k(t) = \int_{x_{k-1}}^{x_k} c(x,t)dx = D_k \; c(x_{k-1},x_k;t) \,. \tag{51}$$

The number of vehicles crossing point x_k between the times t_{i-1} and t_i:

$$n(x_k,t_i) = \int_{t_{i-1}}^{t_i} q(x_k,t)dt = T_i \; q(x_k;t_{i-1},t_i) \,. \tag{52}$$

Using the aggregate variables we can easily write the continuity equation for section k (see Figure 3):

$$\begin{aligned} &\frac{1}{T_i} [c(x_{k-1},x_k;t_i) - c(x_{k-1},x_k;t_{i-1})] \\ &\quad = - \frac{1}{D_k} [q(x_k;t_{i-1},t_i) - q(x_{k-1};t_{i-1},t_i)] \\ &\qquad + \frac{1}{l_k D_k} [r_k(t_{i-1},t_i) - g_k(t_{i-1},t_i)] \end{aligned} \tag{53}$$

where

l_k = number of lanes in section k.

$r_k(t_{i-1},t_i)$ = number of vehicles entering the highway through the on-ramp in section k within $t_{i-1} \leq t \leq t_i$;

$g_k(t_{i-1}, t_i)$ = number of vehicles exiting the highway through the off-ramp in section k within $t_{i-1} \leq t \leq t_i$.

It is easy to see [29] that as $t_{i-1} \to t_i$ and $x_{k-1} \to x_k$, the equation in Eq. (53) will approach the continuity equation in the continuum model in Eq. (4), in the absence of ramps, or it will have the general form of Eq. (5) in the presence of ramps.

We will denote Eq. (53) as the <u>aggregate continuity equation</u>. It could be regarded as one of the possible discretization forms of the general continuity equation, Eq. (5). Equation (3) of the continuum model was also used by Payne in his model [29-34]. The car-following model (Section II.B) was also utilized by Payne [29] in the derivation of the following differential equation for the vehicles velocity u:

$$\frac{du}{dt} = -\frac{1}{T}\left[u - v(c) - \frac{1}{2c}\,\frac{dv(c)}{dc}\,\frac{\partial c}{\partial x}\right] \tag{54}$$

where T is an average driver reaction time, and $v(c)$ is the speed derived in the car-following model. For instance, in the <u>linear</u> car-following model, it is given by Eq. (24):

$$v(c) = v_f(1 - c/c_j)\,. \tag{55}$$

A somewhat more elaborate model was used by Payne in [30]:

$$v(c) = 107 - 2.31c + 0.0215c^2 - 7.4 \times 10^{-5}c^3\,. \tag{56}$$

Equation (56) was obtained empirically [30]. Introducing

$$\nu = \frac{1}{2}\,\frac{dv(c)}{dc}\,. \tag{57}$$

Equation (54) becomes:

$$\frac{du}{dt} = -\frac{1}{T}\left[u - v(c) - \frac{\nu}{c}\,\frac{\partial c}{\partial x}\right]. \tag{58}$$

It should be noted that since $u = u(x,t)$, we actually have on the left-hand side of Eq. (58):

$$\frac{du}{dt} = \frac{\partial u}{\partial t} + u \frac{\partial u}{\partial x} . \tag{59}$$

If we neglect for the moment the third term on the right-hand side of Eq. (58), we will have:

$$\frac{\partial u}{\partial t} + u \frac{\partial u}{\partial x} = - \frac{u - v(c)}{T} . \tag{60}$$

It is obvious that Eq. (60) is of the same form as Eq. (13) in the Prigogine-Herman model (Section II.A). In analogy, we can interpret $v(c)$ as the desired speed the drivers wish to obtain (u is their actual speed). The reaction time T is a personal attribute of each driver and it is different for different persons. In this model, an average reaction time is adopted.

The third term in Eq. (58), $\frac{\nu}{cT} \frac{\partial c}{\partial x}$, can be interpreted as the driver's anticipation term [31]. The drivers observe the change in density, $\partial c/\partial x$, ahead of them, and adjust their speeds accordingly. The coefficient ν can be interpreted here as a sensitivity factor [31].

The aggregate variables, defined above, were substituted into Eq. (58), which was time- and space-discretized. The time axis was divided into equal sampling intervals Δt, and all variables were considered at discrete times $t = 0, \Delta t, 2\Delta t, \ldots, i\Delta t, \ldots$. The following notation is introduced:

$$c_k(i) = c(x_k, x_{k+1}; i\Delta t)$$

$$q_k(i) = q(x_k; (i - 1)\Delta t, i\Delta t)$$

$$u_k(i) = u(x_k, x_{k+1}; i\Delta t)$$

$$r_k(i) = \frac{1}{\Delta t} \int_{(i-1)\Delta t}^{i\Delta t} r_k(t)dt$$

$$g_k(i) = \frac{1}{\Delta t} \int_{(i-1)\Delta t}^{i\Delta t} g_k(t)dt \ .$$

Using this notation, the discretized form of Eq. (58) becomes:

$$u_k(i+1) = u_k(i) - \Delta t \left\{ 2u_k(i) \ \frac{u_k(i) - u_{k-1}(i)}{x_{k+1} - x_{k-1}} \right.$$

$$\left. + \frac{1}{T} \left[u_k(i) - v(c_k(i)) - \frac{2\nu}{c_k(i)} \cdot \frac{c_{k+1}(i) - c_k(i)}{x_{k+1} - x_k} \right] \right\} . \qquad (61)$$

Similarly, Eqs. (3) and (53) are rewritten as:

$$q_i(i+1) = c_{k-1}(i) \ u_{k-1}(i) \qquad (62)$$

$$c_k(i+1) = c_k(i) + \frac{\Delta t}{D_k} \{q_{k-1}(i+1) - q_k(i+1)$$

$$+ [r_k(i+1) - g_k(i+1)]/l_k\} \ . \qquad (63)$$

Equations (61)-(63) are the basic system equations of Payne's aggregate variables model [29,30]. These equations have been used in the optimization and simulations studies reported by Payne and Isaksen [30-34].

If we consider a reduced aggregate model, drawing on Eq. (63) only, a linear programming optimization problem can be formulated [35,36]. It is obvious that if the speed u_k is unknown, Eq. (62) is quadratic, Eq. (61) is nonlinear in any case. For obvious reasons, when real-time control is contemplated, only a linear programming formulation can guarantee a fast solution [19,20]. In case of a nonlinear programming problem, we do not have any guarantees. Even if convergence to the optimal solution is attained, it may take a prohibitive amount of time, rendering the solution impractical for real-time implementation. Therefore, every effort should be made to use linear or linearized equations in a model intended for real-time computer control of highway traffic [35,36]. Indeed, Payne and Isaksen have used a linearized form of their model yielding a linear, quadratic cost, optimal control problem [31,32]. This approach will be discussed in the next paragraph. As an alternative, we could use the linearized form of Eqs. (61), (62) along with the linear Eq. (63) in a linear programming problem. This solution is contemplated in the near future. As a preliminary step, only Eq. (63), was used in [35,36].

We introduce a <u>nominal density</u> c_n, with the understanding that whenever the density on the road falls below c_n, no ramp control action is necessary. If $c \geq c_n$, the procedure, described below, is applied. For the sake of convenience, we introduce the following notation:

$$\begin{aligned} y_k(i) &= c_k(i) - c_n \\ a_k &= \Delta t / D_k \\ q_{ki} &= q_k(i) - q_{k-1}(i) \, . \end{aligned}$$

Substituting this into Eq. (63) we obtain:

$$y_k(i) = y_k(i-1) - a_k q_{ki} + \frac{a_k}{l_k} [r_k(i) - g_k(i)] \, . \tag{63a}$$

From Eq. (63a) we can easily derive:

$$y_k(i) = y_k(0) - a_k \sum_{j=1}^{i} \{q_{kj} - [r_k(j) - g_k(j)]/l_k\} \qquad k = 1,..,K \, . \tag{63b}$$

It should be noted that q_{kj} and $g_k(j)$ are measurable entities while a_k and l_k are constants. (a_k is a constant if we agree to work with a fixed sampling time period.) The on-ramp rates, r_k, are the unknown variables of the linear programming problem, or the control variables in the traffic control problem. In the problem formulated here, we regulate the permitted on-ramp flow rates so that we satisfy the demand and at the same time minimize the excess density $(c - c_n)$ above the nominal value. Therefore, our objective function is:

$$\text{minimize} \quad z = \sum_{k=1}^{K} y_k(N_p) \tag{64}$$

and the constraints:

$$0 \leq y_k(N_p) \leq y_{max}; \qquad k = 1,\ldots,K \tag{65}$$

$$r_k(i) \geq d_k \quad ; \qquad k = 1,\ldots,K, \quad i = 1,\ldots,N_p \tag{66}$$

where

N_p = total number of sampling periods considered
d_k = demand rate at on-ramp k
$y_k(N_p) = y_k(i)$, $i = N_p$, calculated by Eq. (63b)
y_{max} = maximum allowable excess density.

It is obvious that reduction of excess density on the highway, will decrease the travel time. In this respect the problem is similar to that of Wattleworth [17,18], although a different system model is used. Computations using the LP model in Eqs. (64)-(66), applied to a section of Interstate Route 91 in the Greater Hartford, Connecticut area, were performed by Huerta and Tabak [37,38]. Eq. (62) and a linearized version of Eq. (63) were also implemented.

4. The Linear Regulator Approach

The first formulation to the best of the author's knowledge of a highway traffic problem as a linear regulator problem, was given by Yuan and Kreer in 1968 [22]. The density and on-ramp rate were expressed as sums of a steady-state and a perturbation value:

$$\underline{c}(t) = \underline{c}_s + \underline{e}(t) \tag{67}$$

$$\underline{r}(t) = \underline{r}_s + \underline{w}(t) \tag{68}$$

where

$\underline{c} = [c_1, c_2 \ldots c_k \ldots c_K]^T$ = densities vector

$\underline{r} = [r_1, r_2 \ldots r_k \ldots r_K]^T$ = on-ramp rates vector

$\underline{c}_s$ = densities vector in the steady state

$\underline{r}_s$ = on-ramp rates vector in the steady state

$\underline{e}$ = density perturbation vector

$\underline{w}$ = on-ramp rates perturbation vector.

The $\underline{c}_s$ and $\underline{r}_s$ steady-state values were obtained by a linear programming solution (see paragraph III.A.2). Eq. (39) can also be expressed in vector form:

$$\dot{\underline{c}} = \underline{f}(\underline{c}) + B\underline{r} . \tag{69}$$

Substituting Eqs. (67) and (68), expanding $\underline{f}(\underline{c}_s + \underline{e})$ in a Taylor expansion around $\underline{c}_s$, and neglecting second order terms in $\underline{e}$, we obtain

$$\begin{aligned} \dot{\underline{e}}(t) &= A\underline{e}(t) + B\underline{w}(t) \\ \underline{e}(0) &= \underline{c}(0) - \underline{c}_s \end{aligned} \tag{70}$$

where

$$A = [a_{ij}]; \qquad i,j = 1,\ldots,K$$

$$a_{ii} = \frac{l_i v_{fi}}{D_i}(1 - 2c_{is}/c_{ji})$$

$$a_{i,i-1} = \frac{l_i v_{fi}}{D_i}(1 - f_{i-1})\left(1 - \frac{2c_{i-1,s}}{c_{j,i-1}}\right)$$

$$a_{ij} = 0, \quad \text{for } j \neq i,\ i-1$$

$$B = \mathrm{diag}[1/D_1, 1/D_2, \ldots, 1/D_K] .$$

The performance index is:

$$\text{minimize} \quad J = \frac{1}{2} \int_0^\infty (\underline{e}^T Q\underline{e} + \underline{w}^T R\underline{w})dt \tag{71}$$

where Q is a positive semidefinite and R a positive definite matrix. The solution to the linear regulator problem, defined by Eqs. (70), (71) is well known [39-41].

A similar approach has also been adopted by Payne and Isaken [31,32]. The concentration, speed and flow of vehicles on the highway are expressed as a sum of nominal and perturbed quantities:

$$c_k(i) = c_{kn}(i) + c_{kp}(i) \tag{72}$$

$$u_k(i) = u_{kn}(i) + u_{kp}(i) \tag{73}$$

$$q_k(i) = q_{kn}(i) + q_{kp}(i) \tag{74}$$

where the subscript n identifies the nominal and p the perturbed quantities. It is assumed that the perturbed quantities are much smaller than the corresponding nominal ones. In other words, only small disturbances around nominal conditions are considered. The on- and off-ramp flows are expressed as follows:

$$r_k(i) = \alpha_k(i)q_k(i) + m_k(i-1) \tag{75}$$

$$g_k(i) = f_{k-1}(i)q_{k-1}(i) \tag{76}$$

where

α_k = priority parameter for on-ramp k in nominal conditions.

f_{k-1} = exit fraction in section k.

m_k = additional control exerted at on-ramp k.

Substituting Eqs. (72)-(76) into the space-discretized model, we obtain the following set of linearized state equations [31]:

$$\dot{c}_k(t) = a_{k,k-2}c_{k-1}(t) + a_{k,k-1}u_{k-1}(t) + a_{k,k}c_k(t) + a_{k,k+1}u_k(t) + m_k(t) \tag{77}$$

$$\dot{u}_k(t) = b_{k,k-2}u_{k-1}(t) + b_{k,k-1}c_k(t) + b_{k,k}u_k(t) + b_{k,k+1}c_{k+1}(t) \tag{78}$$

where

$$a_{k,k-2} = \frac{l_{k-i}}{l_k}(1-f_{k-1})\frac{u_{k-1,n}}{D_{k+1}}$$

$$a_{k,k-1} = \frac{l_{k-1}}{l_k}(1-f_{k-1})\frac{c_{k-1,n}}{D_{k+1}}$$

$$a_{k,k} = (\alpha_k - 1)\frac{u_{kn}}{D_{k+1}}$$

$$a_{k,k+1} = (\alpha_k - 1)\frac{c_{kn}}{D_{k+1}}$$

$$b_{k,k-2} = \frac{2u_{kn}}{x_{k+1} - x_{k-1}}$$

$$b_{k,k-1} = \frac{1}{T}\left.\frac{\partial v(c_k)}{\partial c_k}\right|_{c_k=c_{kn}} + \frac{\nu}{T}\cdot\frac{2\,c_{k+1,n}}{c_{kn}^2(x_{k+1} - x_{k-1})}$$

$$b_{k,k} = \frac{2(u_{k-1,n} - 2u_{kn})}{x_{k+1} - x_{k-1}} - \frac{1}{T}$$

$$b_{k,k+1} = \frac{-2\nu c_{kn}}{T(x_{k+1} - x_{k-1})}\,.$$

It should be noted that the variables c_k, u_k, appearing in Eqs. (77), (78) are the <u>perturbed</u> ones. The p subscript was dropped for the sake of convenience. Introducing the combined vectors

$$\underline{z} = [c_1, u_1, c_2, u_2, \ldots, c_K, u_K]^T$$
$$\underline{m} = [m_1, m_2, \ldots, m_M]^T$$

where M is the total number of on-ramps where the additional control $\underline{m}$ is exerted, the system equations can be reformulated in the general linear form:

$$\dot{\underline{z}} = A\underline{z} + B\underline{m} \tag{79}$$

$\underline{z}$ being the $2K \times 1$ state vector, $\underline{m}$ - $M \times 1$ control vector, A - $2K \times 2K$ matrix and B - $2K \times M$ matrix.

The control objective is to maintain nominal conditions (that is, to minimize the perturbations) on the highway at a minimal expense of control energy. Therefore the following performance criterion was formulated:

$$\text{minimize} \quad J = \int_0^\infty (\underline{z}^T Q \underline{z} + \underline{m}^T R \underline{m}) dt \tag{80}$$

where the matrices Q and R are defined in the same manner as in Eq. (71). The problem defined by Eqs. (79) and (80) is obviously a linear regulator problem, of the same form as that in Eqs. (70) and (71). The solution to the linear regulator problem is given by

$$\underline{m}^0 = -R^{-1} B P \underline{z} \tag{81}$$

where P is the unique, positive definite solution of the nonlinear matrix algebraic equation [39-41]

$$PA - A^T P - PB^T R^{-1} BP + Q = 0 \; . \tag{82}$$

Details concerning computations performed in the analysis of this model can be found in Isaksen's thesis [31]. It should be noted that the problem is a large-scale one. A partitioning scheme, resulting in the alleviation of some of the computational difficulties, was proposed by Thompson et. al. [33].

5. The Multilevel Approach

A multilevel configuration of ramp control of highway traffic was proposed by Drew and his associates [42,43]. A schematic diagram of the configuration, in a somewhat altered form, is shown in Figure 5. The system consists of:

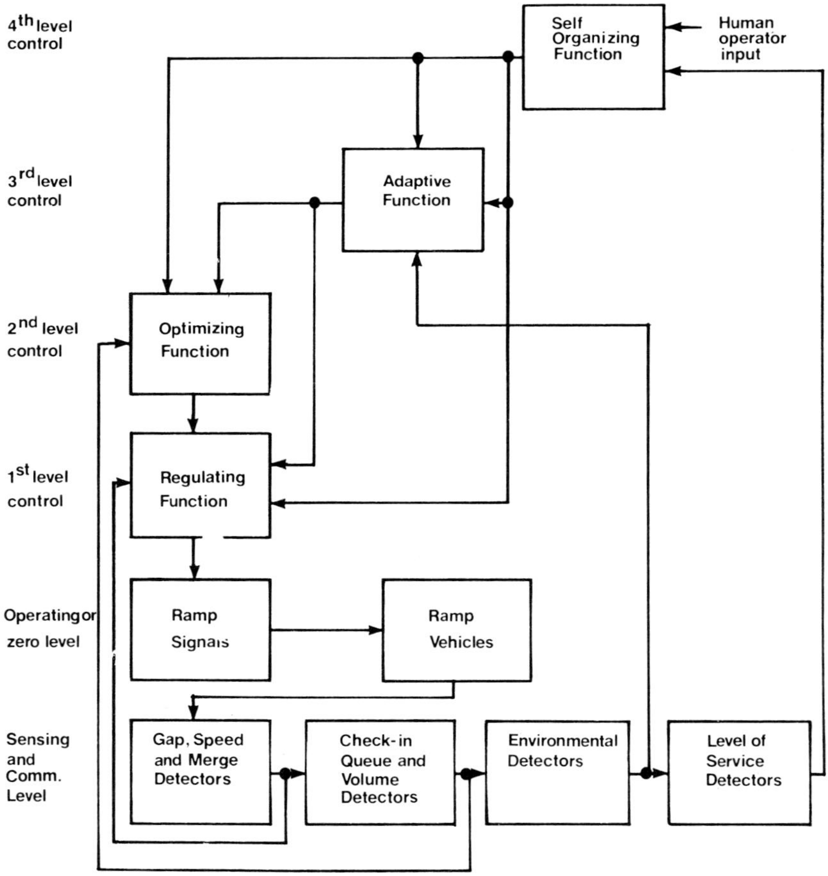

Fig. 5. A Four-Level Ramp Control System.

(a) The sensing and communication to higher levels subsystem.

(b) The operating subsystem which includes the ramp signals and vehicles. It is denoted as the zero-level of operation.

(c) Four hierarchical control levels, denoted as

(1) Regulating function

(2) Optimizing function

(3) Adaptive function

(4) Self-Organizing function.

The regulating function operates directly on the controlling devices — the ramp signals. It translates the decisions of the higher level controllers into direct actions, such as the timely release of ramp vehicles by activating the ramp signal. Another task of this controller is to place as many vehicles as possible into gaps existing in the mainstream of traffic. The gaps are detected by the gap, speed and merge detectors and the information is transmitted to the first level controller.

The optimizing function determines the optimum ramp operating conditions based on a specified performance criterion and on the mathematical model of the process. The speed and volume of the outside lane of the freeway as well as the on-ramp queue are taken directly into account using the check-in, queue and volume detectors.

The adaptive function compensates for the errors in the mathematical model, used in the second level, by adjusting various parameter values in the model. The adjustments are based on additional measurements performed along the freeway.

The self-organizing function acts as a coordinator [44] by supplying coordinating signals to the lower three levels of control. It receives inputs both from the human operator and from highway measurements. It solves a more general optimization problem involving a number of ramps and sections of the highway. A linear programming optimization model, similar to the one proposed by Wattleworth [17,18], was implemented in this case [43].

6. Concluding Remarks

The vast amount of exploratory work concerning ramp control, performed so far, was certainly not exhausted in this section. Only a few cases, where some definite optimization and control methods were directly applied, were selected. Some of the earlier exploratory work was done by May [26] and Gervais [45]. A probabilistic approach to the problem was adopted by Yagoda [46]. Similar results, utilizing queuing theory and multilevel control philosophy, were reported by Shaw [47]. A dynamic programming approach to the ramp-flow assignment problem, was recently proposed by Wang [48].

The highway traffic system is basically a distributed-parameter system, characterized by a partial differential equation, such as Eq. (4). In the approaches described in this section, the partial differential equation of continuity was first converted into an ordinary one through space discretization. There is of course a possibility of treating the system as a distributed-parameter one (which it is) and solve the partial differential equations. This approach was recently undertaken by Sendaula and Knapp [49].

B. Computer Control

Automation and computer control have been penetrating various branches of industry during the past years on an increasing rate basis. Naturally, computerized traffic control has been under development and gradual implementation during the past twelve years. An historic survey, covering most of the activities in this area from the early experiments to date, was recently written by Gazis [50]. In fact, in the seventies, we could not imagine a sophisticated traffic control system which does not utilize a digital computer. All ramp control schemes, described in the preceding paragraph, III.A, can be implemented with a digital computer only. In most cases, the computer would have to work on a real-time basis. For this reason, the schemes utilizing linear programming are much more promising than those using nonlinear programming. In the case of LP we can obtain fast and efficient solutions in a much shorter time, which makes LP much more amenable to real-time implementation. There is also a reliability aspect involved. Using LP we are practically guaranteed to attain convergence in a finite time for most practical cases. In NLP, convergence is never guaranteed. In many cases it is never attained. For this reason it is highly advisable to make a special effort in obtaining a linearized model of the system and formulating linear objective functions. Of course, there is always a danger that the model errors, caused by the linearization, may turn out to be unacceptable. Special care has to be exerted therefore in the linearization process.

Considerable amount of analysis, computer simulation and on-the-road testing has been performed by many researchers in this field. Some of the projects will be briefly described in this paragraph (no claim of completeness is made).

The first project, reporting the use of a digital computer in traffic lights control, was performed in Toronto, in 1959 [50,51]. An IBM 650 computer was used in the control of nine traffic lights. Since then, many more computer control schemes have been implemented.

One of the well-known traffic computer control projects was performed by Gazis and his associates [50,52,53]. The computer control scheme was implemented to vehicle flow through a critical traffic link — a tunnel. Specifically, the New York City Lincoln Tunnel was used as a test site. Appropriate detection devices were placed in selected locations in the tunnel. With the help of the detectors, a vehicle count was conducted. Moreover, the vehicles were identified by a set of coded impulses. Using this information the vehicles length and speed were computed. The control was performed for each of the two existing lanes separately. The tunnel was divided into three sections and the vehicle count in each section was denoted as a state variable. Based on the measurements, the probability of impending congestion was calculated. The ascending sequence of values of this probability was divided into five regions. According to these regions, five control levels were adopted [53].

CONTROL LEVEL	CORRESPONDING SIGNAL
0 (no control)	Green light
1	Flashing amber light
2	Amber light with a sign: "PAUSE HERE THEN GO," and lights flush with the pavement
3	Same signals as in Level 2 plus raised cones to reduce lane width
4	Red light

This control scheme was experimentally implemented in the Lincoln Tunnel using at one time (1967) the IBM 7040 and another time (1969) the IBM 1800 Process Control System [53]. Both experiments were basically successful, however, the 1800 system proved to be more reliable with respect to system stoppages [53].

Another interesting computer traffic control project was

reported by Ross et. al. [54]. It was applied to the control of a critical intersection in an urban network. The scheme calculated predicted traffic delays, taking into account the intersection considered and its neighboring intersections. The optimal timing of traffic lights is calculated by direct minimization of the total delay, measured in vehicles-seconds units. The control scheme was simulated on a digital computer using actual traffic data from San Jose, California [54]. Another scheme of automized, real-time control of traffic lights, utilizing linear programming, was reported by Morris and Yagoda [55] (see Section IV).

A multilevel approach, including a quadratic cost function, to signal control of traffic nets was adopted by McShane, Yagoda et. al. [56] (see Section IV). Dodge and Reiss have proposed a direct computer control scheme for a set of parallel roadways [57]. A scheme for an automated transportation system was recently described by Boyd and Lukas [58].

C. The Oversaturated Intersection

The problem of optimal ramp-control of highway traffic was discussed in Section III.A. The main feature in that case was that the control action, such as traffic lights, was applied to the ramp traffic, but not to the traffic on the mainstream of the highway. On many occasions, especially in populated areas, highways cross each other at the same level, at signalized intersections. As long as the vehicles arriving at the intersection during the red light period, are served during the next green light period, no problem arises. However, particularly during rush hours, when the queues of vehicles at the intersection grow at a faster rate than the throughput, we have the case of the oversaturated intersection. There are many ways of timing the green-red light periods and there is certainly a better way of doing it under specified conditions and a specified performance criterion. In other words, controlling an over saturated intersection can be formulated as an optimization problem. A mathematical model of an oversaturated intersection was developed by Gazis and Potts [59] along with some preliminary analysis of the problem. Subsequently, the problem was solved by direct and graphical methods, as well as optimal control techniques by Gazis [60,61]. Specifically, the Maximum Principle of Pontriagin [62] was used. (To the best of the author's knowledge, Gazis was the first to apply the Maximum Principle in a transportation problem). The results of Gazis will be described in the following.

Two competing flows of vehicles, to be served by the intersection, are taken into account. The arrival rates of the vehicles in the two flows are q_1 and q_2. The maximum throughput rate, or "saturation flow," for both flows is s_1 and s_2 respectively. The effective green light phases in both directions are g_1 and g_2. The cycle time of the light interchange at the intersection will be denoted as t_c.

The average green light time required to serve all the cars arriving during a cycle t_c is:

$$t_{gj} = q_j t_c / s_j ; \qquad j = 1,2 . \tag{83}$$

It is customary in traffic engineering practice to divide the total available green time in proportion to t_{gj}. Thus we have

$$\frac{q_1}{s_1 g_1} = \frac{q_2}{s_2 g_2} . \tag{84}$$

When arrival flows q_1, q_2 increase so that

$$\frac{q_1}{s_1} + \frac{q_2}{s_2} > 1 - \frac{L}{t_c} \tag{85}$$

we have the case of oversaturation. L is the total lost time for acceleration and clearing.

The queue lengths developing at the intersection are designated as the state variables of the problem and denoted as y_1, y_2. The state equations are:

$$\dot{y}_j = q_j - s_j g_j / t_c ; \qquad j = 1,2. \tag{86}$$

The first term on the right hand side of Eq. (86) represents the arriving flow, while the second term represents the flow served. The difference between the two terms is the rate of change of the queue length at the intersection. We have no control over the rate of arrivals, however we can certainly control the timing of the green light, which is expressed in the second term. Therefore we choose this term, for flow 1, as the control variable of the problem:

$$m = s_1 g_1 / t_c ; \qquad 0 \leq m_l \leq m \leq m_u \tag{87}$$

where m_l and m_u are the lower and upper bounds of m, respectively. Without loss of generality we can assume that $s_1 > s_2$. Substituting Eq. (87) into Eq. (86), using Eq. (85) as an equality, and assuming

$$g_1 = t_{g1} = q_1 t_c / s_1$$

we obtain the following form of linear state equations:

$$\begin{aligned} \dot{y}_1 &= q_1 - m \\ \dot{y}_2 &= q_2 - s_2(1 - L/t_c) + \frac{s_2}{s_1} m . \end{aligned} \tag{88}$$

The purpose of the control is the minimization of the queue length over the whole period of time. This goal was expressed by Gazis [60,61] in the following form:

$$\text{minimize} \quad J = \int_0^{t_1} (y_1 + y_2)dt \,. \tag{89}$$

The same goal could of course also be satisfied by a quadratic performance index:

$$\text{minimize} \quad J_q = \int_0^{t_1} (y_1^2 + y_2^2)dt \tag{90}$$

thus obtaining a linear regulator problem. The Hamiltonian of the system is:

$$\begin{aligned} H(\underline{y},\underline{p},m) &= y_1 + y_2 + p_1(q_1 - m) + p_2\left[q_2 - s_2(1 - L/t_c) + \frac{s_2}{s_1} m\right] \\ &= y_1 + y_2 + p_1 q_1 + p_2[q_2 - s_2(1 - L/t_c)] + \left(\frac{s_2}{s_1} p_2 - p_1\right)m \end{aligned} \tag{91}$$

where p_1 and p_2 are the costate variables. The costate equations are:

$$\dot{p}_j = -\frac{\partial H}{\partial y_j} = -1; \qquad j = 1,2 \tag{92}$$

and their solutions

$$p_j(t) = -t + C_j; \qquad j = 1,2 \tag{93}$$

where C_1 and C_2 are integration constants. We will designate the coefficient of m in Eq. (91) as

$$\begin{aligned} Z &= \frac{s_2}{s_1} p_2 - p_1 = \frac{s_2}{s_1}(-t + C_2) + t - C_1 \\ &= (1 - s_2/s_1)t + (s_2 C_2/s_1 - C_1) \,. \end{aligned} \tag{94}$$

By the <u>minimum principle</u> [39,40], we will have for the optimal solution m^0:

$$H(\underline{y}^0,\underline{p}^0,m^0) \leq H(\underline{y}^0,\underline{p}^0,m) . \qquad (95)$$

Taking into account Eqs. (87), (91), (94) and (95) we obtain:

$$m^0 = \begin{cases} m_l; & Z > 0 \\ m_u; & Z < 0 \end{cases} \qquad (96)$$

Defining

$$\text{sgn } Z = \begin{cases} 1; & Z > 0 \\ -1; & Z < 0 \end{cases}$$

we can rewrite Eq. (96) as:

$$m^0 = \tfrac{1}{2}[(1+\text{sgn } Z)m_u + (1-\text{sgn } Z)m_l] . \qquad (97)$$

The <u>switch-over time</u>, t_s, from m_u to m_l occurs when Z passes through the value of zero. Thus we obtain from Eq. (94):

$$t_s = \frac{C_1 - s_2C_2/s_1}{1 - s_2/s_1} . \qquad (98)$$

Further details of the calculations and graphical results can be found in [60,61]. A graphical extension to a system of two oversaturated intersections was also given by Gazis [60]. In some cases one or both of the flows through the signalized intersection come from an exit ramp of a highway. In an oversaturated situation a queue may develop of such lengths as to cause congestion on the highway itself. This is called a <u>spillback</u> on the highway. An analysis of this problem was given by Gazis [63]. A queuing model was developed and direct optimization, utilizing graphical representation was adopted. A queuing model for computer control of critical signalized intersections was also developed by Ross et. al. [54]. A dynamic programming solution, utilizing a probabilistic model, was proposed by Martin-Lof [64] for the same class of problems.

A linear programming model for optimal control of a signalized intersection was recently formulated by Wattleworth [65]. An intersection, where four flows of vehicles converge, is considered. Each of the four flows is divided into three sub-flows, according to their subsequent destination:

(1) Right turn
(2) Continue straight
(3) Left turn.

Each of these sub-flows is considered as a separate variable V_i, so that we have a total of 12 $(i = 1, \ldots, 12)$ variables. The goal of the optimization problem is to maximize the overall volume, that is, the objective function is

$$\text{maximize} \quad z = \sum_{i=1}^{12} V_i \, . \qquad (99)$$

The constraints of the problem are formulated according to the capacity of the destination of various sub-flows. For instance, if sub-flows of volumes V_4, V_6 and V_{10} are going into a destination route whose maximum capacity is C, the appropriate inequality constraint of the LP problem would be:

$$V_4 + V_6 + V_{10} \leq C \, . \qquad (100)$$

A number of constraints of the type shown in Eq. (100) is formulated according to the particular structure of the intersection. A case study utilizing this model, was conducted on a conventional diamond interchange in Orlando, Florida [65]. The IBM MPS system was used to solve the linear programming problem.

D. THE STRING OF MOVING VEHICLES PROBLEM ON AUTOMATED HIGHWAYS

The congestion on the nation's highways has been steadily on the increase during the past years, along with the accidents rate. One of the possible solutions to the problem is the creation of high-speed automated highways. Vehicles would travel in strings along the highways at high speed. They will be automatically controlled by a central computer system. Considerable amount of analysis and optimization studies has been dedicated to this problem during the past decade both by transportation and control professionals [58,66-84]. An automated system where the vehicles are automatically controlled at all times is denoted as a single-mode system. A system where the vehicles are automatically controlled on specially designated guideways, while the driver resumes control anywhere else, are called dual-mode systems [79-83].

One of the earliest projects in this area, performed by control professionals, was reported in 1966 by Levine and Athans [66] of M.I.T. Similar projects have been also conducted in parallel at Ohio State University by Cosgriff, Bender and Fenton [75,76,82,84] and at Stanford Research Institute by Hajdu and his associates [83], although the formal journal publications, describing these projects, appeared somewhat later. As will be seen in the subsequent discussion, other important contributions followed these initial steps.

Levine and Athans [66] proposed a model of a string of vehicles traveling along a single lane. Each vehicle is denoted by an index $k(k=1,\ldots,N)$, index 1 representing the leading vehicle and subsequent indices representing the ordering of other vehicles in the string. A uniform desired string velocity and a desired distance between each pair of cars, are specified.

For each vehicle there is a fixed force applied to overcome the drag force. The drag force is a nonlinear function of the vehicle's velocity. The deviations from the desired distances between the cars $d_k (k = 1, \ldots, N-1)$, and the deviations from the desired string velocity, $u_k (k = 1, \ldots, N)$, are the state variables of the model. The deviations of the forces from the fixed force, denoted as $f_k (k = 1, \ldots, N)$, are the control variables. The state equations are:

$$\dot{d}_k = u_k - u_{k+1} \quad ; \qquad k = 1, \ldots, N-1 \qquad (101)$$

$$\dot{u}_k = -\frac{\alpha_k}{m_k} + \frac{1}{m_k} f_k \; ; \qquad k = 1, \ldots, N \qquad (102)$$

where m_k = mass of vehicle k

α_k = partial derivative of the fixed force with respect to velocity at the desired velocity.

By lumping the state variables into a single $(2N-1)$ dimensional state vector

$$\underline{x} = [u_1 d_1 u_2 d_2 \ldots u_{N-1} d_{N-1} u_N]^T$$

the linear state equations in Eqs. (101), (102) can be expressed in the form:

$$\dot{x} = A\underline{x} + B\underline{u} . \qquad (103)$$

Where

$$\underline{u} = [f_1 f_2 \ldots f_N]^T$$

and A is a $(2N-1) \times (2N-1)$ and B a $(2N-1) \times N$ matrices. A quadratic performance index was chosen:

$$\text{minimize} \quad J = \frac{1}{2} \int_0^\infty (\underline{x}^T Q \underline{x} + \underline{u}^T R \underline{u}) dt . \qquad (104)$$

Where Q and R are constant, diagonal matrices, R is positive definite, while Q is positive semidefinite. It is easy to see that the optimal control problem defined by Eqs. (103) and (104) is a linear regulator problem whose goal is to minimize, and possibly reduce to zero the deviations of the vehicles from the desired interdistances and velocity, at a minimal expense of energy. The solution is given by [39-41]

$$\underline{u} = -R^{-1}B^{T}P\underline{x}$$

where P is a real symmetric positive definite matrix satisfying

$$-PA - A^{T}P + PBR^{-1}B^{T}K - Q = 0 \; .$$

The matrix P can be computed by solving the Riccati matrix differential equation [39-41]

$$\dot{P}_1 = P_1A + A^{T}P_1 - P_1BR^{-1}B^{T}P_1 + Q \tag{105}$$

with the initial condition $P_1(0) = 0$. Matrix P is a steady state solution of Eq. (105), that is

$$P = \lim_{t\to\infty} P_1(t) \; . \tag{106}$$

The numerical solution of the problem for a small number of vehicles is relatively easy. However, as the number of vehicles increases the solution becomes more complicated. This is particularly due to the fact that the control function, f_k, for each vehicle, is influenced by the position and velocity deviations of <u>all</u> vehicles [66]. This would require a very large amount of communications links. For instance for a string of $N = 50$ vehicles (99 state variables, 50 control variables) we would need a total of $99 \times 50 = 4950$ links.

This may be inhibited by economic considerations. A decomposition scheme to alleviate this problem was proposed by Pearson [67], however, no computational experimentation was reported. Continuous control was originally proposed in [66].

Subsequently the design was extended to discrete time control by Levis and Athans [69] and by Melzer and Kuo [71].

An important modification of the model used in [66] was proposed by Peppard and Gourishankar [70]. Instead of considering f_k as the control variables, they propose to use its time derivative, $\dot{f}_k$, as such. The $\dot{f}_k^2$ also appear in the performance index. Minimizing a weighted time derivative of force deviations would tend to reduce and smooth jerks of the vehicles, caused by a sudden change in force. This is an important contribution to improving passenger safety and a smoother operation of the vehicles could be attained.

A different approach to the computational solution of the problem in Eqs. (103) and (104), considerably alleviating the dimensionality difficulties, was proposed by Porter and Crossley [77]. Instead of the traditional solution of the Riccati equation, it uses the so called modal control theory [85]. The same dynamic model as in Eqs. (101) and (102) is used. The overall state vector, $\underline{x}$, is defined using a different ordering of variables:

$$\underline{x} = [u_1 u_2 \ldots u_N \; d_1 d_2 \ldots d_{N-1}]^T .$$

The control vector $\underline{u}$ is the same, and the state equations in condensed form are

$$\dot{\underline{x}} = A\underline{x} + B\underline{u} . \tag{107}$$

Naturally, the ordering of rows and columns in the A, B matrices in Eq. (107) will be different from that in Eq. (103), although identical matrix elements appear in both cases. The vector $\underline{x}$ and the matrices A and B are partitioned as follows:

$$A = \begin{bmatrix} A_{11} & A_{12} \\ A_{21} & A_{22} \end{bmatrix} \qquad B = \begin{bmatrix} B_1 \\ B_2 \end{bmatrix} \qquad \underline{x} = \begin{bmatrix} \underline{u} \\ \underline{d} \end{bmatrix}$$

where

$$A_{11} = \text{diag}[-\alpha_1/m_1, -\alpha_2/m_2, \ldots, -\alpha_N/m_N]$$

$$A_{12} = 0; \quad A_{22} = 0$$

$$A_{21} = \begin{bmatrix} 1 & -1 & 0 & 0 & \ldots & 0 & 0 \\ 0 & 1 & -1 & 0 & \ldots & 0 & 0 \\ \cdot & \cdot & \cdot & \cdot & & \cdot & \cdot \\ \cdot & \cdot & \cdot & \cdot & & \cdot & \cdot \\ \cdot & \cdot & \cdot & \cdot & & \cdot & \cdot \\ 0 & 0 & 0 & 0 & \ldots & 1 & -1 \end{bmatrix}$$

$$B_1 = \text{diag}[1/m_1, 1/m_2, \ldots, 1/m_N]; \qquad B_2 = 0 .$$

The following control law was proposed and derived in [77]:

$$\left.\begin{aligned} f_1 &= K_1 \underline{v}_1^T \underline{x} \\ f_k &= K_k \underline{v}_k^T \underline{x} + K_{N+k-1} \underline{v}_{N+k-1}^T \underline{x}; \qquad k = 2,3,\ldots,N \end{aligned}\right\} \quad (108)$$

$$K_1 = (\rho_1 - \lambda_1)/p_{11}$$

$$\left.\begin{aligned} K_k &= (\rho_k - \lambda_k)(\rho_{N+k-1} - \lambda_k)/p_{kk}(\lambda_{N+k-1} - \lambda_k); \ k = 2,3,\ldots,N \\ K_{N+k-1} &= (\rho_k - \lambda_{N+k-1})(\rho_{N+k-1} - \lambda_{N+k-1})/p_{N+k-1,k}(\lambda_k - \lambda_{N+k-1}); \\ & \qquad k = 2,3,\ldots,N . \end{aligned}\right\} \quad (109)$$

Where

$\underline{v}_k$ = eigenvectors of A^T

λ_k = eigenvalues of A^T and A, that is

$$\left.\begin{aligned} A\underline{u}_k &= \lambda_k \underline{u}_k \\ A^T \underline{v}_k &= \lambda_k \underline{v}_k \end{aligned}\right\} \qquad k = 1,2,\ldots,2N-1$$

$\underline{u}_k$ = eigenvectors of A

ρ_k = eigenvalues of matrix C given by

$$C = A + K_1 \underline{b}_1 \underline{v}_1^T + \sum_{k=2}^{N} (K_k \underline{b}_k \underline{v}_k^T + K_{N+k-1} \underline{b}_{N+k-1} \underline{v}_{N+k-1}^T)$$

$\underline{b}_k$ = kth column of matrix B

$p_{ij} = \underline{v}_i^T \underline{b}_j$; $\quad i = 1,2,\ldots,2N-1; \quad j = 1,2,\ldots,N$.

The control law expressed in Eqs. (108), (109) constitutes a closed form solution, which is easily computable. Considerable insight can be gained by examining the structure of the eigenvectors $\underline{v}_k$ which are columns of the modal matrix:

$$V = [\underline{v}_1 \underline{v}_2 \cdots \underline{v}_k \cdots \underline{v}_{2N-1}] = \begin{bmatrix} V_{11} & V_{12} \\ V_{21} & V_{22} \end{bmatrix}$$

$V_{11} = I_N$; $\quad V_{21} = 0$; $\quad V_{22} = I_{N-1}$

$$V_{12} = \begin{bmatrix} m_1/\alpha_1 & 0 & & \cdots & 0 & 0 \\ -m_2/\alpha_2 & m_2/\alpha_2 & & \cdots & 0 & 0 \\ 0 & -m_3/\alpha_3 & m_3/\alpha_3 & \cdots & 0 & 0 \\ \vdots & \vdots & \vdots & & \vdots & \vdots \\ 0 & 0 & 0 & & -m_{N-1}/\alpha_{N-1} & m_{N-1}/\alpha_{N-1} \\ 0 & 0 & 0 & & 0 & -m_N/\alpha_N \end{bmatrix}$$

The control force, f_k, acting on each vehicle is a function of only one state variable for the first vehicle and of only three state variables for each of the remaining vehicles [77]. So for a string of 50 vehicles only $3 \times 49 + 1 = 148$ communication links are needed (as opposed to the 4950 in [66]). The modal control approach has also been extended to discrete-time control [78].

A somewhat different form of a closed form solution to the string of vehicles problem, was proposed by Melzer and Kuo [72]. A closed form solution of the Riccati equation for the matrix P is obtained in terms of a matrix Φ and λ_k $(k = 1,\ldots,N)$, which satisfy:

$$\Phi^T Q \Phi = \mathrm{diag}[\lambda_1 \lambda_2 \ldots \lambda_N]$$

and

$$\Phi^T R \Phi = I \quad \text{or} \quad \Phi\Phi^T = R^{-1} .$$

The computational realization of the solution is quite simple [72].

A completely different model of a string of moving vehicles was adopted by Cosgriff [75]. The actual positions of the vehicles $x_1(t), x_2(t), \ldots, x_N(t)$ were represented in Laplace transform $X_1(s), X_2(s), \ldots, X_N(s)$. A transfer function (or transmission function) was defined as follows:

$$G(s) = \frac{X_k(s)}{X_{k-1}(s)} ; \qquad k = 2,3,\ldots,N . \quad (110)$$

Consequently:

$$X_k(s) = G^{k-1}(s) X_1(s); \quad k = 2,3,\ldots,N . \quad (111)$$

The same transfer function is assumed between each pair of vehicles. Cosgriff used the asymptotic approximation in the detailed evaluation of $G(s)$ [75]. If the peak value of $G(j\omega)$ occurs at $\omega = 0$, then the following asymptotic limit exists:

$$\left[\frac{G(j\omega)}{G(0)}\right]^k \xrightarrow[k\to\infty]{} \exp[(-j\omega\tau - \omega^2\gamma)k] \ . \tag{112}$$

Both τ and γ are the basic constants characterizing $G(j\omega)$. The mathematical validation of the asymptotic approximation is identical to that of the <u>central limit theory</u> in probability and statistics [86]. If we express $G(s)$ as the following ratio:

$$G(s) = \frac{\Sigma\, b_k s^k}{\Sigma\, c_k s^k} \xrightarrow[s\to 0]{} \exp(-\tau s + \gamma s^2) \tag{113}$$

and then expand both the ratio and the exponential approximation in Eq. (113) in a Taylor series, we obtain:

$$\left.\begin{array}{l} b_0 = c_0 \\ \tau = (c_1 - b_1)/c_0 \\ \gamma = (b_2 - c_2)/c_0 + (c_1^2 - b_1^2)/2c_0^2 \end{array}\right\} \tag{114}$$

Note that only the c_0, b_0, c_1, b_1 coefficients are needed to evaluate τ and γ.

Cosgriff has shown [75] that in order to achieve <u>asymptotic stability</u> the following requirement is to be met:

$$\left.\begin{array}{l} |G(j\omega)| < 1; \quad \omega \neq 0 \\ G(0) = 1 \end{array}\right\} \tag{115}$$

Bender and Fenton [76] have also used a frequency domain transfer function between two subsequent vehicles in a string. They have expressed the transfer function between the velocities $V_1(s)$, $V_2(s)$ of two adjacent vehicles (V_1 leading) as follows:

$$G_v(s) = \frac{V_2(s)}{V_1(s)} = \frac{[(k_1 - k_2 k_3)/k_2]s + 1}{s^2/k_2 + [(k_1 + k_2 k_4)/k_2]s + 1} =$$

$$= \frac{\omega_n}{\beta} \cdot \frac{s + \beta\omega_n}{s^2 + 2\zeta\omega_n s + \omega_n^2} \tag{116}$$

where k_1, k_2, k_3, k_4 and β are positive constants, ω_n is the undamped natural frequency and ζ is the damping ratio. We also have:

$$(k_1 - k_2k_3)/k_2 = 1/\beta\omega_n$$

$$k_2 = \omega_n^2$$

$$(k_1 + k_2k_4)/k_2 = 2\zeta/\omega_n .$$

The performance of the system is characterized by the three parameters (β, ζ, ω_n). A necessary condition for asymptotic stability, utilizing this model, was shown to be [76].

$$|G_v(j\omega)| \leq 1 \quad \text{for all} \quad \omega \tag{117}$$

which is satisfied when

$$[\zeta^2 - (1/2\beta)^2]^{1/2} \geq 0.707 . \tag{118}$$

As long as the string of vehicles travels along a guideway the main control problem is to keep the vehicle spacing and speed at prescribed values.

When two separate guideways merge into one, a new problem arises. When two strings of vehicles arrive at the junction at about the same time, there is a problem of <u>merging</u> them into a single guideway, without loosing time unnecessarily and assuring maximum safety to the passangers. An optimal control solution to the merging problem was proposed by Athans [73], using the same mathematical model of string dynamics as in [66]. Two basic questions were posed by Athans [73]:

(1) Given a set of two strings of vehicles converging through two guideways into a third. A merging sequence of the two strings is also specified.

What should be the control applied to the strings so that the specified merging sequence takes place in a safe and orderly manner?

(2) Is there a <u>best</u>, an <u>optimal</u> (under a specific performance criterion) merging sequence?

The first problem was solved using the linear regulator approach in analogy to the solution in [66]. Athans demonstrated that the matrices G and K, apearing in the control law

$$\underline{u}^0 = -G\underline{x}; \qquad G = R^{-1}B^TK$$

are independent of the merging sequence. A particular merging sequence establishes the structure of the state and control vectors only. The best merging sequence was established by direct enumeration of all possible sequences and comparison of the values of the optimal performance index for each one. It is obvious that for large strings of vehicles this approach may present some computational problems. For instance, if we have n vehicles in one string and m in the other, the total number of merging sequences is [73]:

$$M = \frac{(n+m)!}{n!m!} .$$

If $n = m = 5$, we have $M = 10!/5!5! = 252$ merging possibilities. If we double the number of vehicles in the two strings, $n = m = 10$ (but we still have a modest number of vehicles in each), the total number of possible sequences shoots up to $M = 20!/10!10! = 1,847,560$.

A different approach to the vehicle merging problem was recently proposed by Whitney [74]. The guideways at the approach to the merging junction are subdivided into discrete slots. The presence (or absence) of a vehicle in each slot is indicated by a binary number: "1" for presence, "0" for absence. The problem of establishing merging policies is solved using transition flow diagrams. The model of Whitney is based on the moving cells merging model, proposed by Godfrey [87]. In this model, the vehicles traveling along a guideway are allocated into hypothetical individual cells that move at a constant velocity. The cell's length is chosen to meet spacing requirements and other safety considerations. An optimal control scheme applied to the control of vehicles traveling in the moving cells formation was proposed by Wilkie [79] The dynamic model used in [79] is expressed by the following state equations:

$$\left.\begin{aligned} \dot{z}_1 &= z_2 \\ \dot{z}_2 &= z_3 \\ \dot{z}_3 &= -az_3 + \dot{f}/m \end{aligned}\right\} \tag{119}$$

Where

z_1 = position error

z_2 = velocity error

z_3 = acceleration error

m = mass of the vehicle

f = the control force

$$a = \frac{1}{m}\frac{dg(u)}{du}(u)\bigg|_{u=v}$$

u = speed

v = desired speed; $z_2 = u - v$

$g(u)$ = drag force .

It should be noted that the state equations in Eq. (119) involve the state variables and a control variable associated with a single vehicle. This will eliminate the need of multiple communication links, needed in other models, described previously. The state equations can also be expressed as

$$\underline{\dot{z}} = A\underline{z} + \underline{b}\dot{f} \tag{120}$$

where

$$\underline{z} = [z_1 z_2 z_3]^T ; \quad \underline{b} = [0 \quad 0 \quad 1/m]^T$$

$$A = \begin{bmatrix} 0 & 1 & 0 \\ 0 & 0 & 1 \\ 0 & 0 & -a \end{bmatrix} .$$

The performance index was chosen to be:

$$\begin{aligned} \text{minimize} \quad J &= 1/2 \int_0^\infty (q_{11}z_1^2 + q_{22}z_2^2 + q_{33}z_3^2 + r\dot{f}^2)dt \\ &= 1/2 \int_0^\infty (\underline{z}^T Q\underline{z} + r\dot{f}^2)dt \end{aligned} \tag{121}$$

where

$$Q = \text{diag}[q_{11}, q_{22}, q_{33}] .$$

Again, we have a linear regulator problem, the solution of which is well known:

$$\dot{f}^0 = r^{-1}\underline{b}^T P\underline{z}$$

where P satisfies

$$-PA - A^T P + P\underline{b}r^{-1}\underline{b}^T P - Q = 0 .$$

The numerical solution of a three-state variable — one control variable case is relatively simple. The potential implementation of this control law is much more promising from the economic standpoint in view of the absence of numerous communication links between the vehicles. Details of results of general computational examples can be found in [79]. The merging problem is handled by this approach by formulating a modified performance index [79]:

$$J = 1/2\underline{x}^T F\underline{x} + \int_0^T (\underline{x}^T Q\underline{x} + rf^2)dt$$

where $\underline{x}$ is the actual position-velocity-acceleration state vector and T is a prescribed merging period. This approach has been extended to dual-mode networks by Stefanek and Wilkie [80,81].

The list of references, mentioned in conjunction with the string of moving vehicles problem is by far not exhaustive. Additional references can be found in a comprehensive survey of Fenton [82]. Some basic design considerations, including constraints and performance criteria applicable to automated ground transportation systems were presented by Hajdu and his associates [83]. An alternate solution to the automated highways problem has been developed at the TRW Systems Group in Washington, D. C. area. It is denoted as the Synchronous Longitudinal Guidance (SLG) approach, recently described by Boyd and Lukas [58].

E. ESTIMATION PROBLEMS OF HIGHWAY TRAFFIC

In order to perform effective control of the flow of vehicles on the highway, we need information about the flow rates at various locations, concentration of vehicles and their speeds. Current technology allows us to measure directly the cumulative number of vehicles by presence detection and their speeds by using the effective length of vehicles and the time interval it takes them to pass in front of a detector [30,31,33, 88]. The rest of the data needed are to be estimated. Even the speed measurement is by far not perfect and post-measurement estimation has to be adopted to improve the precision [31,33]. Considerable amount of work in developing and analyzing various estimation techniques, applied to highway traffic, has been performed by transportation professionals [16]. No attempt will be made to cover this area. The main purpose of this section is to discuss work where modern control methods were used.

Sequential estimation has been frequently used by control professionals in many areas of applications. In particular, methods based on the Kalman filter [89,90], received considerable attention in the past ten years [91,92]. The Kalman filter has been applied to a practical problem of estimating traffic densities by Gazis and Knapp [88]. The work included actual computations based on measurements performed in the Lincoln Tunnel of New York City.

Gazis and Knapp considered a section of the roadway located within the interval $x_1 \leq x \leq x_2$. The system is considered at discrete time instants $t_0, t_1, t_2, \ldots, t_i, \ldots,$ uniformly separated. Let n_i denote the total number of vehicles in the section at $t = t_i$ and let n_{1i} and n_{2i} represent the number of cars crossing x_1 and x_2, respectively, within the time interval $t_{i-1} \leq t \leq t_i$. Taking n_i as the variable to be estimated, we have the following system dynamic equation:

$$n_i = n_{i-1} + \Delta n_i + w_i \tag{122}$$

where

$$\Delta n_i = n_{1i} - n_{2i}$$

$$w_i = \text{model error} .$$

The measurement equation is:

$$z_i = n_i + v_i \tag{123}$$

where v_i is the measurement error. The following statistical properties are assumed:

mean values:

$$E[w_i] = E[v_i] = 0$$

covariances:

$$E[w_i w_j] = Q\delta_{ij}$$

$$E[v_i v_j] = R\delta_{ij}$$

$$E[v_i w_j] = 0 \; .$$

An optimal estimate $\hat{n}_i$ was calculated at each sampling time $t = t_i$, minimizing the mean square criterion

$$J_i = E[(n_i - \hat{n}_i)^2] \; . \tag{124}$$

The problem is a standard single dimensional Kalman filter.

The optimal estimate is given by [88]:

$$\hat{n}_i = \hat{n}_{i-1} + \Delta n_i + K_i(z_i - \hat{n}_{i-1} - \Delta n_i) \tag{125}$$

$$K_i = 1 - R(R + Q + J_{i-1})^{-1} \tag{126}$$

$$J_i = K_i R \tag{127}$$

$$\hat{n}_0 = E[n_0] \tag{128}$$

$$J_0 = E\{[n_0 - E(n_0)]^2\} \; . \tag{129}$$

Computations using the Kalman filter in Eqs. (125)-(129) were performed on actual data taken in the Lincoln Tunnel. The estimated results were very close to the results obtained by an exact count of vehicles. In fact, during over 99% of the time, the estimation error was below 10% in all sections of the tunnel [88]. Such accuracy has never before been obtained by other methods of estimation which used flow and speed data [88]. This work has subsequently been extended and additional computations performed by Knapp [93] and Szeto and

Gazis [94].

An extended Kalman-Bucy filtering procedure for estimating the aggregate variables and identification of other highway parameters (ν, T), was proposed by Payne and his associates [30,33]. However, no computational results of recursive estimation were reported in these references.

IV. CONTROL AND SYNCHRONIZATION OF SIGNALIZED URBAN INTERSECTIONS

Optimal control of traffic signals timing at an over-saturated intersection was discussed in Section III.C. The control and synchronization of multiple signalized intersections, in an urban environment, will be discussed in this section. One of the first works, where a systematic optimization technique, was directly applied to the solution of this type of problem, was performed by Little [95]. A mixed-integer linear programming approach was adopted. The model of Little will be described in detail in the following.

A schematic space-time diagram of an urban street with N signalized intersections $S_1, S_2, \ldots, S_N$, is shown in Figure 6. The ordinate, x, represents the distance along the street measured in meters. The abscissa represents time, t, measured in <u>cycles</u> of prescribed length. The zigzag lines represent possible trajectories of vehicles, passing along the street, without stopping at a red light. It will be agreed that the positive x direction will be denoted as direction 1, while the opposite direction — as direction 2. The set of all possible non-stop trajectories form a so called <u>green band</u>, whose horizontal width is denoted as the <u>band width</u> for a particular direction (b_1 and b_2). Although in general, the value of the bandwidth may change along the street, it will be assumed as fixed in this model. The following notation is introduced (see Figure 6):

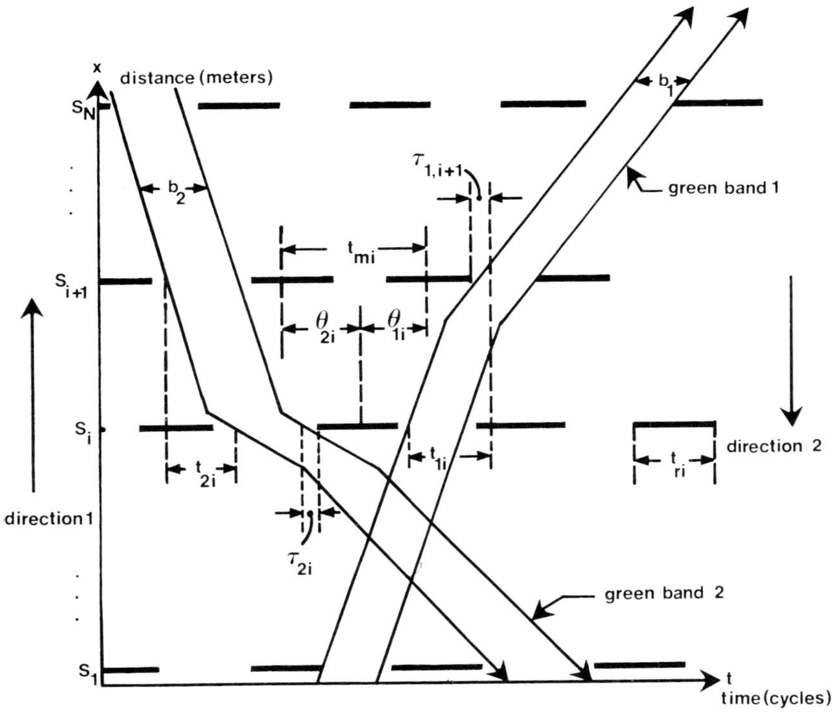

Fig. 6. Signal Band Configuration.

t_{ri} = red time of signal S_i [cycles]

t_{1i} = travel time from S_i to S_{i+1} in direction 1 [cycles]

t_{2i} = travel time from S_{i+1} to S_i in direction 2 [cycles]

θ_{1i} = time from the center of red at S_i to the center of red at S_{i+1}, both immediately to the <u>left</u> of green band 1 [cycles]

θ_{2i} = time from the center of red at S_{i+1} to the center of red at S_i, both immediately to the <u>right</u> of green band 2 [cycles]

τ_{1i} = time from the right side of S_i's red to green band 1 [cycles]

τ_{2i} = time from the left side of S_i's red to green band 2 [cycles]

$t_{mi} = \theta_{1i} + \theta_{2i}$.

The purpose of the optimization procedure is to maximize the bandwidth, b, using τ_{1i}, τ_{2i}, and t_{mi} as design variables. The t_{mi} variables are assumed to be integer. It is also assumed that $b_1 = b_2 = b$, that is, the same bandwidth in both directions. Based on these assumptions and on geometric considerations from Figure 6 the following mixed-integer programming problem is formulated:

maximize b
subject to:

$$\left.\begin{aligned} \tau_{1i} + b &\le 1 - t_{ri} \\ \tau_{2i} + b &\le 1 - t_{ri} \end{aligned}\right\} \qquad i = 1,\ldots,N$$

$$(\tau_{1i} + \tau_{2i}) - (\tau_{1,i+1} + \tau_{2,i+1}) - t_{mi} = (t_{r,i+1} - (t_{1i} + t_{2i}) \qquad i = 1,\ldots,N-1$$

$$\left.\begin{aligned} b;\ \tau_{1i}, \tau_{2i} &\ge 0 \\ t_{mi} &= \text{integer} \end{aligned}\right\} \qquad i = 1,\ldots,N\,.$$

The problem was solved numerically by a branch and bound algorithm [96,97]. The model was also extended to problems where vehicles' speeds and signal periods are also variables, as well as to networks of intersections [95]. A similar problem involving minimization of traffic delay in a signalized network, was solved by Allsop, using a graph-theoretic approach [98].

A very simple and efficient algorithm for traffic signal control of an urban intersection, was proposed by Morris and

Yagoda [55]. The following linear programming problem was formulated for each intersection separately:

$$\text{minimize} \quad z = \sum_{j=1}^{p} t_j$$

subject to:

$$\sum_{j=1}^{p} X_{ij}\, t_j \geq d_i; \qquad i = 1,\ldots,n$$

$$t_j \geq 0\ ; \qquad j = 1,\ldots,p$$

where

t_j = time during which phase j is in effect

p = total number of phases

d_i = demand of flow i (vehicles or pedestrians)

n = total number of flows

$$X_{ij} = \begin{cases} 1; & \text{if vehicular flow } i \text{ is permitted during the jth phase, or if none of the vehicular flows in phase j prohibits movement of pedestrian flow } i. \\ 0; & \text{otherwise} \end{cases}$$

As we can see, it is an easy LP problem. A single computer can control a large number of intersections simultaneously, using this model [55]. A comprehensive discussion of models of traffic networks and basic design considerations can be found in [56].

V. INDIVIDUAL VEHICLE CONTROL PROBLEMS

In the previous sections the vehicles were treated as "points" traveling along highways and urban streets. The main concern was to control the flow of a large number of vehicles, without addressing the problem of control of each individual vehicle. This problem will be touched on in this section. Vehicle control is a large problem and there exists a vast amount of "know-how" with the automobile manufacturers as well as a vast amount of literature published on this subject. However, the main goal of this section is to survey work on the subject where modern control techniques were applied. Several works utilized a frequency domain model to describe the dynamics of an individual automobile [75,99]. A stability analysis of lateral stability characteristics of automobiles, using the Ritz-Galerkin method, was given by Mahig [100]. One of the first works where the maximum principle of Pontriagin was applied to one of the aspects of individual vehicle control, was reported by Kaufman and Larson [101]. In particular, the problem of crashworthiness of a passenger vehicle was considered. The basic dynamic model, adopted by Kaufman and Larson, is expressed by the following state equations:

$$\left.\begin{aligned} \dot{x}_1 &= x_2 \\ \dot{x}_2 &= a \\ \dot{x}_3 &= x_4 \\ \dot{x}_4 &= -\frac{K}{M}(x_3 - x_1) \end{aligned}\right\} \qquad (130)$$

where

x_1 = position of the vehicle
x_2 = velocity of the vehicle
x_3 = position of the passenger
x_4 = velocity of the passenger
a = vehicle acceleration, the forcing function
K = spring constant of the passenger seat belt
M = passenger mass

The initial conditions are (in view of the fact that the control problem starts at the time of impact):

$$x_1(0) = x_3(0) = 0$$
$$x_2(0) = x_4(0) = v_0 .$$

The terminal conditions are

$$x_1(T) = d_c ; \qquad x_2(T) = 0$$

where

d_c = crushable distance or the amount of crumpling permissible in the vehicle front end following a collision

T = duration of the collision.

During the time interval $0 \leq t \leq T$ the purpose of the control action is to minimize the relative dislocation of the passenger with respect to the vehicle. That is, the performance index in phase 1 is:

$$\text{minimize} \quad J_1 = K \int_0^T (x_3 - x_1)^2 dt . \tag{131}$$

In the second phase, only the passenger may be in motion, which can be described by the following set of equations:

$$\begin{aligned} \dot{x}_3 &= x_4 \\ \dot{x}_4 &= - \frac{K}{M} (x_3 - d_c) \end{aligned} \tag{132}$$

with the initial conditions

$$\begin{aligned} x_3(0) &= x_3(T) \\ x_4(0) &= x_4(T) . \end{aligned}$$

This set of equations can be easily solved and as a result we can obtain the peak square displacement of the passenger at $t > T$:

$$h(T) = \frac{M}{K} x_4^2(T) + [x_3(T) - d_c]^2 . \tag{133}$$

Since we are interested in minimizing this term, it is added to the performance index. In addition, a term minimizing the integral of a^2 is added. The final form of the performance index is:

$$\text{minimize} \quad J = wh(T) + \int_0^T [K(x_3 - x_1)^2 + Ra^2]dt \qquad (134)$$

where w and R are weighting factors.

As we can see, Eqs. (130) and (134), define a four-state variables, single control variable, linear dynamics, quadratic cost function, optimal control problem. The approach to the solution of this problem using the minimum principle is a standard procedure [40]. This model was also extended to a case where the restrictive belt force applied to the passenger was considered in two dimensions. This resulted in a non-linear dynamic model [101].

The idea of an electrically driven car became very popular during the past few years in view of the increasing air-pollution awareness of the society. This has not escaped the attention of control professionals. Kokotovic and Singh [102] have applied the maximum principle to the problem of minimum energy control of a vehicle with a dc traction motor.

They used the following model:

$$\begin{aligned} \dot{x} &= v \\ \dot{v} &= -k(v) + \mu i(E - ri)/v \end{aligned} \qquad (135)$$

where

$$\mu = \alpha/b$$

x = distance

v = vehicle speed

k(v) = vehicle resistance force per unit mass

E = voltage applied to the dc motor

b = back emf constant

i = armature current

r = armature circuit resistance

α = a constant

x, v are the state and i is the control variable, which is constrained

$$|i| \leq I . \tag{136}$$

The performance index of minimum energy is formulated as follows:

$$\text{minimize} \quad J = \int_{t_0}^{t_f} Eidt . \tag{137}$$

A complete numerical solution to the problem was given in [102], using the maximum principle approach. A dynamic programming solution of a minimum energy control problem of an electrically driven vehicle, taking into account stochastic disturbances, was reported by Sahinkaya and Sridhar [103]. A Wiener-Hopf approach to active control of vehicle air cushion suspendions was recently proposed by Hullender et. al. [104]. A detailed analysis of a magnetically levitated high speed ground vehicle was reported by Wilkie [105].

VI. TRANSPORTATION PLANNING

The notion of transportation planning is an extremely wide one and certainly deserves an article dedicated just to that subject. Only a few general remarks will be made here. Optimization techniques can be (and should be) used in any planning operation, and transportation planning is no exception. We can find uses of linear and nonlinear programming even in textbooks in this area [106,107]. A dynamic programming approach for planning regional road investments was recently proposed by Chapman [108]. A linear programming solution of a combined traffic distribution-assignment problem, was reported by Tomlin [109]. A nonlinear programming approach to the precise determination of equilibrium in travel forecasting problems, was recently proposed by Wilkie and Stefanek

[110]. A general discussion of transportation planning in connection with the Northeast Corridor Transportation Project, was given by Shuldiner and Nutter [111]. Several alternatives for public transporation planning were given in [112-114].

VII. CONCLUDING REMARKS

A survey of various types of transportation problems, where modern control and optimization practices have been applied, was given in this survey article. No attempt has been made to cover the transportation literature, a vast quantity of which exists. However, some basic references were given, which should lead the interested reader to many others. A particular effort has been made to include work published by professionals specializing in the control field. It is believed that most of the pertinent literature in this category has been identified and referred to in this article.

Perhaps, in many cases discussed in this article, the author has been more brief than the reader may want him to be. It should be remembered that this is basically a survey article with limited space, rather than a textbook and therefore, many details had to be omitted. Sufficient references are given though, which will permit each reader to look up the details that he may be interested in.

There is a very important aspect of transportation systems analysis that has not been touched in this article, namely: Simulation of transportation systems, although quite a number of different models have been presented. An interesting discussion on modeling of transportation systems was recently given by Zames and Kovatch [115]. A significant amount of work in the area of simulation of transportation systems has been performed and it could serve as a subject of a separate survey article. As stated in the introduction, another future subject for a survey article is application of modern control and optimization techniques in Air Traffic Control. To the best of the author's knowledge, there are very few publications of this type. For the benefit of the reader, we will mention a few [116,117].

ACKNOWLEDGMENTS

During the writing of the article, the author was supported by Rensselaer Polytechnic Institute of Connecticut, Hartford, Connecticut and The Travelers Insurance Company, Hartford, Connecticut, with which the author spent the summer of 1972 on a corporate fellowship.

The author appreciates the courtesy of Professors J. B. Kreer (Michigan State University) and J. A. Wattleworth (University of Florida) as well as their publishers Pergamon Press and The Highway Research Board for kindly permitting him to reproduce computational and graphical results from their publications. Valuable information was transmitted to the author by Professor H. J. Payne (University of Southern California).

The drawings in the article have been performed by Mr. J. W. Ashmore (Travelers).

LIST OF SYMBOLS

a_{kj} = decimal fraction of vehicles entering at on-ramp k which pass through partition j of the highway

a_k = $\Delta t/D_k$

b = constant of integration

b_j = capacity at the highway partition j

b_1, b_2 = bandwidth (IV)

c = concentration of vehicles [vehicles/mile]

c_j = jam concentration

c_n = nominal concentration

C = maximum capacity

d_k = hourly demand at on-ramp k;
 = deviations from desired position in the string of moving vehicles

D_k = length of section k [miles]

e_k = sum of waiting times of vehicles at on-ramp k

$\underline{e}$ = density perturbation vector

f = distribution function of vehicles

f_0 = desired speed distribution

f_k = fraction of q_k leaving through the off-ramp in section k + 1;
 = force deviation in the string of moving vehicles

g_k = off-ramp flow rate in section k [vehicles/hour]

g_1, g_2 = effective green light phases at an intersection

i = discrete time index

j = $\sqrt{-1}$

k = highway sections index

K = total number of highway sections considered

l_k = number of lanes in section k

L = total lost time for acceleration and clearing at an intersection

m_k = additional control exerted at on-ramp k
 = mass of kth vehicle

M = total number of highway partitions in the LP model

N_p = total number of sampling periods considered

p = probability of a car passing another;

$\underline{P}$ = costate vector

q = flow of vehicles [vehicles/hour]

q_1, q_2 = arrival flow rates at an intersection

r_k = on-ramp flow rate in section k [vehicles/hour]

s = spacing

s_j = jam spacing

s_1, s_2 = saturation flows at an intersection

T = relaxation time;

= driver reaction time lag;

T_i = discrete time interval = $t_i - t_{i-1}$

$_c$ = cycle time at an intersection

u = speed of individual vehicles [miles/hour]

$\bar{u}$ = average speed

u_k = speed deviation of kth vehicle in the string

v = space-mean speed of vehicles

v_f = free speed

V_i = sub-flow volume at an intersection

w = weighting factor

$\underline{w}$ = on-ramp rates perturbation vector

x = distance; space coordinates

$y_k = c_k - c_n$ = excess density in section k

α_k = priority parameter for on-ramp k in nominal conditions;

= partial derivative of drag force with respect to velocity for vehicle k in a string

γ = parameter of $G(jw)$ in asymptotic approximation (III.D)

η_k = queue length at the on-ramp in section k

λ = sensitivity coefficient

λ_k = eigenvalue of matrix A (III.D)

ν = sensitivity factor

ρ_k = eigenvalues of matrix C (III.D)

τ = parameter of $G(jw)$ in (III.D)

ω = angular frequency [rad/sec]

REFERENCES

1. D. C. GAZIS and L. C. EDIE, "Traffic Flow Theory," Proce. IEEE, 56, pp. 458-471, April 1968.
2. I. PRIGOGINE and R. HERMAN, "Kinetic Theory of Vehicular Traffic," American Elsevier, N. Y., 1971.
3. M. J. LIGHTHILL and G. B. WHITHAM, "On Kinematic Waves, II. A. Theory of Traffic Flow on Long Crowded Roads," Proc. Roy. Soc. (London), 229A, pp. 317-345, May 1955.
4. P. I. RICHARDS, "Shock Waves on the Highway," Operations Research, 4, pp. 42-51, 1956.
5. F. Haight, "Mathematical Theories of Traffic Flow," Academic Press, N. Y., 1963.
6. D. R. DREW, "Traffic Flow Theory and Control," McGraw Hill, N. Y., 1968.
7. I. PRIGOGINE, "A Boltzmann-like Approach to the Statistical Theory of Traffic Flow," in Theory of Traffic Flow, R. Herman, ed., pp. 158-164, Elsevier, Amsterdam, 1961.
8. F. P. PREPARATA, "Analysis of Traffic Flow on a Signalized One-Way Artery," Transportation Science, 6, pp. 32-51, Feb. 1972.
9. R. E. FRANKLIN, "On the Flow-Concentration Relationship for Traffic," Proc. 2nd International Symposium on the Theory of Traffic Flow, London, 1963.
10. R. HERMAN, E. W. MONTROLL, R. B. POTTS, and R. ROTHERY, "Traffic Dynamics: Analysis of Stability in Car-Following," Operations Research, 7, pp. 86-106, 1959.
11. D. C. GAZIS, R. HERMAN and R. B. POTTS, "Car-Following Theory of Steady-State Traffic Flow," Operations Research, 7, pp. 499-505, 1959.
12. D. C. GAZIS, R. HERMAN and R. W. ROTHERY, "Nonlinear Follow the Leader Models of Traffic Flow," Operations

Research, 9, pp. 545-567, 1961.

13. R. HERMAN, and R. B. POTTS, "Single-Lane Traffic Theory and Experiment," in Theory of Traffic Flow, R. Herman, ed., pp. 120-146, Elsevier, Amsterdam, 1961.

14. D. C. GAZIS, R. HERMAN and R. W. ROTHERY, "Analytical Methods in Transportation: Mathematical Car-Following Theory of Traffic Flow," Proc. ASCE, J. Eng. Mechanics Div., 89, pp. 29-46, Dec. 1963.

15. R. HERMAN, "Theoretical Research and Experimental Studies in Vehicular Traffic," Proc. Australian Road Research Board, 3, pp. 26-56, 1966.

16. J. A. WATTLEWORTH, "Estimation of Demand at Freeway Bottlenecks," Traffic Engineering, 35, pp. 21-26, February, 1965.

17. J. A. WATTLEWORTH, and D. S. BERRY, "Peak-Period Analysis and Control of a Freeway System-Some Theoretical Investigations," Highway Research Record, No. 89, pp. 1-25, 1965.

18. J. A. WATTLEWORTH, "Peak-Period Analysis and Control of a Freeway System," Highway Research Record, No. 157, pp. 1-21, 1967.

19. G. HADLEY, "Linear Programming," Addison-Wesley, Reading, Mass., 1962.

20. S. I. GASS, "Linear Programming," 3rd edition, McGraw Hill, N. Y., 1969.

21. C. I. CHEN, J. B. CRUZ, and J. G. PAQUET, "Expressway Entrance Ramp Control for Flow Maximization," 1972 Princeton Conference of Information Sciences and Systems, Princeton, N. J., March 1972.

22. L. S. YUAN, and J. B. KREER, "An Optimal Control Algorithm for Ramp Metering of Urban Freeways," Proc. 6th Annual Allerton Conference on Circuit and System Theory, Allerton, Illinois, October, 1968.

23. L. S. YUAN, and J. B. KREER, "Adjustment of Freeway Ramp Metering Rates to Balance Entrance Ramp Queues," Transportation Research, 5, pp. 127-133, 1971.

24. A. KAYA, "On the Optimization of Traffic Flow for Urban Transportation Systems," 1971 JACC, pp. 884-892, St. Louis, Mo., August 1971.

25. A. KAYA, "Computer and Optimization Techniques for Efficient Utilization of Urban Freeway Systems," 1972 IFAC Conference, Paris, France, June 1972.

26. A. D. MAY, "Improving Network Operations with Freeway Ramp Control," 43rd Annual Meeting of the Highway Research Board, Washington, D. C., January 1964.

27. M. FRANK, and P. WOLFE, "An Algorithm for Quadratic Programming," Naval Research Logistics Quarterly, 3, pp. 95-110, 1956.

28. G. HADLEY, "Nonlinear and Dynamic Programming," Addison Wesley, Reading, Mass., 1964.

29. H. J. PAYNE, "Models of Freeway Traffic and Control," 1970 National Invitational Seminar on Advanced Simulation, San Diego, California, Sept. 24-26, 1970, and in Simulation Council Proceedings, Mathematical Models of Public Systems, 1, No. 1, 1971.

30. H. J. PAYNE, "Systems Problems in Freeway Traffic Control and Surveillance," Joint ASCE-ASME Transportation Engineering Meeting, Seattle, Washington, July 26-30, 1971. (Paper No. 1511).

31. L. ISAKSEN, "Suboptimal Control of Large-Scale Systems with Application to Freeway Regulation," Ph.D. thesis, Dept. of Electrical Eng. University of Southern California, Los Angeles, California, 1971.

32. H. J. PAYNE, and L. ISAKSEN, "Suboptimal Control of Large Scale Systems with Application to Freeway Traffic

Regulation," 4th Hawaii International Conference on System Sciences, pp. 332-334, Honolulu, Hawaii, Jan. 1971.

33. W. THOMPSON, H. J. PAYNE, and L. ISAKSEN, "Design of a Traffic Responsive Control System for a Los Angeles Freeway," Proc. 1972 International Conference on Cybernetics and Society, IEEE Systems, Man and Cybernetics Society, pp. 461-468, Washington, D. C., October, 1972.

34. L. ISAKSEN, and H. J. PAYNE, "Simulation of Freeway Traffic Control Systems," Simulation Council Proceedings, The Mathematics of Large-Scale Simulation, Paul Brock, ed., 2, No 1, 1972.

35. D. TABAK, "Real-Time Computer Control of Highway Traffic," 1971 IEEE Conference on Decision and Control, pp. 484-485, Miami, Florida, December, 1971.

36. D. TABAK, "A Linear Programming Model of Highway Traffic Control," 6th Annual Princeton Conference on Information Sciences and Systems, Princeton, N. J., March, 1972.

37. J. HUERTA, "Real-Time Ramp Control by Linear Programming using the Aggregate Model," M. S. project in Systems Engineering, RPI of Connecticut, Hartford, Connecticut, September, 1972.

38. J. HUERTA, and D. TABAK, "Simulation of a Linear Programming Ramp Traffic Computer Procedure," 5th IFIP Conference on Optimization Techniques, Rome, Italy, May, 1973.

39. M. ATHANS, and P. L. FALB, "Optimal Control," McGraw Hill, N. Y., 1966.

40. D. E. KIRK, "Optimal Control Theory; An Introduction," Prentice-Hall, Englewood Cliffs, New Jersey, 1970.

41. B. D. O. ANDERSON, and J. B. MOORE, "Linear Optimal Control," Prentice-Hall, Englewood Cliffs, New Jersey, 1971.

42. D. R. DREW, K. A. BREWER, J. H. BUHR, and R. H. WHITSON, "Multilevel Approach to the Design of a Freeway Control System," Highway Research Record, No. 279, pp. 40-55, 1969.

43. K. A. BREWER, J. H. BUHR, D. R. DREW, and C. J. MESSER, "Ramp Capacity and Service Volume as Related to Freeway Control," Highway Research Record, No. 279, pp. 70-86, 1969.

44. M. D. MESAROVIC, D. MACKO, and Y. TAKAHARA, "Theory of Multi-Level Hierarchical Systems," Academic Press, New York, 1970.

45. E. F. GERVAIS, "Optimization of Freeway Traffic by Ramp Control," Highway Research Record, No. 59, 1964.

46. H. N. YAGODA, "The Dynamic Control of Automotive Traffic at a Freeway Entrance Ramp," Automatica, 6, pp. 385-393, May, 1970.

47. L. SHAW, "On Optimal Ramp Control of Traffic Jam Queues," 1971 IEEE Conference on Decision and Control, pp. 479-483, Miami Beach, Florida, December, 1971.

48. C. F. WANG, "On a Ramp-Flow Assignment Problem," Transportation Science, 6, pp. 114-130, May, 1972.

49. H. M. SENDAULA, and C. H. KNAPP, "Optimal Control of Hyperbolic Distributed Parameter Systems with Applications to Traffic Bottlenecks," 1972 IEEE SMC Conference, pp. 473-474, Washington, D. C., October 1972.

50. D. C. GAZIS, "Traffic Control: From Hand Signals to Computers," Proc. IEEE, 59, pp. 1090-1099, July, 1971.

51. S. CASS, and L. CASCIATO, "Centralized Traffic Signal Control by a General Purpose Computer," Proc. Inst. Traffic Eng., pp. 203-211, 1960.

52. D. C. GAZIS, and R. S. FOOTE, "Surveillance and Control of Tunnel Traffic by an On-Line Digital Computer,"

Transportation Science, 3, pp. 255-275, August, 1969.

53. B. C. BLACK, and D. C. GAZIS, "Real-Time Traffic Flow Optimization," IBM Systems J., 10, No. 3, pp. 217-231, 1971.

54. D. W. ROSS, R. C. SANDYS, and J. L. SCHLAEFLI, "A Computer Control Scheme for Critical Intersection Control in an Urban Network," Transportation Science, 5, pp. 141-160, May, 1971.

55. A. A. MORRIS, and H. N. YAGODA, "A New Design Procedure for Traffic Signal Control Policies," IEEE Trans. on Systems Science and Cybernetics, SSC-6, pp. 330-336, October, 1970.

56. W. R. MCSHANE, H. N. YAGODA, L. J. PIGNATARO, and K. W. CROWLEY, "Control Considerations and Smooth Flow in Vehicular Traffic Nets," Highway Research Record, No. 334, pp. 8-22, 1970.

57. K. W. DODGE, and R. A. REISS, "Surveillance, Control and Simulation for a System of Parallel Roadways," Proc. 1971 IEEE Systems, Man and Cybernetics Conference, pp. 219-224, Anaheim, California, October, 1971.

58. R. K. BOYD, and M. P. LUKAS, "How to Run an Automated Transportation System," IEEE Trans. on Systems, Man and Cybernetics, SMC-2, pp. 331-341, July, 1972.

59. D. C. GAZIS, and R. B. POTTS, "The Over-Saturated Intersection," Proc. 2nd. International Symposium on the Theory of Traffic Flow, pp. 221-237, London, 1963.

60. D. C. GAZIS, "Optimum Control of a System of Oversaturated Intersections," Operations Research, 12, pp. 815-831, 1964.

61. D. C. GAZIS, "Control Problems in Automobile Traffic," IBM Scientific Computing Symposium: Control Theory and Applications, pp. 171-185, T. J. Watson Research Center, Yorktown Heights, New York, October, 1964.

62. L. S. PONTRIAGIN, V. G. BOLTIANSKII, R. V. GAMKRELIDZE, and E. F. MISHCHENKO, "The Mathematical Theory of Optimal Processes," Wiley, New York, 1962.

63. D. C. GAZIS, "Spillback from an Exit Ramp of an Expressway," Highway Research Record, No. 89, pp. 39-46, 1965.

64. A. MARTIN-LOF, "Computation of an Optimal Control for a Signalized Traffic Intersection," Transportation Science, 1, pp. 1-5, February, 1967.

65. J. A. WATTLEWORTH, "A Capacity Analysis Technique for Highway Junctions," Traffic Engineering, June, 1972, pp. 30-36; 68-69.

66. W. S. LEVINE, and M. ATHANS, "On the Optimal Error Regulation of a String of Moving Vehicles," IEEE Trans. on Automatic Control, AC-11, pp. 355-361, July, 1966.

67. J. D. PEARSON, "On Controlling a String of Moving Vehicles," IEEE Trans. on Automatic Control, AC-12, pp. 328-329, June, 1967.

68. M. MASAK, and J. D. PEARSON, "Comment on Controlling a String of Moving Vehicles," IEEE Trans. on Automatic Control, AC-13, pp. 113-114, February, 1968.

69. A. H. LEVIS, and M. ATHANS, "On the Optimal Sampled Data Control of Strings of Vehicles," Transportation Science, 2, pp. 362-382, November, 1968.

70. L. E. PEPPARD, and V. GOURISHANKAR, "Optimal Control of a String of Moving Vehicles," IEEE Trans. on Automatic Control, AC-15, pp. 386-387, June, 1970.

71. S. M. MELZER, and B. C. KUO, "The Optimal Regulation of a String of Moving Vehicles Through Difference Equations," 1970 JACC, pp. 175-180, Atalanta, Georgia, June 22-26, 1970.

72. S. M. MELZER, and B. C. KUO, "A Closed-Form Solution for the Optimal Error Regulation of a String of Moving

Vehicles," IEEE Trans. on Automatic Control, AC-16, pp. 50-52, February, 1971.

73. M. ATHANS, "A Unified Approach to the Vehicle-Merging Problem," Transportation Research, 3, pp. 123-133, April, 1969.

74. D. E. WHITNEY, "A Finite State Approach to Vehicle Merging," Trans. ASME, J. of Dynamic Systems, Measurement and Control, Series G, 94, pp. 147-151, June, 1972.

75. R. L. COSGRIFF, "The Asymptotic Approach to Traffic Dynamics," IEEE Trans. on Systems Science and Cybernetics, SSC-5, pp. 361-368, October, 1969.

76. J. G. BENDER, and R. E. FENTON, "On the Flow Capacity of Automated Highways," Transportation Science, 4, pp. 52-63, February, 1970.

77. B. PORTER, and T. R. CROSSLEY, "Modal Control of Cascaded-Vehicle Systems," Int. J. Systems Science, 1, pp. 111-122, 1970.

78. T. R. CROSSLEY, and B. PORTER, "Direct Digital Modal Control of Cascaded-Vehicle Systems," Int. J. Systems Science, 1, pp. 323-329, 1971.

79. D. F. WILKIE, "A Moving Cell Control Scheme for Automated Transportation Systems," Transportation Science, 4, pp. 347-364, November, 1970.

80. R. G. STEFANEK, and D. F. WILKIE, "Control Aspects of a Dual-Mode Transportation System," 1971 JACC, St. Louis, Missouri, August 11-13, 1971.

81. D. F. WILKIE, "Where Does the Control Theorist Fit into the Transportation Picture?" 1970 IEEE Symposium on Adaptive Processes: Decision and Control, University of Texas at Austin, December 7-9, 1970.

82. R. E. FENTON, "Automatic Vehicle Guidance and Control-A State of the Art Survey," IEEE Trans. on Vehicular

Technology, VT-19, pp. 153-161, February, 1970.

83. L. P. HAJDU, K. W. GARDINER, H. TAMURA, and G. L. PRESSMAN, "Design and Control Considerations for Automated Ground Transportation Systems," Proc. IEEE, 56, pp. 493-513, April, 1968.

84. J. G. BENDER, and R. E. FENTON, "A Study of Automatic Car Following," IEEE Trans. on Vehicular Technology, VT-18, pp. 134-140, November, 1969.

85. B. PORTER, "Synthesis of Dynamical Systems," Nelson, London, 1969.

86. H. CRAMER, "Mathematical Methods of Statistics," Princeton University Press, Princeton, N. J., 1946.

87. M. B. GODFREY, "Merging in Automated Transportation Systems," Ph.D. thesis, Dept. of Civil Engineering, M.I.T., June, 1968.

88. D. C. GAZIS, and C. H. KNAPP, "On-Line Estimation of Traffic Densities from Time-Series of Flow and Speed Data," Transportation Science, 5, pp. 283-301, August, 1971.

89. R. E. KALMAN, "A New Approach to Linear Filtering and Prediction Problems," ASME Trans., Series D, J. Basic Eng., 82, pp. 34-45, 1960.

90. H. W. SORENSON, "Kalman Filtering Techniques," Advances in Control Systems, C. T. Leondes, ed., Vol. 3, pp. 219-292, Academic Press, N. Y., 1966.

91. A. P. SAGE, and J. L. MELSA, "Estimation Theory with Applications to Communications and Control," McGraw Hill, N. Y., 1971.

92. A. H. JAZWINSKI, "Stochastic Processes and Filtering Theory," Academic Press, N. Y., 1970.

93. C. H. KNAPP, "Traffic Estimation and Control at Bottlenecks," 1972 IEEE Systems, Man and Cybernetics Conference,

pp. 469-472, Washington, D. C., October 9-12, 1972.

94. M. W. SZETO, and D. C. GAZIS, "Surveillance and Control of Traffic Systems by Application of Kalman Filtering Techniques," IBM Research Report, RC 3690, January 17, 1972.

95. J. D. C. LITTLE, "The Synchronization of Traffic Signals by Mixed-Integer Linear Programming," Operations Research, 14, pp. 568-594, 1966.

96. D. R. PLANE, and C. MCMILLAN, "Discrete Optimization," Prentice-Hall, Englewood-Cliffs, N. J., 1971.

97. H. GREENBERG, "Integer Programming," Academic Press, N. Y., 1971.

98. R. E. ALLSOP, "Selection of Offsets to Minimize Delay to Traffic in a Network Controlled by Fixed-Time Signals," Transportation Science, 2, pp. 1-13, February, 1968.

99. D. H. WEIR, and D. T. MCRUER, "Dynamics of Driver Vehicle Steering Control," Automatica, 6, pp. 87-98, January, 1970.

100. J. MAHIG, "Method for Determining Effects of Oscillation Amplitude on Lateral Stability Characteristics of Automobiles," 1970 JACC, pp. 160-168, Atlanta, Georgia, June 22-26, 1970.

101. H. KAUFMAN, and D. B. LARSON, "Application of Optimal Control Theory to the Crashworthiness of a Passenger Vehicle Model," IEEE Trans. on Systems Science and Cybernetics, SSC-5, pp. 251-256, July, 1969.

102. P. KOKOTOVIC, and G. SINGH, "Minimum-Energy Control of a Traction Motor," IEEE Trans. on Automatic Control, AC-17, pp. 92-95, February, 1972.

103. Y. E. SAHINKAYA, and R. SRIDHAR, "Minimum-Energy Control of a Class of Electrically Driven Vehicles," IEEE Trans. On Automatic Control, AC-17 pp. 1-6, February, 1972.

104. D. A. HULLENDER, D. N. WORMLEY, and H. H. RICHARDSON, "Active Control of Vehicle Air Cushion Suspensions," ASME Trans., J. of Dynamic Systems, Measurement and Control, 94, pp. 41-49, March, 1972.

105. D. F. WILKIE, "Dynamics, Control and Ride Quality of a Magnetically Levitated High Speed Ground Vehicle," Transportation Research (to appear).

106. M. WOHL, and B. V. MARTIN, "Traffic System Analysis for Engineers and Planners," Chapter 6, McGraw Hill, N. Y., 1967.

107. W. R. BLUNDEN, "The Land-Use/Transport System," Pergamon Press, Oxford, England, 1971.

108. L. D. CHAPMAN, "A Planning Method for Regional Road Investments," 1971 IEEE Systems, Man and Cybernetics Society Conf., pp. 213-218, Anaheim, California, October 25-27, 1971.

109. J. A. TOMLIN, "A Mathematical Programming Model for the Combined Distribution-Assignment of Traffic," Transportation Science, 5, pp. 122-140, May, 1971.

110. D. F. WILKIE, and R. G. STEFANEK, "Precise Determination of Equilibrium in Travel Forecasting Problems using Numerical Optimization Techniques," Highway Research Record, No. 369, pp. 239-252, 1971.

111. P. W. SHULDINER, and R. D. NUTTER, "Strategy and a Tactic for Generation of Transportation Alternatives," IEEE Trans. on Systems Science and Cybernetics, SSC-5, pp. 257-260, July, 1969.

112. K. W. HOLLANDER, "A Multimode Hybrid Vehicle in an Urban Network for Integrated Traffic," 1972 IEEE Systems, Man and Cybernetics Conf., pp. 447-448, Washington, D. C., October 9-12, 1972.

113. W. RAITHEL, "The Morgantown PRT System and its Automation,"

1972 IEEE Systems, Man and Cybernetics Conf., pp. 449-450, Washington, D. C., October 9-12, 1972.

114. A. VATSKY, and D. TABAK, "PERBUS-A Mass Transit System Alternative," 1972 IEEE Systems, Man and Cybernetics Conf., pp. 451-452, Washington, D.C., October 9-12, 1972.

115. G. ZAMES, and G. KOVATCH, "Modeling Transportation Systems: An Overview," 1971 IEEE Conf. on Decision and Control, pp. 294-304, Miami Beach, Florida, December 15-17, 1971.

116. M. ATHANS, "System Aspects of Air Traffic Control," 1971 Hawaii Conference, January, 1971.

117. D. K. SCHMIDT, and R. L. SWAIM, "An Optimal Control Approach to Terminal Area Air Traffic Control," 1972 JACC, Stanford, California, August 16-18, 1972.

Integrated System Identification and Optimization

YACOV Y. HAIMES
Systems Research Center,
Case Western Reserve University,
Cleveland, Ohio

I. INTRODUCTION

The modern systems approach in handling large scale problems includes the concepts of system identification and optimization. The coupling relationship between these two concepts is inherent in the nature of the desired "optimal solution." Any mathematical model consists of unknown variables and "known" parameters characterizing the system. In general, these parameters are not known exactly, but rather are estimated or determined under non-optimal conditions.

Accordingly, the solution generated from such system models may be non-optimal.

Clearly, the identification of the system's parameters, often referred to as system modeling, is essential in order to obtain an optimal control policy. Unfortunately, system identification and system optimization are generally treated as two separate problems in the literature, and hence the co-ordination between the above two concepts is often overlooked.

In the following sections the coupling relationship between these two concepts is investigated and analytical tools and methods for tackling the joint problem are introduced.

A. Modeling, Identification and Parameter Estimation

Mathematical models, which aim at representing real physical systems in quantitative form, have become important tools in the design, synthesis, analysis, operation and control of complex systems. A mathematical model is a set of relations, often equations, that describes and represents the real system. This set of equations uncovers the various aspects of the real system, establishes measures of effectiveness and constraints and indicates what data should be collected in order to deal with the problem quantitatively.

The nature of the physical system under consideration determines which class of mathematical models will closely represent it. The following classifications of mathematical models may be considered:

(a) Linear and nonlinear models
(b) Static and dynamic models
(c) Deterministic and stochastic models
(d) Lumped parameter and distributed parameter models.

If the response of both the real system and the mathematical model to the same signal input is identical (ideally), then the mathematical model simulating the system is considered to be indeed "perfect." In general, however, these two responses are not identical, and an error exists. Thus the purpose in system modeling is to construct a mathematical model such that the above difference error is minimized.

System parameters of many processes are unknown or only approximately known in the design or operational stages. Often they are empirically obtained and thus induce a great deal of uncertainty to the system model. Furthermore, system parameters can be time varying and may depend on the system's state. In this case a "truly" optimal control policy could be obtained only when these parameters are exactly known at each instant of time. The dynamic behavior of these systems is generally described by sets of linear or nonlinear ordinary differential equations (O.D.E.).

The system identification problem may be defined as: given a set of inputs and corresponding outputs from a system, find a set of parameters which minimize a norm of errors between the mathematical model and the system.

Thus system identification is a procedure to determine a mathematical model which represents a physical system. Such important factors as the kind and degree of equations, the computational complexity, and the data available determine how well the model approximates the physical system.

Often the term "system identication" refers to the determination of a process or system topology; that is the type and form of the best representative model relations or equations. The term parameter estimation on the other hand refers to the determination of the parameter values of the process assuming the topology is known. Systems modeling in general includes both system identification and parameter estimation.

The following problems arise in system identification of dynamic models represented by differential equations. The form of the differential equations are known, that is the order and the degree of the differential equations are known, however, the coefficients (parameters) and/or the initial conditions are unknown. More generally, the form, the order, the coefficients and/or the initial conditions of the differential equations are unknown. In the following section, O.D.E. with unknown coefficients and unknown initial conditions are considered. For static system models, the O.D.E. with the proper initial conditions will degenerate to a set of algebraic equations.

Extensive research has been conducted in the field of system identification in the last decade. Astrom and Eykhoff [1] give extensive references in the field and also discuss the basic concepts of system modeling for various classes of processes. Cuenod and Sage [2] in a survey paper present a discussion and comparison of various numerical techniques used for process identification. Balakrishnan and Peterka [3] in a survey paper discuss the development of the field of system identification from 1966 to 1969 and present an extensive literature survey. Bellman and Kalaba [4] contributed much to the theory and computational aspects of system identification, especially in developing the applications of quasilinearization to system identification. Bellman and Astrom [5] introduce a new concept of structural identifiability where certain identifiable structures are discussed.

B. Joint System Identification and Optimization

The optimal control is determined from the mathematical model and is only optimal for a system represented by that model. Thus the optimal control is only as good as the mathematical model. In reality an "optimal solution" to a problem often involves questions relating to the cost of modeling the system as well as the benefit received from optimal control. If the cost of additional improvement in the mathematical model is not compensated by an additional benefit from the control over a suitable time period, then the solution to the combined problem of identification and optimization can hardly be considered optimal. Donoghue and Lefkowitz [6] considered the cost of additional improvement in the mathematical model as part of the optimization problem.

Despite the great degree of interaction, problems of system modeling and system optimization are generally treated separately in the literature. This chapter will address itself to the modeling problem where the form and order of the model are given and the model parameters are to be determined. In general these parameters may be time-varying and the model will usually be represented by a set of differential equations. For purposes of terminology, this problem is called one of dynamic identification and dynamic optimization. While the general solution of this latter problem is at present unknown, however, some important special cases are also of interest.

Joint problems can be classified as follows:

(a) Static identification and static optimization,

(b) Static identification and dynamic optimization,

(c) Dynamic identification and static optimization,

(d) Dynamic identification and dynamic optimization.

In the above classification, the problem is termed as static identification when the parameters are constants, otherwise it is termed dynamic identification. Similarly, the problem is termed static optimization when the state variables are constant, otherwise the problem is termed dynamic optimization. Static identification and static optimization fall more into the domain of mathematical programming rather than optimal control since the system model is represented by algebraic or transcendental equations rather than by differential equations. The remaining special case of dynamic identification-static optimization is of little practical interest and is not considered here. In the following sections theoretical analysis as well as computational procedures for solving the static and dynamic problems are developed and demonstrated by example problems.

C. Example Problem

There are many important physical problems which lend themselves to the joint system identification and optimization formulation. Consider for example the problem of optimal conjunctive use of ground and surface water, (Haimes [7,8]). Extensive research has been devoted to this problem. About 60 percent of the United States' supply of water comes from ground water sources. Aquifers are capable of storing most of the cumulative excess water runoff or of being recharged with water from other sources for use in water shortage periods. This eliminates the need for constructing expensive large dams and reservoirs. Hence, an optimal management of the conjunctive use of ground water and surface water facilities is desired. Once the physical properties and characteristics of an aquifer are known, it is possible to apply appropriate physical laws and to predict the response of the aquifer system to demands placed on it. Without this "identification," prediction is impossible.

Key to the optimal management of this resource, for any of the multiple beneficial uses, is an ability to predict with reasonable accuracy the response of the system to decisions affecting recharge and withdrawals. This in turn is dependent upon the values of the important parameters such as permeability, specific yield, compressibility, transmissivity, etc. as these are distributed over the aquifer formations in space and time. The identification of aquifer system parameters, e.g. transmissivity and storage functions, as part of system modeling, has traditionally been considered and treated separately from the optimization procedure. However, since the unknown aquifer parameters are used in determining the optimal decision variables, and part of these parameters, i.e. the transmissivity is a nonlinear function of the water level which in turn is a function of the decision variables, recharge and withdrawals, a joint solution is desirable.

II. PROBLEM FORMULATION

Generally, mathematical programming problems are posed in terms of a single unknown vector, say $\underline{X}$. However, in this chapter the concepts of system state variables $\underline{X}$ and control variables $\underline{U}$ will be adopted for the static system models [7]. This notation simplifies the analogy between the static and dynamic system problems to be developed later.

A. The Static Identification and Static Optimization Problem

Consider the problem:

$$\min_{\underline{U}} f(\underline{X},\underline{U},\underline{\alpha}) \tag{1}$$

subject to the constraints

$$\begin{aligned} g_i(\underline{X},\underline{U},\underline{\alpha}) &= 0 \quad i = 1,2,\ldots,n \\ g_i(\underline{X},\underline{U},\underline{\alpha}) &\leq 0 \quad i = n+1,\ldots,n+r \end{aligned} \tag{2}$$

where:

$\underline{X}$ is an n-dimensional state vector

$\underline{U}$ is an l-dimensional control vector

$\underline{\alpha}$ is a k-dimensional vector of unknown parameters

f, g_i are functions of class C^2.

If $(\underline{X}^\circ, \underline{U}^\circ)$ solves (2) and all $\nabla g_i(\underline{X}^\circ, \underline{U}^\circ, \underline{\alpha})$ are linearly independent, then the Implicit Function Theorem guarantees existence of solution $\underline{X}(\underline{U})$ for all $\underline{U}$ near $\underline{U}^\circ$. With this justification, (2) is now rewritten as

$$g_i(\underline{X}, \underline{U}, \underline{\alpha}) \leq 0 \qquad i = 1,2,\ldots,m \qquad (3)$$

where, $m = 2n + r$. Note that each equality constraint $g_i = 0$ is equivalently replaced by two inequality constraints $g_i \leq 0$ and $g_i \geq 0$. The p-dimensional vector $\underline{Y}$ represents the system output given by

$$\underline{Y} = \underline{H}(\underline{X}, \underline{U}, \underline{\alpha}) \qquad (4)$$

where $\underline{H}$ is a p-dimensional vector of functions of class C^2. It is assumed that J observations of the output denoted by

$$\hat{\underline{Y}}^j, \qquad j = 1,2,\ldots,J$$

are made.

The parameter identification problem is

$$\min_{\underline{\alpha}} \sum_{j=1}^{J} [\underline{Y}^j - \hat{\underline{Y}}^j]^T W^j [\underline{Y}^j - \hat{\underline{Y}}^j] \qquad (5)$$

where W^j are $p \times p$ positive semidefinite weighting matrices, $j = 1,2,\ldots,J$. By combining (4) and (5), the parameter identification problem becomes

$$\min_{\underline{\alpha}} G(\underline{\alpha})$$

subject to

$$g_i(\hat{\underline{U}}^j, \underline{\alpha}) \leq 0 \qquad i = 1,2,\ldots,m; \quad j = 1,2,\ldots,J \qquad (6)$$

where:

$$G(\underline{\alpha}) = \sum_{j=1}^{J} [\underline{H}(\hat{\underline{X}}^j, \hat{\underline{U}}^j, \underline{\alpha}) - \hat{\underline{Y}}^j]^T W^j [\underline{H}(\hat{\underline{X}}^j, \hat{\underline{U}}^j, \underline{\alpha}) - \hat{\underline{Y}}^j]$$

and $\hat{\underline{X}}^j$ and $\hat{\underline{U}}^j$ are known values at the jth time.

The combined problem can be written

$$\left.\begin{aligned} &\min_{\underline{U},\underline{\alpha}} \; [f(\underline{X},\underline{U},\underline{\alpha}), G(\underline{\alpha})] \\ &\text{subject to} \\ &g_i(\underline{X},\underline{U},\underline{\alpha}) \leq 0 \qquad i = 1,2,\ldots,m\,. \end{aligned}\right\} \quad (7)$$

B. The Bicriterion Problem

Few have considered the interaction between the identification and optimization problems. Bellman and Astrom [5] studied the concept of structural identifiability. Durbeck [9] studied the problem of developing simplified models of a process. Such models are essential in on-line control applications. In determining the parameters for the simplified model, the solution to a specialized form of the joint identification and optimization problem was found. Here, the identification criterion was that of matching the real system and model gradients. Holloway [10] developed algorithms using the solution strategies of inner and outer linearization followed by relaxation and restriction (for a review of these techniques see Geoffrion [11]). These algorithms were then used for solving, on-line, the joint problem when both the identification and the optimization problems are convex. However, the approach was essentially that of repeated identification followed by optimization. Orr's [12] approach, using a multilevel approach to the identification and optimization problems, was essentially the same where the coupling between the two problems was not considered.

Haimes, et. al. [7, 13-15] studied the coupling of the two problems in detail and considered the more general class of non-linear static and dynamic systems, and introduced the bicriterion formulation (7) to the joint identification and optimization problem. It has been found in the above cited work that the system identification and optimization problems are frequently coupled. This is particularly apparent when the model parameters are control dependent and when the system identification problem does not possess a unique solution. Furthermore, as pointed out in [16], the effects of measurement noise, computational inaccuracies, and inexact modeling often result in many equally valid solutions to the identification problem (6). In such cases the parameters chosen should be those which, when used in the system optimization problem will result in the best possible system performance.

Viewing the joint problem as a bicriterion optimization provides a unified approach for a precise definition of the optimal solution to the integrated problem. Furthermore, past work in the area of multicriterion optimization can then be applied to develop techniques for finding the optimal solution. The works by Geoffrion [17] and Everett [18] are of importance with this respect.

C. <u>The Efficient Point Concept</u>

The solution to the bicriterion minimization problem posed in (7) lies in the efficient set. Koopmans [19] introduced the concept of efficient point for problems of multiobjective functions in economics. He defined an efficient point as:

> A possible point in the commodity space is called efficient whenever an increase in one of its coordinates (the net output of one good) can be achieved only at the cost of a decrease in some other coordinate (the net output of another good).

Kuhn and Tucker [20] extended the theory of nonlinear programming for one objective function to a vector minimization problem and introduced necessary and sufficient conditions for "proper" solution.

DEFINITION 1. A point, $(\underline{X}^*, \underline{U}^*, \underline{\alpha}^*)$, is said to be an efficient point of (7) if and only if there does not exist another point, $(\underline{X}, \underline{U}, \underline{\alpha})$, such that $G(\underline{\alpha}) \leq G(\underline{\alpha}^*)$ and $f(\underline{X}, \underline{U}, \underline{\alpha}) \leq f(\underline{X}^*, \underline{U}^*, \underline{\alpha}^*)$, with strict inequality holding for at least one of the expressions.

Consequently, any point at which neither the identification nor the performance can be improved without causing a degradation in the other criterion is an efficient point of (7). This obviously is not a unique point. It is easily verified that, every point in the interval between the minima of the two component function of Figure 1 is an efficient point of (7). The optimum solution to the joint problem must be in this set of efficient points. To arrive at a complete definition of the optimal solution to the joint problem a second criterion must be specified. For this, the intuitive definition is used. Namely, the optimal solution is the efficient point of (7) which results in the best possible process identification. The following definition is therefore made.

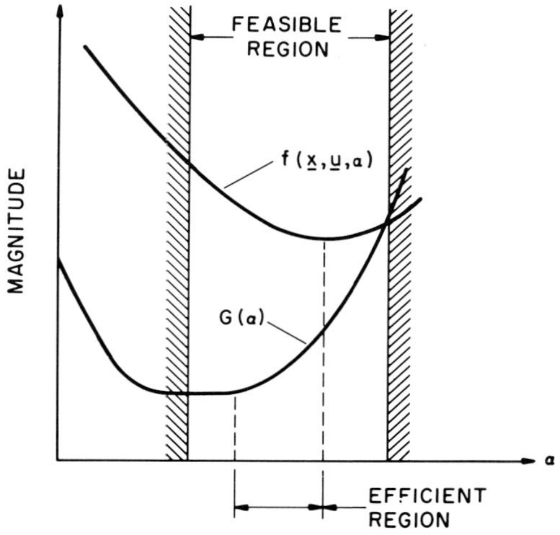

Fig. 1. The Efficient Region

DEFINITION 2. $(\underline{X}^*, \underline{U}^*, \underline{\alpha}^*)$ is an optimal solution to the integrated identification and optimization problem if (1) $(\underline{X}^*, \underline{U}^*, \underline{\alpha}^*)$ is an efficient point of (7); (2) $\underline{\alpha}^*$ minimizes $G(\underline{\alpha})$ over all $\underline{\alpha}$ such that $(\underline{X}^*, \underline{U}^*, \underline{\alpha}^*)$ is efficient.

Note that separate identification and optimization will, in general, not result in an efficient solution of (7). In such an approach the solution to the identification problem is independent of the performance criterion. While the identification problem may have several solutions, only one will in general be efficient. Thus an integrated approach is often necessary.

D. Approaches for Integrated Problem Formulation

Several approaches for the solution of (7) are now proposed:

a. Parametric Approach

The bicriterion problem (7) can be replaced by a parametric formulation

$$\min_{\underline{U}, \underline{\alpha}} \theta G(\underline{\alpha}) + (1 - \theta) f(\underline{X}, \underline{U}, \underline{\alpha}) \tag{8}$$

$$0 \leq \theta \leq 1$$

subject to

$$\underline{g}(\underline{X}, \underline{U}, \underline{\alpha}) \leq 0$$

where $\underline{g}(\underline{X}, \underline{U}, \underline{\alpha})$ is an m-dimensional vector of functions $g_i(\underline{X}, \underline{U}, \underline{\alpha})$, $i = 1, 2, \ldots, m$.

b. A two-step Approach

The combined problem can be considered as two separate problems and solved recursively. The first step would solve the identification problem

$$\min_{\underline{\alpha}} G(\underline{\alpha}) \tag{9}$$

subject to

$$g_i(\underline{X}^\circ, \underline{U}^\circ, \underline{\alpha}) \leq 0 \qquad i = 1,2,\ldots,m$$

where $\underline{X}^\circ, \underline{U}^\circ$ represent estimates of $\underline{X}^*$, $\underline{U}^*$, the optimal solution. If this procedure yields a solution $\underline{\alpha}^\circ$, the next step would solve

$$\min_{\underline{U}} f(\underline{X}, \underline{U}, \underline{\alpha}^\circ) \tag{10}$$

subject to

$$g_i(\underline{x}, \underline{U}, \underline{\alpha}^\circ) \leq 0 \qquad i = 1,2,\cdots,m .$$

If this solution is $\underline{X}^1$, $\underline{U}^1$, these values replace $\underline{X}^\circ$, $\underline{U}^\circ$ in step 1 and the cycle is repeated. The conditions under which convergence is obtained must of course be examined.

This two-step approach, which is probably a more common one, has been shown [16,21] to be inferior to the joint approach when $\underline{\alpha}$ is nct unique, and therefore is of no interest here. Note that a solution to the joint approach is always contained in the efficient set where a solution to the separate approach may not be so.

c. The ε-constraint Approach

This method involves replacing one of the objective functions by a constraint. For instance

$$G(\underline{\alpha}) \leq \varepsilon$$

$$\varepsilon \geq \delta > 0$$

where ε is the maximum tolerated error and δ is an arbitrary small positive scalar. Now the combined problem becomes

$$\left.\begin{aligned} &\min_{\underline{U},\underline{\alpha}} f(\underline{X},\underline{U},\underline{\alpha}) \\ &\text{such that} \\ &g_i(\underline{X},\underline{U},\underline{\alpha}) \leq 0 \qquad i = 1,2,\ldots,m \\ &G(\underline{\alpha}) \leq \varepsilon \qquad \varepsilon \geq \delta > 0 \,. \end{aligned}\right\} \quad (11)$$

By considering ε as a variable in (11), the identification can be made as accurate as desired. In fact minimizing ε implies minimizing $G(\underline{\alpha})$. Olagundoye [22] studied the relationship between vector minimization and this approach.

d. The Minimax Approach

A conservative alternative to the previous procedures is the minimax problem. Assume that "nature" will choose the "worst" $\underline{\alpha}$ after the best $\underline{U}$ is found. The minimax problem is therefore:

$$\left.\begin{aligned} &\min_{\underline{U}} \{\max_{\underline{\alpha}\in\underline{A}} f(\underline{X},\underline{U},\underline{\alpha})\} \\ &\text{such that} \\ &g_i(\underline{X},\underline{U},\underline{\alpha}) \leq 0 \qquad i = 1,2,\ldots,m \\ &\text{where} \\ &\underline{A} = \{\underline{\alpha} \mid G(\underline{\alpha}) \leq \varepsilon, \ \varepsilon \geq \delta > 0 \\ &\text{and} \\ &g_i(\underline{X},\underline{U},\underline{\alpha}) \leq 0 \,, \qquad i = 1,2,\ldots,m\} \,. \end{aligned}\right\} \quad (12)$$

As in the other approaches, this version is important when the Set $\underline{A}$ contains more than one element, consequently the solution to the identification problem is not unique.

Let

$$\underline{A} = \{\underline{\alpha}_1, \underline{\alpha}_2, \ldots, \underline{\alpha}_k\} \,.$$

In this case the interest lies not only in $\underline{\alpha} \in \underline{A}$, but in

$\underline{\alpha}_k \in \underline{A}$, such that the objective function $f(\underline{X},\underline{U},\underline{\alpha})$ is minimized, or in min/max language, it is desired to minimize the worst $f(\underline{X},\underline{U},\underline{\alpha})$ introduced by "nature."

E. The Dependency of $\underline{\alpha}$ on the System Variables

In this chapter, the vector of system model parameters, $\underline{\alpha}$, appears as an argument in the formulation of the objective function for the following reasons:

(a) The objective function $f(\underline{X},\underline{U},\underline{\alpha})$ may indeed include $\underline{\alpha}$ explictly. As an example, consider a manufacturing process where $\underline{\alpha}$ is the probability of defect in the outputs due to defect in the input materials and/or the process. In this case $\underline{\alpha}$ does appear explicitly in the objective function (e.g., more raw material would be purchased as the value of $\underline{\alpha}$ increases).

(b) Even though $\underline{\alpha}$ may appear explicitly only in the system model constraints and not in the objective function, the minimum of the objective function is still affected (implicitly) by $\underline{\alpha}$ via the course of satisfying the system model constraints.

For the above reasons $\underline{\alpha}$ will be included as an argument in the objective function in order to stress the role that the identification of $\underline{\alpha}$ plays in the overall system model minimum. In the preceeding discussion $\underline{\alpha}$ has been considered to be either a constant, $\underline{\alpha}$, or a function of time, $\underline{\alpha}(t)$. In some cases, however, $\underline{\alpha}$ may be a function of the control variable vector $\underline{\alpha}(\underline{U})$ or a function of both time and control, $\underline{\alpha}(t,\underline{U})$. The objective function, f, and the identification norm, G, can be classified for different cases of $\underline{\alpha}$ as depicted in Figure 2.

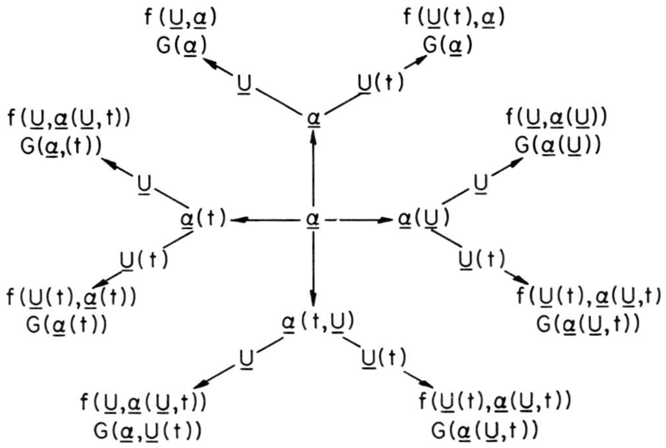

Fig. 2. On the Dependency of $\underline{\alpha}$.

F. The Dynamic Identification and Dynamic Optimization Problem

The above discussion of the static identification and static optimization problem facilitates the formulation of the more general integrated problem that is, the dynamic identification and dynamic optimization problem. The formulation of the remaining two classes of problems: dynamic identification-static optimization; static identification-dynamic optimization will be obtained as special cases.

System models for which both the parameter vector $\underline{\alpha}$ and the control vector $\underline{U}$ are time-varying introduce serious computational and theoretical difficulties. Such system models are generally represented by a system of differential equations with variable coefficients. The notation previously introduced will be adopted here with the following generalization:

Let

$$\underline{U} = \underline{U}(t)$$
$$\underline{X} = \underline{X}(t)$$

$$\underline{Y} = \underline{Y}(t)$$
$$\underline{\alpha} = \underline{\alpha}(t)$$

where t is the time variable, $t_0 \leq t \leq t_f$.

The optimization problem is

$$\min_{\underline{U} \in \mathcal{U}} \left\{ \rho = \int_{t_0}^{t_f} f(\underline{X}(t), \underline{U}(t), \underline{\alpha}(t), t) dt \right\} \tag{13}$$

subject to the constraints:

$$\left. \begin{aligned} & \underline{g}(\underline{X}(t), \underline{U}(t), \underline{\alpha}(t), t) \leq \underline{0} \\ & t_0 \leq t \leq t_f \\ & \underline{\dot{X}}(t) = \underline{F}(\underline{X}(t), \underline{U}(t), \underline{\alpha}(t), t) \\ & \underline{X}(t_0) = \underline{X}_0 \end{aligned} \right\} \tag{14}$$

where $\underline{F}$ is an $\underline{M}$-vector of functions F_i and $\mathcal{U}$ is the feasible set of $\underline{U}$.

$$\underline{Y}(t) = \underline{H}(\underline{X}(t), \underline{U}(t), \underline{\alpha}(t), t)$$

where $\underline{H}$ is a p-vector of output functions H_i.

The vector $\underline{Y}(t)$, $t_0 \leq t \leq t_f$, is the output vector of the system and $\underline{H}$ is assumed to be a vector of single valued functions of the arguments. Also note that no system of differential equations in $\underline{\alpha}$ is introduced here, since it is assumed that these relationships are generally not known.

J discrete time observations are made and the notation

$$\underline{U}^j \triangleq \underline{U}(t_j);\ \underline{H}^j \triangleq \underline{H}(t_j)$$
$$\underline{Y}^j \triangleq \underline{Y}(t_j);\ \underline{X}^j \triangleq \underline{X}(t_j)$$

$j = 1, 2, \ldots, J$ is adopted.

Then the system identification problem can be formulated as follows:

$$\min_{\underline{\alpha}(t)} G(\underline{\alpha}(t))$$

subject to (14), where

$$G(\underline{\alpha}(t_1),\ldots,\underline{\alpha}(t_J)) = \sum_{j=1}^{J} [\underline{\hat{Y}}^j - \underline{H}^j(\underline{X}^j,\underline{\hat{U}}^j,\underline{\alpha}(t_j))]^T W^j [\underline{\hat{Y}}^j - \underline{H}^j(\underline{X}^j,\underline{\hat{U}}^j,\underline{\alpha}(t_j))]$$

for the discrete observation case, and

$$G(\underline{\alpha}(t)) = \int_{t_0}^{t_f} [\underline{\hat{Y}} - \underline{H}(\underline{X},\underline{\hat{U}},\underline{\alpha})]^T W(t) [\underline{\hat{Y}} - \underline{H}(\underline{X},\underline{\hat{U}},\underline{\alpha})]dt$$

for the continuous observation case, where

$$\underline{U}^j_{obs} \triangleq \underline{\hat{U}}^j$$
$$\underline{Y}^j_{obs} \triangleq \underline{\hat{Y}}^j$$
$$\underline{Y}^j_{cal} \triangleq \underline{H}^j(\underline{\hat{X}},\underline{\hat{U}}^j,\underline{\alpha}(t_j)) \qquad j = 1,2,\ldots,J .$$

Only the ε-constraint approach will be applied to this class of problems.

The integrated dynamic problem is thus:

$$\min_{\underline{U},\underline{\alpha}} \int_{t_0}^{t_f} f(\underline{X}(t),\underline{U}(t),\underline{\alpha}(t),t)dt$$

subject to the constraints

$$\left.\begin{aligned}
&\underline{g}(\underline{X}(t),\underline{U}(t),\underline{\alpha}(t),t) \le 0\\
&t_0 \le t \le t_f\\
&\underline{\dot{X}}(t) = \underline{F}(\underline{X}(t),\underline{U}(t),\underline{\alpha}(t),t)\\
&\underline{X}(t_0) = \underline{X}_0\\
&\underline{Y}(t) = \underline{H}(\underline{X}(t),\underline{U}(t),\underline{\alpha}(t),t)\\
&G(\underline{\alpha}(t)) \le \varepsilon\\
&\varepsilon \ge \delta > 0.
\end{aligned}\right\} \quad (15)$$

G. The Static Identification and Dynamic Optimization Problem

This special class of problems is of particular interest. Since many system models can be adequately represented by a system of O.D.E. with constant coefficients, this class is general enough to cover many systems while being computationally tractable. The present advanced theory and applications of optimal control and mathematical programming [23-27] can be readily applied to this class of static identification and dynamic optimization problems in conjunction with the experience gained in the field of system identification [1,2,3, 28].

In applying the ε-constraint approach, the integrated static identification and dynamic optimization problem is

$$\left.\begin{aligned}
&\min_{\underline{U},\underline{\alpha}} \int_{t_0}^{t_f} f(\underline{X},\underline{U},\underline{\alpha},t)\,dt\,, \\
&\text{subject to the constraints} \\
&\underline{g}(\underline{X},\underline{U},\underline{\alpha}) \le 0\,, \\
&\dot{\underline{X}} = \underline{F}(\underline{X},\underline{U},\underline{\alpha},t) \\
&X(t_0) = \underline{X}_0 \\
&t_0 \le t \le t_f \\
&G(\underline{\alpha}) \le \varepsilon \\
&\varepsilon \ge \delta > 0\,.
\end{aligned}\right\} \qquad (16)$$

H. The Dynamic Identification and Static Optimization Problem

This class is subject to limited realistic cases where both the system state and control variables are time invariant but the system parameters are time variant. Therefore it will not be further discussed in the following sections.

III. CHARACTERISTICS OF THE JOINT PROBLEM

A. The Equivalence Theorem

This theorem provides the linkage between formulations (7) and (8):

Define the following two Problems A and $B(\varepsilon)$ [7,13].

Problem A:

Find an efficient point for the vector minimization problem

$$\left.\begin{aligned} &\min_{\underline{U},\underline{\alpha}} [f(\underline{X},\underline{U},\underline{\alpha}),G(\underline{\alpha})] \\ &\text{subject to} \\ &g(\underline{X},\underline{U},\underline{\alpha}) \leq 0 . \end{aligned}\right\} \tag{17}$$

Problem $B(\varepsilon)$:

Consider

$$\left.\begin{aligned} &\min_{\underline{U},\underline{\alpha}} f(\underline{X},\underline{U},\underline{\alpha}) \\ &\text{subject} \\ &G(\underline{\alpha}) \leq \varepsilon \\ &\underline{g}(\underline{X},\underline{U},\underline{\alpha}) \leq 0 . \end{aligned}\right\} \tag{18}$$

Let

$$\underline{\beta} = (\underline{X},\underline{U},\underline{\alpha}) . \tag{19}$$

Equivalence Theorem

Let $\varepsilon \geq \min G(\underline{\alpha})$, let $\underline{\beta}^*$ solve Problem $B(\varepsilon)$, and assume that, if $\underline{\beta}^*$ is not unique, then $\underline{\beta}^*$ is an optimal solution of Problem B(ε) with minimal $G(\underline{\alpha})$ value. Then $\underline{\beta}^*$ solves Problem A.

Proof. Assume $\underline{\beta}^*$ does not solve Problem A. Then there is a $\hat{\underline{\beta}} = (\hat{\underline{X}}, \hat{\underline{U}}, \hat{\underline{\alpha}})$ satisfying (17) such that either

$$f(\hat{\underline{\beta}}) < f(\underline{\beta}^*),\ G(\hat{\underline{\alpha}}) \leq G(\underline{\alpha}^*) \qquad (20)$$

or

$$G(\hat{\underline{\alpha}}) < G(\underline{\alpha}^*),\ f(\hat{\underline{\beta}}) \leq f(\underline{\beta}^*)\ . \qquad (21)$$

But (20) contradicts the fact that $\underline{\beta}^*$ solves Problem B(ε) while (21) contradicts the hypothesis that $\underline{\beta}^*$ is an optimal solution of Problem B(ε) with smallest $G(\underline{\alpha})$. Hence the theorem is proved.

Consider now solving Problem B(ε) with the minimal value of ε such that Problem B(ε) is feasible. Let this ε be ε^*. Clearly, if the original solution of the least square model fitting problem

$$\min G(\underline{\alpha})$$

(let this solution be $\underline{\alpha}^*$) is unique and there exists a vector $(\underline{X},\underline{U})$ such that $g(\underline{X},\underline{U},\underline{\alpha}^*) \leq 0$, then

$$\varepsilon^* = G(\underline{\alpha}^*)$$

and Problem B(ε) may be solved by first minimizing $G(\underline{\alpha})$ and then solving the optimization problem

$$\min f(\underline{X},\underline{U},\underline{\alpha}^*) \qquad (22)$$

subject to

$$\underline{g}(\underline{X},\underline{U},\underline{\alpha}^*) \leq 0 \qquad (23)$$

that is, the optimization and identification problems may be decomposed. Hence we need consider an integrated formulation (i.e., solving Problem B(ε)) only if $\underline{\alpha}^*$ is not unique or if it is not feasible, i.e., if there exists no $(\underline{X},\underline{U})$ such that $\underline{g}(\underline{X},\underline{U},\underline{\alpha}^*) \leq 0$, since then the optimization problem (22) and (23) has no feasible solution.

B. Parametric Approach

Consider the parametric formulation presented in (8). Since the following characteristics of the joint problem relate to the static identification-static optimization case, it is convenient to make the following augmentation:

Let

$$\underline{\beta} = (\underline{X},\underline{U},\underline{\alpha}) \, . \tag{24}$$

Accordingly, the parametric formulation (8) can be simplified as follows:

$$\left.\begin{aligned} &\min_{\underline{\beta}}\{\theta G(\underline{\alpha}) + (1-\theta)\, f(\underline{\beta})\} \\ &\text{subject to} \qquad \underline{g}(\underline{\beta}) \leq \underline{0},\ 0 \leq \theta \leq 1 \end{aligned}\right\} \tag{25}$$

Form the Lagrangian of (25), L:

$$L = \theta G(\underline{\alpha}) + (1-\theta)\, f(\underline{\beta}) + \underline{\mu}^T \underline{g}(\underline{\beta}) \tag{26}$$

where $\underline{\mu}$ is an m-dimensional vector of Kuhn-Tucker multipliers.

LEMMA 1. Assume the constraints qualifications [20] hold. If for some $0 < \theta < 1$, $(\underline{\beta}^\circ,\underline{\mu}^\circ)$ is a saddle point of (26), then $(\underline{\beta}^\circ)$ is an efficient point of (7). A proof of this Lemma and its converse is given by Karlin [29] under convexity assumptions.

While the parametric approach may be used to determine efficient points of the vector minimization, it is not obvious that it can be used to find the optimum solution to the joint problem. Note that finding a saddle point of (26) for $\theta = 1$ does not guarantee that an efficient point will be found since Lemma 1 is valid only over the open interval $(0,1)$. It will be proven that in the limit, as θ approaches one, the parametric solution approaches the integrated identification and optimization problem solution [16,21]. Define the feasible set, B, to problem (25) as follows:

$$B = \{\underline{\beta} \mid g_i(\underline{\beta}) \leq 0, \quad i = 1,2,\ldots,m\}. \quad (27)$$

Define the set of optimal solutions to (25) for a fixed value of the parameter θ as $B(\theta)$. It is assumed that the sets B and $B(\theta)$ are not empty and that $G(\underline{\beta})$, $f(\underline{\beta})$, and $\underline{g}(\underline{\beta})$ are of class C^2.

DEFINITION 3. Any $\underline{\beta}(\theta) \in E^{k+\ell}$ on $(0,1)$ that satisfies $\underline{\beta}(\theta) \in B(\theta)$ for every θ on $(0,1)$ is called an optimal function for (25).

For notational ease, denote $G(\underline{\beta}(\theta))$ and $f(\underline{\beta}(\theta))$ by $G(\theta)$ and $f(\theta)$ respectively. Note that $G(\theta)$ and $f(\theta)$ are defined only over the feasible set B.

THEOREM 1. If $\underline{\beta}(\theta)$ is an optimal solution function for (25), then $f(\theta)$ and $G(\theta)$ are unimodal for θ over the open interval $(0,1)$. The proof for this theorem is given in Appendix A.

THEOREM 2. If $\underline{\beta}(\theta)$ is an optimal solution function for (25), then $f(\theta)$ is non-decreasing and $G(\theta)$ is non-increasing in the open interval $(0,1)$, [16,21].

Proof. By definition $\underline{\beta}(\theta)$ solves (25). Clearly, for $\theta = 0$, $\underline{\beta}(0)$ is at the global minimum of $f(\underline{\beta})$. By Theorem 1, $f(\theta)$ is unimodal over $(0,1)$ and therefore $f(\theta)$ is non-decreasing in this same interval. Likewise, at $\theta = 1$, $\underline{\beta}(1)$ is at the global minimum of $G(\underline{\beta})$. And, by Theorem 1, $G(\theta)$ must be non-increasing over $(0,1)$.

The result of this theorem is shown graphically in Figure 3.

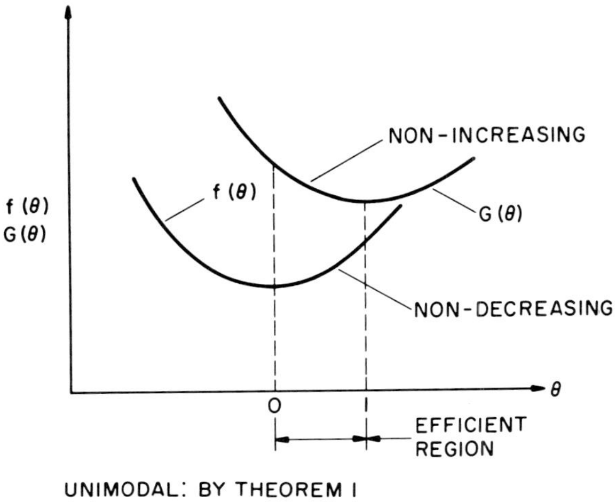

Fig. 3. Illustration of Monotonicity Theorem.

THEOREM 3. For an optimal solution function $\underline{\beta}(\theta)$ of (25), the $\lim_{\theta\to 1-} \underline{\beta}(\theta)$ solves the joint problem in the sense of Definition 2, [16,21].

Proof. By the continuity of $\underline{\beta}(\theta)$, $\lim_{\theta\to 1-} \underline{\beta}(\theta)$ is in B(1) (where θ approaches 1- from below). Suppose that $\underline{\beta}(1-)$ is not efficient. Then there exists a point $\underline{\beta} \in B$ such that

$$f(\underline{\beta}') < f[\underline{\beta}(1-)] \tag{28}$$

$$G(\underline{\beta}') = G[\underline{\beta}(1-)] . \tag{29}$$

Equation (29) must be true since $\underline{\beta}(1)$ solves (25) at $\theta = 1$. Thus, $\underline{\beta}' \in B(1)$. Let $\tilde{\theta} < 1$ be such that $f(\underline{\beta}') < f(\underline{\beta}(\tilde{\theta}))$. This contradicts the monotonicity of $f(\underline{\beta}(\theta))$. Consequently, $\underline{\beta}(1-)$ minimizes $G(\underline{\beta})$ and is efficient.

The above theorem implies that the solution to the joint problem may be obtained by choosing the parameters, $\underline{\alpha}$, as

$$\lim_{\theta\to 1-} \underline{\alpha}(\theta)$$

The unimodality of $f(\theta)$ and $G(\theta)$ for $0 < \theta < 1$ guarantee that this is the joint solution even if the optimum parameters for the separate identification problem are not unique.

IV. THE MULTILEVEL APPROACH

The concept of the multilevel approach, Mesarovic et. al. [30] is based on the decomposition of large scale and complex systems, into independent subsystems. This decentralized approach, by utilizing the concepts of strata, layers and echelons, enables the system analyst to analyze and comprehend the behavior of small parts of the system, called subsystems, in a lower level and to transmit the information obtained to fewer subsystems of a higher level. Whenever more decentralization is needed, the system is further decomposed. This decomposition is accomplished by introducing new variables, which will be called pseudo variables, into the system. Then each subsystem is separately and independently optimized. This is called a first level solution. The subsystems are joined together by coupling variables which are manipulated by second level controllers in order to arrive at the optimal solution of the whole system. This coordination is called the second level solution. One way to achieve subsystem independence is by relaxing one or more of the necessary conditions for optimality and then satisfying these conditions with the second level controller. Mesarovic and Eckman [31], Macko [32], and Macko and Pearson [33] formally define the concept of the multilevel system within a general systems theory framework.

Dantzig [34] in his Decomposition Principle for linear programming extends the theory of linear programming to a special class of linear systems with high dimensionality. In particular he makes use of duality properties. Mesarovic et. al. [35] extend the decomposition idea to dynamic optimization models. Bauman [36] outlines two methods for the decomposition of systems of high dimension which are known as feasible and nonfeasible decomposition. In the feasible method the subsystems' coupling constraints are satisfied at any iteration, but optimality is not achieved until the last iteration. In the nonfeasible method, the subsystems' coupling constraints at any iteration are not satisfied, but the subsystems are always at optimum. Lefkowitz [37] extends the theory and applications of the multilevel approach to control system design. In this respect, he presents four layers in the hierarchy: the regulator or direct control function, the optimizing control function, the adaptive control function, and the self-organizing control function. Lasdon [24] and Lasdon and Schoeffler [38] discuss the applications of these techniques to processes where the input of each subsystem is determined by the combined outputs of other subsystems. Wismer [25] and other contributors present the state of the art of this growing field. Wismer [39] applies these techniques to the optimization of a class of distributed parameter systems. Haimes [8] applies the multilevel approach to the system identification of aquifer parameters. In addition, Haimes et. al. [40-44] discuss the application of the hierarchical multilevel approach to the modeling and mangement of water resource systems.

As was previously noted, the joint problem of optimization and identification can be hampered by high dimensionality. This dimensionality problem can be nicely handled by the multilevel approach. Unlike the two step approach previously discussed in Section II D-b, in using the multilevel approach, the interactions between the subsystems are taken into account by the second level controller which influences the individual subsystems. The goal of the second level is thus to coordinate the actions of the first level units so that the solution to the integrated problem is obtained. Conceptually the multi-level approach has the advantages of both the two-step approach (low dimensionality) and the integrated approach (considers coupling). While solving the integrated problem as a single problem may require excessive computer time and storage, decomposition into coordinated subproblems can be computationally (and economically) feasible, provided the subsystems can be solved efficiently.

A. The ε-Constraint Approach for Static Identification and Static Optimization Problem

A solution procedure to the ε-constraint approach for the joint static identification and static optimization problem is presented below. It is convenient for this purpose to separate the constraints $\underline{g}_i(\underline{X},\underline{U},\underline{\alpha}) \leq \underline{0}$, $i = 1,2,\ldots,m$ given by (3) into three groups based on the depencency of the functions $g_{ji}(\underline{X},\underline{U},\underline{\alpha})$, $i = 1,\ldots,m_j$, $j = 1,2,3$ on their arguments as follows:

$$g_{1i}(\underline{X},\underline{U},\underline{\alpha}) \leq 0 \qquad i = 1,2,\ldots,m_1 \tag{30}$$

$$g_{2i}(\underline{X},\underline{U}) \leq 0 \qquad i = 1,2,\ldots,m_2 \tag{31}$$

$$g_{3i}(\underline{\alpha}) \leq 0 \qquad i = 1,2,\ldots,m_3 \tag{32}$$

$$m_1 + m_2 + m_3 = m . \tag{33}$$

Now form the Lagrangian, $\overline{L}$:

$$\overline{L} = f(\underline{X},\underline{U},\underline{\alpha}) + \underline{\lambda}_1^T\underline{g}_1(\underline{X},\underline{U},\underline{\alpha}) + \underline{\lambda}_2^T\underline{g}_2(\underline{X},\underline{U}) + \underline{\lambda}_3^T\underline{g}_3(\underline{\alpha}) + \mu_1[G(\underline{\alpha}) - \varepsilon] \tag{34}$$

where

$$\underline{g}_1(\underline{X},\underline{U},\underline{\alpha}) = [g_{1i}(\underline{X},\underline{U},\underline{\alpha}),\ldots,g_{1m_1}(\underline{X},\underline{U},\underline{\alpha})]^T \tag{35}$$

$$g_2(\underline{X},\underline{U}) = [g_{21}(\underline{X},\underline{U}),\ldots,g_{2m_2}(\underline{X},\underline{U})]^T \tag{36}$$

$$g_3(\underline{\alpha}) = [g_{31}(\underline{\alpha}),\ldots,g_{3m_3}(\underline{\alpha})]^T \tag{37}$$

$\underline{\lambda}_i$ is an m_i-dimensional Lagrange multiplier $i = 1,2,3.$

μ_1 is a one-dimensional Lagrange multiplier.

In order to uncouple the terms in (34) with respect to $\underline{\alpha}$ and thus to be able to decompose the Lagrangian into independent subsystems, it is convenient to introduce the pseudo variable vector, $\underline{\sigma}$, and to require that:

$$\underline{\sigma} = \underline{\alpha}\,. \tag{38}$$

The additional equality constraint (38) will be satisfied by adding it to the Lagrangian (34) as follows:

$$L = \bar{L} + \underline{\lambda}_4^T\,[\underline{\sigma} - \underline{\alpha}] \tag{39}$$

where $\underline{\lambda}_4$ is a k-dimensional Lagrange Multiplier Vector.
Note that

$$L = L(\underline{X},\underline{U},\underline{\alpha},\varepsilon,\underline{\sigma},\underline{\lambda}_1,\underline{\lambda}_2,\underline{\lambda}_3,\underline{\lambda}_4,\mu_1)\,.$$

By considering the $\underline{\sigma}$ as a vector of known parameters in some subsystems (as will be shown later), the Lagrangian (39) is readily decomposed into two independent subsystems:

$$L = \sum_{i=1}^{2} L_i \tag{40}$$

where

$$L_1(\underline{X},\underline{U},\underline{\sigma},\underline{\lambda}_1,\underline{\lambda}_2,\underline{\lambda}_4) = f(\underline{X},\underline{U},\underline{\sigma}) + \underline{\lambda}_1^T\underline{g}_1(\underline{X},\underline{U},\underline{\sigma}) + \underline{\lambda}_2^T\underline{g}_2(\underline{X},\underline{U}) + \underline{\lambda}_4^T\underline{\sigma} \tag{41}$$

$$L_2(\underline{\alpha} : \underline{\lambda}_3\underline{\lambda}_4,\mu_1) = \underline{\lambda}_3^T\underline{g}_3(\underline{\alpha}) + \mu_1[G(\underline{\alpha}) - \varepsilon] - \underline{\lambda}_4^T\underline{\alpha}\,.$$

The above decomposition is termed nonfeasible decomposition, (Lasdon and Durbeck [45] and Bauman [36]).

a. The Gauss-Seidel Type Second-Level Controller

At the first level the Psuedo variable vector, $\underline{\sigma}$, is considered to be known in subsystem 1 and to be determined at the second-level optimization. In addition the Lagrange multiplier vector, $\underline{\lambda}_4$, is also assumed to be known at the first level for both subsystems and to be determined at the second-level optimization. The Lagrange multipliers, $\underline{\lambda}_1$ and $\underline{\lambda}_2$, appearing in subsystem 1, and $\underline{\lambda}_3$ and μ_1, appearing in subsystem 2, are determined at the first level in each corresponding subsystem. A schematic description of the Gauss-Seidel type second-level controller is given in Figure 4, (Haimes [7], Wismer [25,39]).

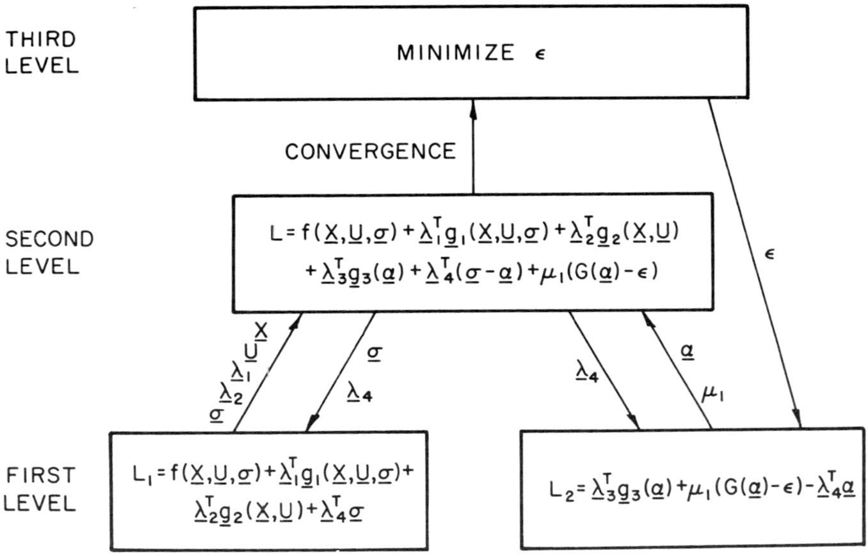

Fig. 4. Gauss-Seidel Type Second Level Controller.

The following Kuhn-Tucker stationarity conditions must be

satisfied at the first level for subsystems 1 and 2 respectively:

1. Subsystem 1:

$$\frac{\partial L_1}{\partial \underline{U}} = 0; \quad \frac{\partial L_1}{\partial \underline{X}} = 0 \tag{43}$$

$$\frac{\partial L_1}{\partial \underline{\lambda}_i} \leq 0; \quad \underline{\lambda}_i^T \frac{\partial L_1}{\partial \underline{\lambda}_i} = 0; \quad \underline{\lambda}_i \geq 0, \qquad i = 1,2 . \tag{44}$$

2. Subsystem 2:

$$\frac{\partial L_2}{\partial \underline{\alpha}} = 0 \tag{45}$$

$$\frac{\partial L_2}{\partial \underline{\lambda}_3} \leq 0; \quad \underline{\lambda}_3^T \frac{\partial L_2}{\partial \underline{\lambda}_3} = 0; \quad \underline{\lambda}_3 \geq 0 \tag{46}$$

$$\frac{\partial L_2}{\partial \mu_1} \leq 0; \quad \mu_1 \frac{\partial L_2}{\partial \mu_1} = 0; \quad \mu_1 \geq 0 . \tag{47}$$

At the second level the following necessary conditions for stationarity must be satisfied:

$$\frac{\partial L}{\partial \underline{\sigma}} = 0 \tag{48}$$

$$\frac{\partial L}{\partial \underline{\lambda}_4} = 0 . \tag{49}$$

Carrying out the differentiation of the Lagrangian as given in Eq. (39) yields:

$$\frac{\partial L}{\partial \underline{\sigma}} = \frac{\partial f(\underline{X},\underline{U},\underline{\sigma})}{\partial \underline{\sigma}} + \underline{\lambda}_1^T \frac{\partial \underline{g}_1}{\partial \underline{\sigma}} (\underline{X},\underline{U},\underline{\sigma}) + \underline{\lambda}_4 = 0 \tag{50}$$

$$\frac{\partial L}{\partial \underline{\lambda}_4} = \underline{\sigma} - \underline{\alpha} = 0 . \tag{51}$$

From the above two equations the Gauss-Seidel second-level

controller provides the following values for $\underline{\lambda}_4$ and $\underline{\sigma}$:

$$\underline{\lambda}_4 = - \frac{\partial f(\underline{X},\underline{U},\underline{\sigma})}{\partial \underline{\sigma}} - \underline{\lambda}_1^T \frac{\partial \underline{g}_1}{\partial \underline{\sigma}} (\underline{X},\underline{U},\underline{\sigma}) \tag{52}$$

$$\underline{\sigma} = \underline{\alpha} . \tag{53}$$

Sufficient conditions for a saddle point to the Lagrangian function, L, are given in Appendix B. Note that ε is assumed to be a known parameter to be determined by the third-level controller as will be shown in Section c.

b. The Gradient Type Second-Level Controller

A second-level controller based on the gradient of the second-level objective function is developed hereafter. At the first level the pseudo variable vector, $\underline{\sigma}$, is considered to be unknown in subsystem 1. The Lagrange multiplier vector, $\underline{\lambda}_4$, is assumed to be known at the first level for both subsystems and to be determined at the second-level optimization. The Lagrange multipliers, $\underline{\lambda}_1$ and $\underline{\lambda}_2$, appearing in subsystem 1, and $\underline{\lambda}_3$ and μ_1 appearing in subsystem 2, are determined at the first level in each corresponding subsystem. A schematic description of the gradient second-level controller is given in Figure 5. Equations (43)-(48), developed for the Gauss-Seidel type second-level controller, constitute the necessary conditions for first level stationarity. Note that, unlike the Gauss-Seidel method, Eq. (48) is satisfied here at the first level.

At the second level $\underline{\lambda}_4$ is the only unknown vector of parameters, and a necessary condition for stationarity at the second level is thus:

$$\frac{\partial L}{\partial \underline{\lambda}_4} = 0$$

or

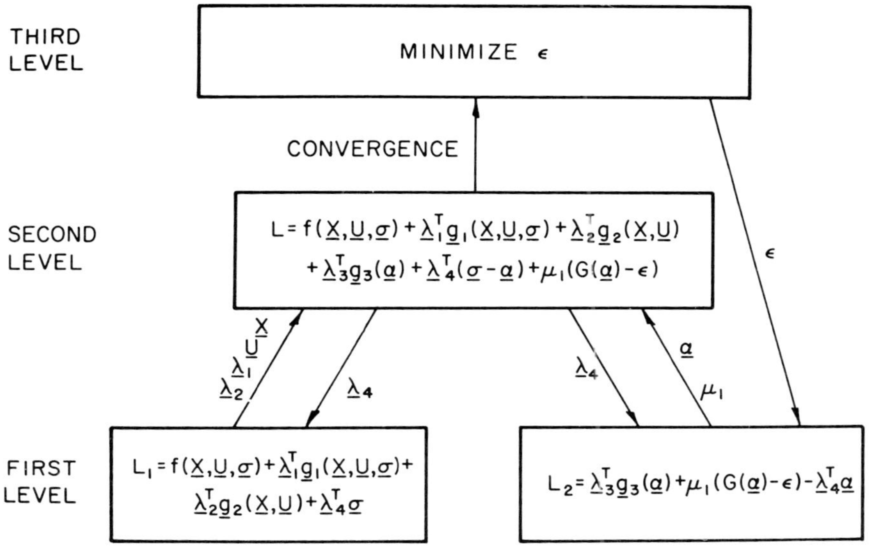

Fig. 5. Gradient Type Second-Level Controller.

$$\frac{\partial L}{\partial \underline{\lambda}_4} = \underline{\sigma}(\underline{\lambda}_4) - \underline{\alpha}(\lambda_4) = 0 \, . \tag{54}$$

A gradient type second-level controller as proposed by Lasdon and Schoeffler [46] is adopted here. A set of sufficient conditions for the gradient controller to converge to a local minimum is that for all $\underline{\lambda}_4$ in some neighborhood of $\underline{\lambda}_4^{\circ}$ at the minimum, L_1 and L_2 have a local minimum with respect to the other variables. If the subsystems meet these requirements, then L has a maximum with respect to $\underline{\lambda}_4$ (i.e., L has a saddle point at the minimum to the original problem $\overline{L}$. Suppose L possesses a maximum with respect to $\underline{\lambda}_4$; then $dL > 0$ as $\underline{\lambda}_4$ moves towards the optimum. Consider now the motions along the negative gradient direction (steepest descent path):

$$\frac{d\underline{\lambda}_4}{d\tau} = - k \frac{\partial L}{\partial \underline{\lambda}_4} \tag{55}$$

where $k > 0$ and τ is a time parameter, related to the process of successive approximations. Since Eq. (55) is derived for a continuous process, while the computational procedure is not, a discretization of $\underline{\lambda}_4$ is introduced:

$$d\underline{\lambda}_4 = \underline{\lambda}_4^{(n+1)} - \underline{\lambda}_4^{(n)} \tag{56}$$

where the superscript (n) represents the nth iteration.

Introducing a step size $\Delta\tau$, Eq. (55) becomes:

$$\underline{\lambda}_4^{(n+1)} = \underline{\lambda}_4^{(n)} - k \frac{\partial L}{\partial \underline{\lambda}_4} \Delta\tau . \tag{57}$$

Let $K = k\Delta\tau$. Then

$$\underline{\lambda}_4^{(n+1)} = \underline{\lambda}_4^{(n)} - K \frac{\partial L}{\partial \underline{\lambda}_4} . \tag{58}$$

Since

$$\frac{\partial L}{\partial \underline{\lambda}_4} = \underline{\sigma}(\underline{\lambda}_4) - \underline{\alpha}(\underline{\lambda}_4) .$$

Then

$$\underline{\lambda}_4^{(n+1)} = \underline{\lambda}_4^{(n)} - K[\underline{\sigma}(\underline{\lambda}_4^{(n)}) - \underline{\alpha}(\underline{\lambda}_4^{(n)})] \tag{59}$$

where $K \geq 0$. This process will converge to an optimum when the problem possesses the appropriate convexity (see Appendix B).

A second order gradient method such as the Fletcher-Powell method [47] has been applied and shown fast convergence [16,21,41].

c. Third-Level Controller for the Determination of ε.

More than one method may be employed for the determination of ε at the third level. Probably the most efficient and straightforward method is to introduce a third-level controller which will assign a certain value for ε to be imposed on the two lower levels. When convergence has been achieved at the two lower levels with a given ε, the third-level controller will decrease that value of ε and repeat the second-level solution. This procedure will be continued until convergence for the second-level optimization becomes impossible. At that point $\varepsilon = \delta$.

B. The ε-Constraint Approach for Static Identification and Dynamic Optimization

Extending the notation employed for the static optimization to the dynamic one, the integrated problem is formulated as follows:

$$\min_{\underline{U}} \left\{ \rho = \int_{t_0}^{t_f} f(\underline{X},\underline{U},\underline{\alpha},t)dt \right\} \tag{60}$$

subject to the constraints

$$\left.\begin{aligned} &\underline{g}_1(\underline{X},\underline{U},\underline{\alpha},t) = 0 \\ &\quad t_0 \leq t \leq t_f \\ &\underline{g}_2(\underline{X},\underline{U},t) \leq 0 \\ &\quad t_0 \leq t \leq t_f \\ &\underline{g}_3(\underline{\alpha}) \leq 0 \end{aligned}\right\} \tag{61}$$

$$\dot{\underline{X}} = \underline{F}(\underline{X},\underline{U},\underline{\alpha},t);\; \underline{X}(t_0) = \underline{X}_0 \tag{62}$$

$$\underline{Y} = \underline{H}(\underline{X},\underline{U},\underline{\alpha},t) \tag{63}$$

$$G(\underline{\alpha}) \leq \varepsilon \geq \delta > 0 \tag{64}$$

$$\underline{\sigma} = \underline{\alpha}\,. \tag{65}$$

Form the Hamiltonian, $\mathcal{H}$, as follows:

$$\left.\begin{aligned} \mathcal{H} = {} & f(\underline{X},\underline{U},\underline{\sigma},t) + \underline{\lambda}_1^T \underline{g}_1(\underline{X},\underline{U},\underline{\sigma},t) \\ & + \underline{\lambda}_2^T \underline{g}(\underline{X},\underline{U},t) + \underline{\lambda}_3^T \underline{g}_3(\underline{\alpha}) \\ & + \underline{\lambda}_4^T(\underline{\sigma} - \underline{\alpha}) + \underline{\lambda}_5^T \underline{F}(\underline{X},\underline{U},\underline{\sigma},t) \\ & + \mu_1(G(\underline{\alpha}) - \varepsilon) \end{aligned}\right\} \quad (66)$$

where $\underline{\lambda}_5$ is an M-dimensional Lagrange multiplier vector, and all other Lagrange multipliers $\underline{\lambda}_1$, $\underline{\lambda}_2$, $\underline{\lambda}_3$, $\underline{\lambda}_4$, μ_1 are of a similar dimensionality to the corresponding ones in the static case. The constraint (64) will be satisfied via the computational procedure at the third level. The Hamiltonian Eq. (66) is readily decomposed into two subsystems:

$$\mathcal{H} = \mathcal{H}_1 + \mathcal{H}_2$$

where

$$\left.\begin{aligned} & \mathcal{H}_1(\underline{X},\underline{U},t;\ \underline{\sigma},\underline{\lambda}_1,\underline{\lambda}_2,\underline{\lambda}_4,\underline{\lambda}_5) \\ & \quad = f(\underline{X},\underline{U},\underline{\sigma},t) + \underline{\lambda}_1^T \underline{g}_1(\underline{X},\underline{U},\underline{\sigma},t) \\ & \quad + \underline{\lambda}_2^T \underline{g}_2(\underline{X},\underline{U},t) + \underline{\lambda}_4^T \underline{\sigma} \\ & \quad + \underline{\lambda}_5^T \underline{F}(\underline{X},\underline{U},\underline{\sigma},t) \end{aligned}\right\} \quad (67)$$

and

$$\begin{aligned} \mathcal{H}_2(\underline{\alpha};\ \varepsilon,\underline{\lambda}_3,\underline{\lambda}_4,\mu_1,\mu_2) = {} & \mu_1(G(\underline{\alpha}) - \varepsilon) \\ & + \underline{\lambda}_3^T \underline{g}_3(\underline{\alpha}) - \underline{\lambda}_4^T \underline{\alpha}\,. \end{aligned} \quad (68)$$

a. The Gauss-Seidel Second-Level Controller

The following stationarity conditions must be satisfied at the first level for subsystems 1 and 2 respectively:

1. Subsystem 1:

$$\frac{\partial \mathcal{H}_1}{\partial \underline{U}} = 0; \quad \frac{\partial \mathcal{H}_1}{\partial \underline{X}} = -\dot{\underline{\lambda}}_5; \quad \frac{\partial \mathcal{H}_1}{\partial \underline{\lambda}_5} = \dot{\underline{X}} \quad (69)$$

$$\underline{X}(t_0) = \underline{X}_0; \quad \underline{\lambda}(t_f) = 0 \tag{70}$$

$$\frac{\partial H_1}{\partial \underline{\lambda}_i} \leq 0; \quad \underline{\lambda}_i^T \frac{\partial H_i}{\partial \underline{\lambda}_i} = 0; \quad \underline{\lambda}_i \geq 0; \qquad i = 1,2 . \tag{71}$$

2. Subsystem 2:

$$\frac{\partial H_2}{\partial \underline{\alpha}} = 0; \quad \frac{\partial H_2}{\partial \underline{\lambda}_3} = 0; \quad \underline{\lambda}_3^T \frac{\partial H_2}{\partial \underline{\lambda}_3} = 0; \quad \underline{\lambda}_3 \geq 0 \tag{72}$$

$$\frac{\partial H_2}{\partial \mu_1} \leq 0; \quad \mu_1 \frac{\partial H_2}{\partial \mu_1} = 0; \quad \mu_1 \geq 0 . \tag{73}$$

At the second level the following necessary conditions have to be satisfied:

$$\frac{\partial H}{\partial \underline{\sigma}} = 0; \quad \frac{\partial H}{\partial \underline{\lambda}_4} = 0 . \tag{74}$$

At the third level ε will be determined as in the static case by successively decreasing ε whenever convergence is achieved at the second level, until a further decrease will produce no feasible solution. Example problems will be discussed in the following sections.

C. The Parametric Approach for Static Identification and Static Optimization

Forming the Lagrangian to the parametric formulation presented by (8), the unconstrained problem is

$$\min_{\underline{U},\underline{\alpha},\underline{\mu}} \{\theta G(\underline{\alpha}) + (1-\theta) f(\underline{X},\underline{U},\underline{\alpha}) + \underline{\mu}^T \underline{g}(\underline{X},\underline{U},\underline{\alpha})\} \tag{75}$$

$$0 < \theta < 1$$

where $\underline{\mu}$ is an m-dimensional vector of Kuhn-Tucker multipliers for appending the system constraints.

Decomposition of (75) is achieved by introducing the

pseudo-variables $\underline{\sigma}$ wherever the controls, $\underline{U}$, and model parameters, $\underline{\alpha}$, are coupled. The constraint, $\underline{\sigma} = \underline{\alpha}$ is then forced. Form the new overall system Lagrangian L as follows:

$$L(\underline{X},\underline{U},\underline{\alpha},\sigma,\underline{\mu},\lambda,\theta) = \theta G(\underline{\alpha}) + (1-\theta)f(\underline{X},\underline{U},\underline{\sigma}) + \underline{\mu}^T\underline{g}(\underline{X},\underline{U},\underline{\sigma}) + \underline{\lambda}^T(\underline{\sigma}-\underline{\alpha}) \tag{76}$$

where $\underline{\lambda}$ is a k-dimensional vector of Lagrange multipliers. The decomposition adopted for the parametric approach here is based on the "feasible" decomposition presented by Brosilow et. al. [48]. Other decompositions are possible, for example the "nonfeasible" decomposition proposed by Lasdon and Durbeck [47] or by Bauman [36].

Assume the pseudo-variables, $\underline{\sigma}$, and the parameter θ in the Lagrangian L are known at the first level. Then the system Lagrangian L may be decomposed into the following "Independent" subLagrangians

$$L(\underline{X},\underline{U},\underline{\alpha},\underline{\sigma},\underline{\mu},\underline{\lambda},\theta) = L_1(\underline{\alpha},\underline{\lambda},\underline{\sigma},\theta) + L_2(\underline{X},\underline{U},\underline{\mu},\underline{\sigma},\theta) \tag{77}$$

where

$$L_1(\underline{\alpha},\underline{\lambda};\underline{\sigma},\theta) = \theta G(\underline{\alpha}) + \underline{\lambda}^T(\sigma-\underline{\alpha})$$

and

$$L_2(\underline{X},\underline{U},\underline{\mu},\underline{\sigma},\theta) = (1-\theta)f(\underline{X},\underline{U},\underline{\sigma}) + \underline{\mu}^T\underline{g}(\underline{X},\underline{U},\underline{\sigma}) .$$

Thus the task of the first level is to:

subsystem 1:

$$\min_{\underline{\alpha},\underline{\lambda}} L_1(\underline{\alpha},\underline{\lambda};\underline{\sigma},\theta) \tag{78}$$

subsystem 2:

$$\min_{\underline{U},\underline{\mu}} L_2(\underline{X},\underline{U},\underline{\mu};\underline{\sigma},\theta) . \tag{79}$$

Where $\underline{\sigma}$ and θ are assumed known at the first level and to be determined by higher levels, the following are the necessary conditions for stationarity at the first level:

subsystem 1:

$$\left.\begin{aligned} \frac{\partial L_1}{\partial \underline{\alpha}} &= \frac{\partial G}{\partial \underline{\alpha}} - \underline{\lambda} = 0 \\ \frac{\partial L_1}{\partial \underline{\lambda}} &= \underline{\sigma} - \underline{\alpha} = 0 \end{aligned}\right\} \tag{80}$$

subsystem 2:

$$\frac{\partial L_2}{\partial \underline{U}} = (1 - \theta) \frac{\partial f}{\partial \underline{U}} + \underline{\mu}^T \frac{\partial \underline{g}}{\partial \underline{U}} = 0$$

$$\frac{\partial L_2}{\partial \underline{\mu}} = \underline{g}(\underline{X},\underline{U},\underline{\sigma}) \leq 0; \quad \underline{\mu} \geq 0; \quad \underline{\mu}^T \underline{g}(\underline{X},\underline{U},\underline{\sigma}) = 0 .$$

A second level problem can then be defined to determine the optimum value, $\underline{\sigma}^*$, used in the two first level subproblems. The parameter θ will be determined at the third level. Thus the second level problem is

$$\min_{\underline{\sigma}} \theta G[\underline{\alpha}^*(\underline{\sigma})] + (1 - \theta) f[\underline{X},\underline{U}^*(\underline{\sigma})] + \underline{\lambda}^T[\underline{\sigma} - \underline{\alpha}^*(\underline{\sigma})]$$
$$+ \underline{\mu}^*(\underline{\sigma})^T \underline{g}[\underline{X},\underline{U}^*(\underline{\sigma})] . \tag{82}$$

If analytic solution of the first level problem is possible, i.e., $\underline{U}^*(\underline{\sigma})$, $\underline{\alpha}^*(\underline{\sigma})$, $\underline{\mu}^*(\sigma)$ and $\underline{\lambda}^*(\sigma)$, then the second level controller will easily yield the integrated problem solution. Since this is generally not possible, the solution is then reached by iterating between the first and second levels as illustrated in Figure 6.

The second level controller determines the values of the pseudo-variables, $\underline{\sigma}$, used in the first level optimizations by the following algorithm.

$$\underline{\sigma}^{I+1} = \underline{\sigma}^I - K\underline{D} \tag{83}$$

where I is the iteration number and K is taken to be the

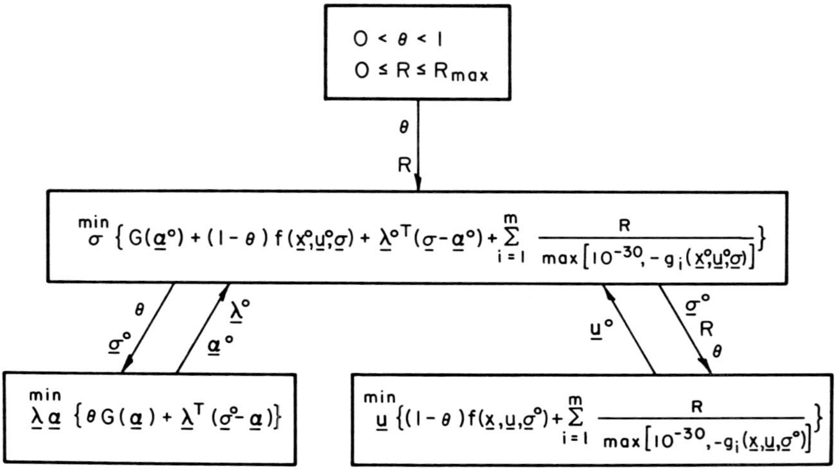

Fig. 6. Feasible Decomposition of Joint Problem.

step size which minimizes the joint problem down the search direction, $\underline{D}$. Brosilow et. al. [48] used a steepest desent technique in which $\underline{D}$ is defined to be the gradient of the system Lagrangian, $(\nabla_{\underline{\sigma}} L)$. More rapid convergence to the joint problem solution can be expected through the use of one of the second order conjugate gradient minimization algorithms. In the example problems to follow, the search direction, $\underline{D}$, is determined by a modified Fletcher-Powell algorithm [47].

It is shown in [48] that convergence to the integrated problem solution when using Eq. (83) as a second level controller requires that each subsystem be minimized to at least a local optimum for any choice of the pseudo variables, $\underline{\sigma}$. This should always be possible for subsystem 2. It is also true for subsystem 1, since it is not an optimization problem, but, has the following unique solution, provided that $\nabla_{\underline{\alpha}} G(\underline{\alpha})$ exists.

$$\left.\begin{array}{l} \underline{\lambda}^{*} = \theta \nabla_{\underline{\alpha}} G(\underline{\alpha}); \quad \underline{\alpha}^{*} = \underline{\sigma} \\ \theta \text{ and } \underline{\sigma} \text{ are given by the higher levels.} \end{array}\right\} \qquad (84)$$

The following Kuhn-Tucker conditions for the Lagrangian L, are difficult to handle computationally.

$$\underline{g}(\underline{X}^\circ,\underline{U}^\circ,\underline{\sigma}^\circ) \leq 0; \quad \underline{\mu} \geq 0; \quad \underline{\mu}^{\circ T}\underline{g}(\underline{X}^\circ,\underline{U}^\circ,\underline{\sigma}^\circ) = 0 .$$

A preferred means of handling system inequality contraints is via penalty functions. In some of the example problems to follow, a modified Fiacco-McCormic penalty function [49] was utilized. This penalty is particularly appropriate for the feasible decomposition since then the solution is always feasible. It has the disadvantage however in that an initial feasible solution must be known. Thus, for the constraint

$$g_i(\underline{X},\underline{U},\underline{\alpha}) \leq 0, \qquad i = 1,2,\ldots,m$$

the penalty function is

$$\frac{R}{\max[10^{-30},-g_i(\underline{X},\underline{U},\underline{\alpha})]}$$

where R is a penalty factor which is decreased to zero. The incorporation of the penalty function is illustrated in the decomposition of Figure 6. The penalty factor, R, is decreased simultaneously as θ is increased at the third level.

At the third level, θ is initialized at 10^{-20} to guarantee that the solution to the joint problem would be approached from within the efficient region. Then θ is increased in steps.

V. QUASILINEARIZATION APPROACH

The ε-constraint formulation for the joint static identification and dynamic optimization problem posed in (16) may be solved via quasilinearization [50,7,14]. The Hamiltonian $\mathcal{H}$ for the system (16) is given by

$$\mathcal{H} = f(\underline{X},\underline{U},\underline{\alpha},t) + \underline{\lambda}_1^T\underline{F}(\underline{X},\underline{U},\underline{\alpha},t) + \underline{\lambda}_2^T\underline{g}(\underline{X},\underline{U},\underline{\alpha}) \qquad (85)$$

where $\underline{\lambda}_1$ is an n-dimensional vector of Lagrange multipliers, $\underline{\lambda}_2$ is an m-dimensional vector of Lagrange multipliers and the constraints

$$G(\underline{\alpha}) \leq \varepsilon; \quad \varepsilon \geq \delta > 0 \tag{86}$$

will be satisfied at a later stage of the optimization process. Necessary conditions for stationarity are

$$\nabla_{\underline{U}} \mathcal{H} = 0, \; \nabla_{\underline{X}} \mathcal{H} = -\dot{\underline{\lambda}}_1, \; \nabla_{\underline{\lambda}_1} \mathcal{H} = \dot{\underline{X}} \tag{87}$$

$$\underline{\lambda}_2 \geq 0, \; \underline{\lambda}_2^T \underline{g}(\underline{X},\underline{U},\underline{\alpha}) = 0, \; \underline{g}(\underline{X},\underline{U},\underline{\alpha}) \leq 0 . \tag{88}$$

The conditions (87)-(88) yield the following two-point boundary-value problem:

$$\begin{aligned} \dot{\underline{X}} &= \underline{\varphi}_1(\underline{X},\underline{U},\underline{\alpha},\underline{\lambda}_1,\underline{\lambda}_2), \; \underline{X}(t_0) = \underline{X}_0 \\ \dot{\underline{\lambda}}_1 &= \underline{\varphi}_2(\underline{X},\underline{U},\underline{\alpha},\underline{\lambda}_1,\underline{\lambda}_2), \; \underline{\lambda}_1(t_f) = 0 \end{aligned} \tag{89}$$

where $\underline{\varphi}_1$ and $\underline{\varphi}_2$ are n-dimensional vectors of functions. At each iteration of the computational procedure, the Kuhn-Tucker stationarity conditions (88) can be satisfied by checking whether the constraints are active or inactive.

The solution of the system (86) and (89) is obtained via quasilinearization technique. Quasilinearization is an effective algorithm which proceeds by solving a sequence of linear initial-value problems such that the sequence of solutions converges to the solution of the nonlinear problem. This process is well suited for high-speed computers. Since $\underline{\alpha}$ is a constant vector of parameters, the ordinary differential equations $\dot{\underline{\alpha}} = 0$ can be added to Eq. (89) with unknown initial conditions $\underline{\alpha}_0$.

It is convenient to introduce the following canonical notation:

$$\underline{Z} = [\underline{X},\underline{\lambda}_1,\underline{\alpha}], \; \underline{\varphi} = [\underline{\varphi}_1,\underline{\varphi}_2,\underline{0}]$$

where $\underline{Z}$ is an M-dimensional vector, $\underline{\varphi}$ is an M-dimensional vector of functions, and $M = 2n + k$. The integrated problem can now be formulated compactly as

$$\dot{\underline{Z}} = \underline{\varphi}(\underline{Z}, \underline{U}, \underline{\lambda}_2), \quad \underline{Z}(t_0) = \underline{Z}_0$$
$$G(\underline{\alpha}) \leq \varepsilon, \quad \varepsilon \geq \delta > 0 . \tag{90}$$

The computational procedure for the integrated problem (90) consists of the following successive steps:

Step (a): Approximation Solution to the Ordinary Differential Equations (89). Generate numerically an approximate time history of $\underline{Z}(t)$, $\overline{\underline{Z}}(t)$. This can be done by either solving a two-point boundary-value problem with unknown $\underline{Z}_0$ and $\underline{\lambda}_1(t_f)$ or by solving an initial-value problem with unknown $\underline{Z}_0$, and then satisfying the boundary condition $\underline{\lambda}_1(t_f) = 0$ iteratively.

Step (b): Linearization of the System. Linearize the system (90) about $\overline{\underline{Z}}$ using Taylor series expansion. Since an iterative procedure is employed, it is convenient to denote the nth time histories of $\underline{Z}(t)$ by $\underline{Z}^{(n)}(t)$. Specifically,

$$\dot{\underline{Z}}^{(n)}(t) = J[\underline{Z}^{(n-1)}(t)]\, \underline{Z}^{(n)} + \underline{\varphi}(\underline{Z}^{(n-1)}(t)) - J[\underline{Z}^{(n-1)}(t)]\, \underline{Z}^{(n-1)}(t) \tag{91}$$

where

$$\underline{Z}^{(0)}(t) = \overline{\underline{Z}}(t)$$

and

$$J[\underline{Z}^{(n-1)}(t)] = [\partial \underline{\varphi}_i / \partial \underline{Z}_j]^{(n-1)}$$

is the Jacobian evaluated at the $(n-1)$st time history.

Step (c): Solution of the Linearized System. The homogeneous solution $\underline{Z}^h$ of the system (91) is obtained by solving the homogeneous part of (91) M times, with

$\underline{Z}_1^h(t_0) = [1,0,\ldots,0]$ the first time and with $\underline{Z}_M^h(t_0) = [0,0,\ldots,0,1]$ the Mth time. This yields the following results:

$$Z_i^h(t) = \sum_{j=1}^{M} Z_j(t_0)Z_{ij}^h(t), \quad i = 1,\ldots,M. \tag{92}$$

The particular solution $\underline{Z}^p(t)$ is obtained by integrating the system (91) for the initial conditions $\underline{Z}(t_0) = 0$. The general solution of the system (91) at the Nth iteration is thus given by

$$[Z_i(t)]^{(n)} = [Z_i^p(t)]^{(n)} + \left[\sum_{j=1}^{M} Z_j(t_0)Z_{ij}^h(t)\right]^{(n)},$$

$$=1,2,\ldots,M. \tag{93}$$

Step (d): Substitution of the Value of $\underline{Z}$ from (93) into (86). It was previously shown that the constraint (86) must be satisfied for a minimum feasible value of ε; that is, ε should be minimized, while the constraint (86) is still satisfied. This goal can be achieved by assuming that the equality in (86) holds for a feasible ε. Then, the minimization of G is equivalent to the minimization of ε. Now, the unknown parameters in G are the initial conditions $\underline{Z}_0$, which include $\underline{\alpha}$. Necessary conditions for minimizing G with respect to $\underline{Z}_0$ are

$$\nabla_{\underline{Z}_0} G = 0. \tag{94}$$

The system (94) yields M algebraic equations in $\underline{Z}_0$. Its solution provides new value for $\underline{Z}_0$.

Step (e): Checking of System Convergence. This can be done by either computing the new value of G and checking whether it is less than some maximum tolerated error or by checking the difference between the new and old values of $\underline{Z}_0$.

Step (f): Optimality. If convergence is achieved, stop the iterations and calculate the optimal control $\underline{U}$ from (87). Otherwise, return to step (b), and linearize the system of ordinary differential equations about a new time history of $\underline{Z}$ as the system initial conditions.

A numerical example is presented in the following section to illustrate the applications of quasilinearization to the joint problem.

VI. APPLICATIONS OF THE JOINT APPROACH

Every system model has its own structure, dimensionality, complexity, and other specific characteristics. Accordingly, the solution procedure to the system model should be carefully chosen. The integrated solution to the identification and optimization problem should not be rigid, but rather flexible and adaptive. In order to demonstrate the calculation procedure for a combined system optimization and identification problem, several example problems are introduced, formulated and solved.

A. A Production Plant: ε-Constraint Apprcach

A production plant is able to process N different types of materials, $U_1, U_2, \ldots, U_N$ into M different types of products $Y_1, Y_2, \ldots, Y_M$. It is assumed that defects in raw material or finished product occur with probabilities α_1 and α_2 respectively. The objective is to maximize the net revenue from the sales of the M types of products.

a. Problem Formulation

Consider the following process relationships:

(1) Normal total processing time is

$$\sum_{i=1}^{N} U_i t_i \tag{95}$$

where t_i is the unit processing time for the ith raw material.

(2) Process maintenance and delay time due to defects in raw material is:

$$\sum_{i=1}^{N} (\alpha_1 + \alpha_2 - \alpha_1\alpha_2)t_i U_i \tag{96}$$

(3) Let X be the time allocated for production by prior scheduling. Then by combining (95) and (96) the constraint on time becomes

$$\sum_{i=1}^{N} (1 + \alpha_1 + \alpha_2 - \alpha_1\alpha_2)t_i U_i \leq X \tag{97}$$

(4) The number of units of the kth product are given by

$$Y_k = [1 - (\alpha_1 + \alpha_2 - \alpha_1\alpha_2)] \sum_{i=1}^{N} a_{ik} U_i \tag{98}$$

where a_{ik} is the number of units of the ith raw material needed to produce one unit of the kth product.

(5) The cost of the ith raw material is

$$b_i = b_{1i} + b_{2i} U_i \tag{99}$$

(6) The additional cost incurred by defective materials is assumed to be

$$C_1 = \alpha_1 \sum_{i=1}^{N} c_{1i} U_i^2 \tag{100}$$

The additional cost incurred by a defective product is assumed to be

$$C_2 = \alpha_2 \sum_{i=1}^{N} C_{2i} U_i^2 \tag{101}$$

where C_{1i} and C_{2i} are given scalars dependent on the process.

Given some observations of the plant, the objective is to find an optimal production for each product such that the net profit is maximized. Since α_1 and α_2 are unknown, these parameters must also be identified. The objective function is denoted by

$$F(X,\underline{U},\underline{\alpha}) = \sum_{k=1}^{M} R_k Y_k - \sum_{i=1}^{N} \{K_1(1 + \alpha_1 + \alpha_2 - \alpha_1\alpha_2)t_i U_i + b_{1i} + b_{2i}U_i + \alpha_1 C_{1i} U_i^2 + \alpha_2 C_{2i} U_i^2\} \tag{102}$$

where R_k is the unit revenue for selling Y_k, and it may be negative if Y_k is waste rather than saleable product. Also, k_1 is the cost of one time unit in the production process. Let Y_{M+1} be the unused (slack) time.

$$Y_{M+1} = X - \sum_{i=1}^{N} (1 + \alpha_1 + \alpha_2 - \alpha_1\alpha_2)t_i U_i \; .$$

For convenience define

$$\begin{aligned} \underline{U} &= [U_1, U_2, \ldots, U_N]^T \\ \underline{\alpha} &= [\underline{\alpha}_1, \underline{\alpha}_2]^T \\ f(X,\underline{U},\underline{\alpha}) &= -F(X,\underline{U},\underline{\alpha}) \\ \underline{Y}^j &= [Y_1^j, Y_2^j, \ldots, Y_{M+1}^j]^T \; . \end{aligned}$$

Denoting the jth observed output vector by $\hat{\underline{Y}}^j$, the identification problem is

$$\min_{\alpha_1,\alpha_2} \sum_{j=1}^{J} [\hat{\underline{Y}}^j - \underline{Y}^j]^T W^j [\hat{\underline{Y}}^j - \underline{Y}^j] \; . \tag{103}$$

Substituting for $\underline{Y}^j$ from Eq. (98), the combined problem becomes

$$\min_{\underline{U},\underline{\alpha}} \sum_{i=1}^{N} \Big[k_1(1 + \alpha_1 + \alpha_2 - \alpha_1\alpha_2)t_iU_i + b_{1i} + b_{2i}U_i + \alpha_1 C_{1i}U_i^2 + \alpha_2 C_{2i}U_i^2 - (1 - \alpha_1 - \alpha_2 + \alpha_1\alpha_2) \cdot \sum_{k=1}^{M} R_k a_{ik} U_i \Big] \tag{104}$$

subject to the constraints

$$\left.\begin{aligned} & G(\underline{\alpha}) \leq \varepsilon \\ & \varepsilon \geq \delta > 0 \\ & \sum_{i=1}^{N} (1 + \alpha_1 + \alpha_2 - \alpha_1\alpha_2)t_iU_i \leq X \\ & 0 \leq \alpha_1 \leq 1 \\ & 0 \leq \alpha_2 \leq 1 \\ & U_i \geq 0 \qquad i = 1,2,\ldots,N \end{aligned}\right\} \tag{105}$$

where

$$G(\underline{\alpha}) = \sum_{j=1}^{J} \left\{ \sum_{k=1}^{M} \Big[\hat{Y}_k^j - (1 - \alpha_1 - \alpha_2 + \alpha_1\alpha_2) \cdot \sum_{i=1}^{N} a_{ik}\hat{U}_i^j \Big]^2 + \Big[\hat{Y}_{M+1}^j - X + \sum_{i=1}^{N} (1 + \alpha_1 + \alpha_2 - \alpha_1\alpha_2)t_i\hat{U}_i^j \Big]^2 \right\} \tag{106}$$

b. Computational Procedure and Results

The ε-constraint approach is a promising one, especially for large scale and nonlinear system models such as the production plant problem. In this case decomposition and multilevel optimization techniques seem a suitable solution procedure. Two different second-level controllers, the Gauss-Seidel [25] and the gradient [24] are employed for the solution of the above problem. The difference in the rate of convergence of the above second level controllers may determine their feasible applicability.

In order to decompose the system into independent parts it is convenient to introduce the pseudo variables σ_1 and σ_2 and require that

$$\begin{aligned} \sigma_1 &= \alpha_1 \\ \sigma_1 &= \alpha_2 \end{aligned} \tag{107}$$

Adding the new constraints (107) to the system (104) and (105) and forming the Lagrangian, L, yields:

$$\begin{aligned} \underline{L}(X,\underline{U},\underline{\alpha},\underline{\sigma},\underline{\lambda},\underline{\mu}) = \sum_{i=1}^{N} \Big[& k_1(1+\sigma_1+\sigma_2-\sigma_1\sigma_2)t_iU_i \\ & + b_{1i} + b_{2i} + U_i + \sigma_1 C_{1i}U_i^2 + \sigma_2 C_{2i}U_i^2 \\ & - (1-\sigma_1-\sigma_2+\sigma_1\sigma_2) \sum_{k=1}^{M} R_k a_{ik} U_i \Big] + \mu_1[G(\underline{\alpha}) - \varepsilon] \\ & + \mu_2 \Big[\sum_{i=1}^{N} (1+\sigma_1+\sigma_2-\sigma_1\sigma_2)t_iU_i - X \Big] \\ & + \lambda_1(\sigma_1-\alpha_1) + \lambda_2(\sigma_2-\alpha_2) \end{aligned}$$

where $\underline{\lambda}$ and $\underline{\mu}$ are the Kuhn-Tucker multipliers. The constraints $0 \le \alpha_1 \le 1$, $0 \le \alpha_2 \le 1$, $U_i \ge 0$, $i = 1,\ldots,N$ will be satisfied computationally by fixing the value of the variable on the boundary whenever the constraint is violated.

Several decomposition schemes are available. The one utilized here is termed the "nonfeasible" one since the constraints Eq. (107) are in general not satisfied unless convergence is achieved [36].

The decomposition of $L(X,\underline{U},\underline{\alpha},\underline{\sigma},\underline{\lambda},\underline{\mu})$ into two subsystems yields

$$L(X,\underline{U},\underline{\alpha},\underline{\sigma},\underline{\lambda},\underline{\mu}) = L_1(X,\underline{U},\underline{\sigma},\mu_2;\underline{\lambda}) + L_2(\underline{\alpha},\mu_1;\underline{\lambda}) \qquad (109)$$

where

$$\left.\begin{aligned} L_1(X,\underline{U},\underline{\sigma},\mu_2;\underline{\lambda}) = \sum_{i=1}^{N} \Bigg[& k_1(1+\sigma_1+\sigma_2-\sigma_1\sigma_2)t_iU_i \\ & + b_{1i} + b_{2i}U_i + \sigma_1 C_{1i}U^2 + \sigma_2 C_{2i}U_i^2 \\ & - (1-\sigma_1-\sigma_2+\sigma_1\sigma_2)\sum_{k=1}^{M} R_k a_{ik}U_i \Bigg] \\ & + \mu_2\left[\sum_{k=1}^{N}(1+\sigma_1+\sigma_2-\sigma_1\sigma_2)t_iU_i - X\right] \\ & + \lambda_1\sigma_1 + \lambda_2\sigma_2 \end{aligned}\right\} \qquad (110)$$

and

$$L_2(\underline{\alpha},\mu_1;\underline{\lambda}) = \mu_1[G(\underline{\alpha}) - \varepsilon] - \lambda_1\alpha_1 - \lambda_2\alpha_2 \,. \qquad (111)$$

In spite of the fact that difficulties in convergence arose in the gradient second-level controller [7], it is introduced here as a possible alternative to the Gauss-Seidel type second-level controller.

Figures 7 and 8 represent block diagrams of the Gradient and the Gauss-Seidel type second level controllers. This example problem was solved for the case having four raw materials and three finished products. Three observations of the system were used, and the third level computation converged to the solution of the identification problem ($\alpha_1 = 0 \cdot 1, \alpha_2 = 0 \cdot 2$) in 9 iterations. The input data are shown in Table 1 and the convergence of the optimization and the identification are given in Figure 9. The calculations were carried out on an IBM 360/91 digital computer.

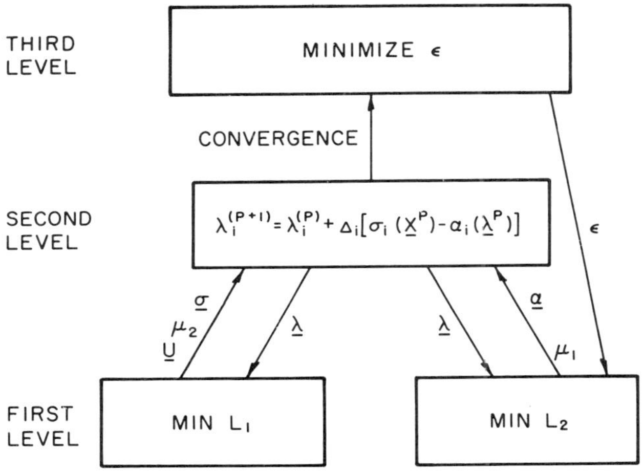

Fig. 7. Gradient Type Second-Level Controller for the Production Plant (at the P Iteration).

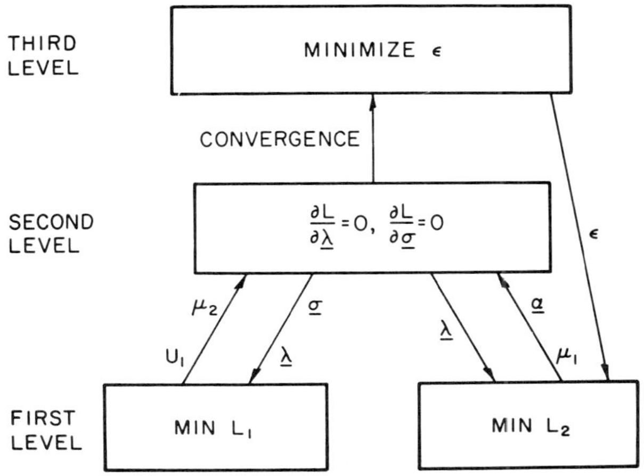

Fig. 8. Gauss-Seidel Type Second-Level Controller for the Production Plant.

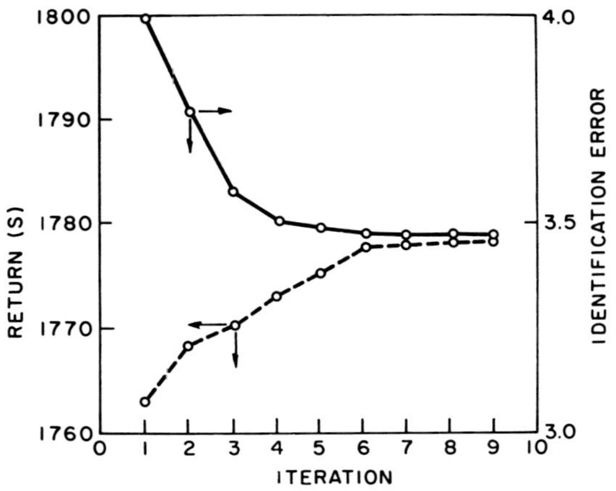

Fig. 9. Solution Convergence of the Integrated Problem.

t_i	1.0	0.9	0.95	0.85
b_{1i}	2.0	2.5	2.3	2.8
b_{2i}	0.1	0.2	0.5	0.18
C_{1i}	1.0	1.5	1.2	1.6
C_{2i}	1.5	1.4	1.5	1.9
a_{i1}	1.0	2.0	1.6	1.1
a_{i2}	1.2	1.5	1.2	0.9
a_{i3}	1.4	1.3	1.3	0.8
R_k	20.0	15.0	17.0	

TABLE 1

Production Plant Data.

B. A Chemical Process: Parametric Approach

This process is typical of many chemical processes and as such has been employed by Burghart [51] as a vehicle for demonstrating control approaches. Consider a stirred tank reactor, Figure 10, where two components X and Y, of concentrations X_0 and Y_0 enter the tank at a flow rate Q (McGrew and Haimes [21]).

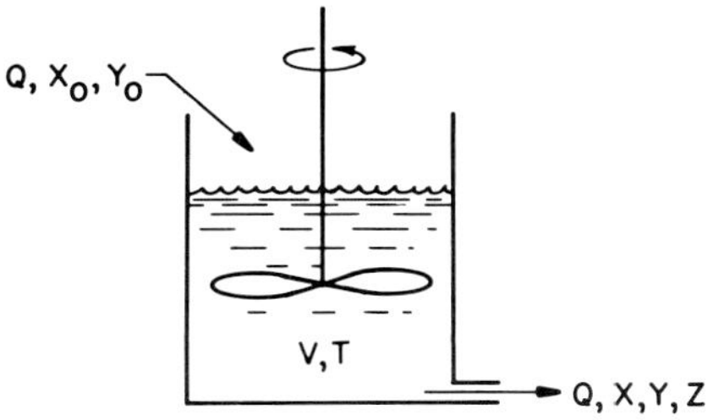

Fig. 10. Stirred Tank Reactor.

Let V be the volume of material in the reactor, T the temperature of the reactor, Q the flow rate of the material through the reactor, X_0 and Y_0 the concentrations of the input stream, and X, Y, Z be the concentrations of the output stream components. The assumption is that the flow rate Q is constant and that the temperature and composition are uniform throughout the reactor. By controlling the volume, V, and the temperature, T, within the reactor, it is desired to maximize the total profit from this process.

The process may be represented symbolically by

$$X \xrightarrow{K_1} Y \xrightarrow{K_2} Z .$$

Where K_1, K_2 are reaction rate coefficients, which in turn are functions of the controls, T and V, as shown later. The steady state relationships are

$$X_0 - X(1 + K_1) = 0 \tag{112}$$

$$Y_0 - Y(1 + K_2) + K_1 X = 0 \tag{113}$$

$$Z - K_2 Y = 0 . \tag{114}$$

Thus the identification problem is to determine the parameters K_1 and K_2 which best describe the process. Discrete observations of Y, V, and T are used for this purpose. From a knowledge of the process, it is assumed that

K_1 and K_2 can be approximated by,

$$K_1 = [A_1 V \exp(-B_1/T)]/Q \tag{115}$$

$$K_2 = [A_2 V \exp(-B_2/T)]/Q \tag{116}$$

and thus the identification problem becomes that of determining the coefficients A_1, A_2, B_1, and B_2. A least squares criterion is used for the identification error function. It is assumed that the identification and optimization must also be subject to the following system constraints:

$$T \geq 250.\ ^\circ K; \quad 0 \leq V \leq 200.\ M^3;$$

$$B_1 \geq 3000.\ ^\circ K; \quad B_2 \geq 2000.\ ^\circ K.$$

Given the following simulated observations in Table 2 for $A_1 = 1400.$; $A_2 = 8.$; $B_1 = 4000.$; $B_2 = 2500.$; $Q = 10.$; $X_0 = 0.9$; $Y_0 = 0.1$:

$\hat{T}(^\circ K)$	$\hat{V}(M^3)$	$\hat{Y}$
250.	65.0	.1066
270.	80.0	.1278
290.	95.0	.1827
310.	110.	.2737
330.	125.	.3563
350.	140.	.3780
370.	155.	.3410
390.	170.	.2793
410.	185.	.2181

TABLE 2

Simulated Observations

The identification problem can be formulated as follows:

$$\min_{A_1, A_2, B_1, B_2} G(A_1, A_2, B_1, B_2) \tag{117}$$

Subject to:

$$\left.\begin{array}{l} T \geq 250. \\ 0 \leq V \leq 200 \\ B_1 \geq 3000 \\ V_2 \geq 2000 \\ Y = [Y_0 + (X_0 + Y_0)K_1]/(1 + K_1)(1 + K_2) \\ K_i = [A_i V \exp(-B_i/T)]/Q \end{array}\right\} \quad (118)$$

Where

$$G(A_1,A_2,B_1,B_2) = \sum_{j=1}^{9} [Y(\hat{V}^j,\hat{T}^j,A_1,A_2,B_1,B_2) - \hat{Y}^j]^2 .$$

To optimize the process, the volume and temperature which result in the maximization of the performance function must be found. That is,

$$\max_{T,V} f(T,V)$$

where

$$f(T,V) = C_1YQ - C_2(Z - C_3)^2 - C_4T^4 - C_5T^4 - C_5T^2Q - (C_6Q - C_7)$$

and

$$C_1 = 2.5\times10^4,\ C_2 = 5.0\times10^6,\ C_3 = 5.0\times10^{-3},$$
$$C_4 = 3.0\times10^{-6},\ C_5 = 5.,\ C_6 = 3590.,\ C_7 = 2.65\times10^4 .$$

The first term of the performance function represents the value of the desired product, Y. The second term represents a loss due to high concentrations of Z. The third term represents heat losses due to radiation. The fourth term represents the cost of heating the mixture, and the last term corresponds to the cost associated with the input stream.

The model, (112)-(114), was simulated using the parameter values of $A_1 = 1400.$, $A_2 = 8.$, $B_1 = 4000.$, $B_2 = 2500.$, and

$Q = 10$, $X_0 = .9$, $Y_0 = .1$, to generate the hypothetical data given above. Through a feasible decomposition [36], the joint problem was solved by multilevel optimization [30], the optimization was also done by a separate treatment, i.e., identification followed by optimization for comparison purposes.

An independent search of the parameter space indicated that the identification problem possesses a unique global minimum at the true parameter values. However, since the output concentration Z was assumed not to be observable, the estimation problem became insensitive to the model parameters near the true optimal solution. Because of computer accuracy, neither the joint nor the separate solutions could find the theoretical optimum solution. This is of course typical of most realistic optimization problems. It was, however, possible for both treatments to minimize the identification norm to less than 10^{-5}.

In both treatments, the separate and the joint, the Fletcher-Powell minimization algorithm was used. The system constraints were handled via Fiacco-McCormic type penalty functions having the form

$$\varphi(\underline{X},\underline{U},\underline{\alpha}) = \frac{R}{\max\{10^{-20}, g_i(\underline{X},\underline{U},\underline{\alpha})\}}$$

where $\varphi(\underline{X},\underline{U},\underline{\alpha})$ is the penalty for violating the constraint $g_i(\underline{X},\underline{U},\underline{\alpha}) \leq 0$ and R is a constant penalty factor. These pnealities were added to the identification and performance criteria.

In the case of separate identification and optimization the constrained optimum solution was arrived at by solving the identification problem and then the optimization problem for a sequence of penalty factors, R. Each successive solution used a smaller value of R. In the integrated approach, R was decreased only when the parameter θ was increased being zero when θ was unity. This approach reduced the computational effort involved in using penalty functions at the expense of not resulting in an "efficient" solution to the joint problem until $\theta = 1$. The joint treatment predicted a better performance than the separate treatment (see Table 5). Furthermore, when the calculated controls of the respective solutions were applied to the exact system model (the parameters previously obtained via simulation are considered to be the "exact" system parameters), the true performance due to a joint treatment was 176.2 (the theoretical optimum is 208.). Separate identification followed by optimization resulted in a performance of only 47.5. Further improvement in the separate solution was not possible by initializing the parameter search at various points, indicating that the primary source of error was due to the limited computer accuracy. It can be expected that under identical conditions (starting point, computer word length, stopping criteria, search technique, etc.) the joint solution of the system model will always predict at least a comparable performance as a separate solution since it is always operating within the efficient region.

While the computation time for the joint treatment was approximately four times that of the separate treatment, the notable improvement in performance makes it a tractable approach for this process. A summary of the result for both approaches is tabulated in Tables 3 and 4, while a comparison is tabulated in Table 5. Convergence to the optimal controls is shown in Figure 11.

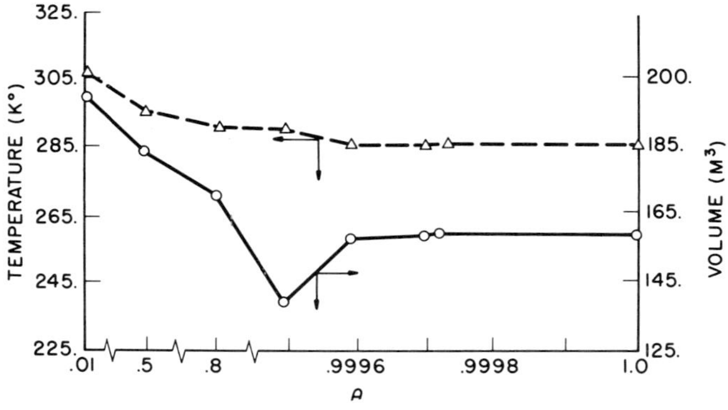

Fig. 11. Convergence to Reactor Optimal Controls.

The predicted performance values, $f(\underline{X}^*,\underline{U}^*,\underline{\alpha}^*)$, of Table 3, as θ is increased, are not monotonic as indicated by Theorem 2. This occurs because the penalty factors are decreased simultaneously as θ is increased.

θ	A_1	B_1	A_2	B_2	V	T	$f(\underline{X}^\circ,\underline{U}^\circ,\underline{\alpha}^\circ)$	$G(\underline{\alpha}^\circ)$
.01	1389.	3002.	.0384	2574.	196.	307.2	1.8×10^6	3.4
.8	1114.	3926.	15.1	2743.	196.	283.	5056.	$.4\times10^{-3}$
.9995	1212.	3950.	7.05	2453.	137.	287.	48.6	$.11\times10^{-4}$
.9996	1219.	3955.	7.65	2485.	155.	284.5	358.	$.46\times10^{-5}$
.99966	1219.	3956.	7.70	2487.	154.6	284.6	377.	$.42\times10^{-5}$
.9997	1224.	3956.	8.19	2511.	156.	284.9	621.	$.166\times10^{-5}$
.99972	1230.	3964.	8.63	2532.	158.2	285.0	814.5	$.773\times10^{-6}$
1.000	1230.	3964.	8.63	2532.	158.8	284.9	814.8	$.772\times10^{-6}$

Note: $f(\underline{X}^\circ,\underline{U}^\circ,\underline{\alpha}^\circ)$ is a predicted performance.

TABLE 3

Joint Identification and Optimization of Reactor

Results of Identification of Reactor Parameters:	Results of Reactor Optimization:
$A_1 = 1074.$	$V = 200.$
$A_2 = 9.5$	$T = 279.3$
$B_1 = 3929.$	$f(\underline{X}^\circ, \underline{U}^\circ, \underline{\alpha}^\circ) = 1610.$
$B_2 = 2574.$	$G(\underline{\alpha}^\circ) = .3 \times 10^{-5}$

Note: $f(\underline{X}^\circ, \underline{U}^\circ, \underline{\alpha}^\circ)$ is a predicted performance.

TABLE 4

Separate Identification and Optimization of Reactor.

	Separate Treatment	Joint Solution	Theoretical Optimum
A_1	1074.	1230.	1400.
A_2	9.5	8.63	8.0
B_1	3929.	3964.	4000.
B_2	2572.	2532.	2500.
V	200.	158.8	145.
T	279.3	284.9	286.
$f(\underline{X}^\circ, \underline{U}^\circ, \underline{\alpha}^\circ)$	1610.	814.8	---
$G(\underline{\alpha}^\circ)$	$.3 \times 10^{-5}$	$.772 \times 10^{-6}$	0
$f(\underline{X}^\circ, \underline{U}^\circ, \underline{\alpha}^*)$	47.5	176.2	208.

Note: $\underline{U}^\circ$, $\underline{\alpha}^\circ$ are estimated (predicted) values and $\underline{U}^*$, $\underline{\alpha}^*$ are exact values.

TABLE 5

Comparison of Separate and Joint Solutions to Reactor Optimization.

C. A Circuit Design Problem: Parametric Approach

Many design problems can be formulated as vector minimization. For example, consider the circuit design problem taken from Calahan [52]. Suppose a filter is to be built having the following gain vs. frequency characteristics, as given by Table 6, (McGrew and Haimes [21]).

$\hat{W}$(RADIANS/SEC)	$\hat{H}(\hat{W})$
1000.	.9901
5000.	.8000
10,000.	.5000
15,000.	.3077
20,000.	.2000

TABLE 6

Gain vs. Frequency Characteristics

Where W is the angular frequency of the input signal and H(W) is the circuit gain. The circuit structure to be used is the RLC circuit given in Figure 12. The circuit gain is given in (119) as follows:

$$H(W) = 1. \Big/ \sqrt{(1 - W^2LC)^2 + (WRC)^2}\,. \tag{119}$$

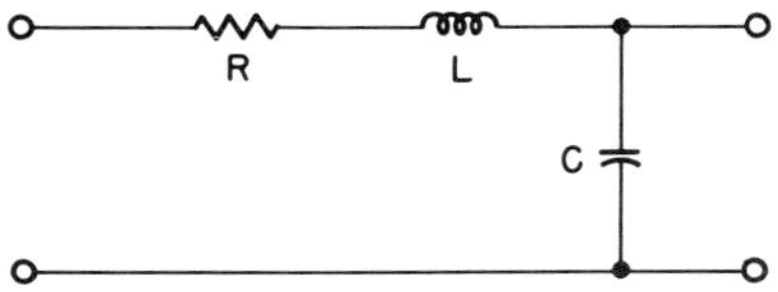

Fig. 12. Electrical Circuit Design.

Quite obviously, there is an infinite number of values of resistance, R, inductance, L, and capacitance, C, which can be used to "match" the given frequency characteristics.

If the filter is to be mass produced, then the values of R, L, C should be chosen to minimize production cost of the circuit since large values of capacitance and inductance tend to be expensive. Likewise, for consumer needs, the source impedance, R, should be kept relatively large to prevent "loading."

Suppose, however, that resistors, capacitors, and inductors in the range

$$10 \text{ Ohms} \leq R \leq 1 \text{ Kohm}$$
$$1 \text{ m henry} \leq L \leq 100 \text{ m henrys}$$
$$.3\ \mu \text{ Farads} \leq C \leq 100\ \mu \text{ Farads}$$

are available and inexpensive. It is desired to design the filter with the above characteristics, but with as large a source impedance as possible. If identification were attempted without simultaneous optimization, only the products LC and RC of (119) could be determined uniquely. The optimization problem then would be to find the largest resistance, R, that satisfies these products and the system constraints. Confidence in a trial and error solution, which is a very inefficient procedure, cannot be had without a reasonably exhaustive search. This problem can be solved by the joint identification and optimization problem below.

$$\min_{R,C,L} \ [-R, G(R,L,C)]$$

subject to:

$$\left.\begin{array}{l} 10 \leq R \leq 1000. \\ .001 \leq L \leq .1 \\ .3 \times 10^{-6} \leq C \leq 1. \times 10^{-4} \end{array}\right\} \quad (120)$$

where $G(R,L,C) = \sum_{j=1}^{5} [H(\hat{W}^j) - \hat{H}(\hat{W}^j)]^2$ represents the identification criterion and (-R) represents the performance criterion.

The bicriterion problem of (120) was solved using the parametric approach to the joint problem. The resulting optimum values were

$$R = 663.\ \text{Ohms}$$
$$L = .033\ \text{Henrys}$$
$$C = .301\ \mu\ \text{Farads}\ .$$

Convergence of the identification and R to their optimum values is shown in Figure 13.

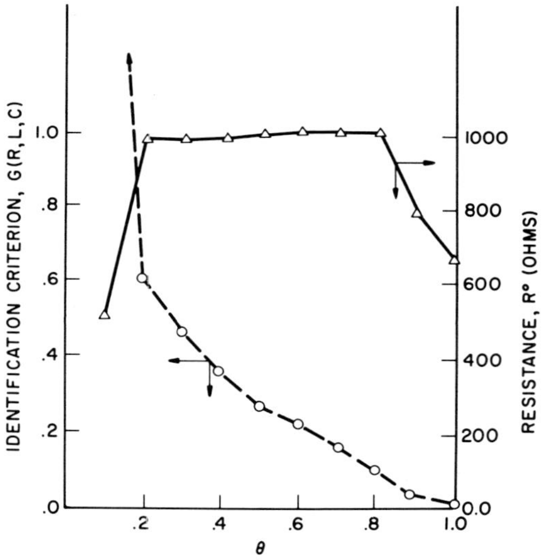

Fig. 13. Design Problem Identification and Performance Criteria Convergence.

D. A Mixing Process: Parametric Approach

The process consists of two mixer-settler type liquid-liquid extraction units [53]. The extractors remove a valuable component, A, from the cross-feed-raffinate by cross-current extraction. The concentration of A in the extract and raffinate phases are denoted by y and x respectively. The value of the extract is directly proportional to the output concentration y. The flows of raffinate and extract solvents to each stage are denoted by q and w_i respectively, see Figure 14. Applying conservation of material at each extractor gives the following relationships

$$\left.\begin{aligned} q(x_f - x_1) + w_1 y_1 &= 0 \\ q(x_1 - x_2) + w_2 y_2 &= 0 \\ y_1 &= \varepsilon_1 x_f \\ y_2 &= \varepsilon_2 x_1 \end{aligned}\right\} \qquad (121)$$

where x_f is the feed concentration and ε_1, ε_2 are the extractor efficiencies. It is assumed that ε_1 and ε_2 are related to the feed flow, q, in the following manner

$$\varepsilon_i = \hat{\varepsilon}_i \frac{q_{max} - q}{q_{max}} \qquad i = 1,2 \qquad (122)$$

where $\hat{\varepsilon}_i$ are static parameters and q_{max} is the maximum allowable feed flow.

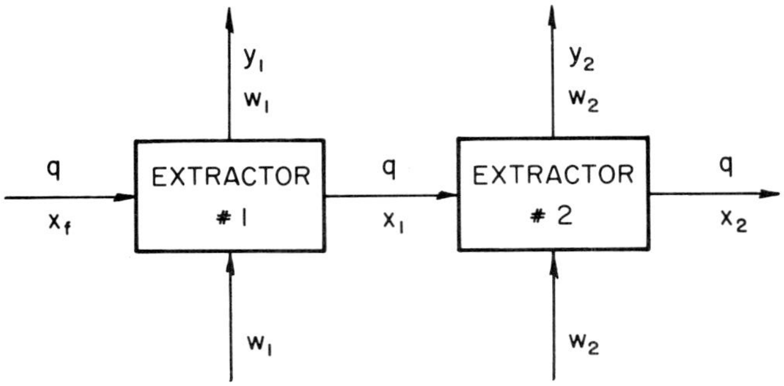

Fig. 14. Extractor Configuration.

For the purposes of optimization it is desired to choose the flows q, w_1, w_2 so as to maximize the criterion below:

$$\max_{q,w_1,w_2} f(q,w_1,w_2,\hat{\varepsilon}_1,\hat{\varepsilon}_2)$$

where

$$f(q,w_1,w_2,\hat{\varepsilon}_1,\hat{\varepsilon}_2) = w_1(y_1 - .025) + w_2(y_2 - .025) - P_f q x_f \,.$$

That is, the profit is equal to the value of the solute in the extract minus the cost of the solute, $.025(w_1 + w_2)$, and the cost of the feed is $p_f q x_f$.

The optimization requires a knowledge of the extractor efficiencies, ε_1 and ε_2. Thus, the parameters $\hat{\varepsilon}_1$ and $\hat{\varepsilon}_2$ must be determined. This is to be done by fitting, in a least squares sense, the observed data given in Table 7 to the process model (121-122).

$\hat{x}_f$	$\hat{q}$	$\hat{w}_1$	$\hat{y}_1$	$\hat{y}_2$
.2	.5	2.5	.0858	0.0
.2	1.0	2.0	.0736	.0199
.2	1.5	1.5	.06134	.0939
.2	2.0	2.5	.04912	.0347
.2	2.5	2.5	.0369	.0307
.2	2.5	2.0	.0369	.0321
.2	3.0	.5	.0247	.0247
.2	3.0	1.0	.0247	.0242
.2	1.0	.5	.0736	.0613

TABLE 7

Observed Data to the Process Model.

The resulting joint identification and optimization problem is

$$\min_{q,w_1,w_2,\hat{\varepsilon}_1,\hat{\varepsilon}_2} [-f(q,w_1,w_2,\hat{\varepsilon}_1,\hat{\varepsilon}_2),G(\hat{\varepsilon}_1,\hat{\varepsilon}_2)] \qquad (123)$$

subject to

$$y_1 = \hat{\varepsilon}_1 \; \frac{q - q_{max}}{q_{max}} \; x_f$$

$$y_2 = \hat{\varepsilon}_2 \; \frac{q - q_{max}}{q_{max}} \; x_1$$

$$q(x_1 - x_f) + w_1 y_1 = 0$$
$$q(x_2 - x_1) + w_2 y_2 = 0$$
$$0 \leq \hat{\varepsilon}_1 \leq 1$$
$$0 \leq \hat{\varepsilon}_2 \leq 1$$
$$0 \leq q \leq 4.01 = q_{max}$$
$$0 \leq w_1 \leq 4.01$$
$$0 \leq w_2 \leq 4.01$$

where

$$G(\hat{\varepsilon}_1, \hat{\varepsilon}_2) = \sum_{j=1}^{q} [(y_1^j - \hat{y}_1^j)^2 + (y_2^j - \hat{y}_2^j)^2] .$$

Analysis of $G(\hat{\varepsilon}_1, \hat{\varepsilon}_2)$ indicated that its minimum is strongly dependent on the ratio $\varepsilon_1/\varepsilon_2$. Thus, from a computational standpoint, there will likely be many equally valid solutions to the identification problem. An integrated approach to the identification and optimization problems is therefore desirable. Such an approach was taken by applying the feasible decomposition of Figure 6. The parameter θ was initialized at zero to guarantee that the solution to the joint problem would be approached from within the efficient region. θ was increased to .8 in steps of .1. The penalty factor, R, was decreased for each new value of θ. A θ equal to unity was then approached more slowly to simulate the limiting procedure of Section IIIB.

The optimization converged to the following solution

$$\hat{\varepsilon}_1^* = .49008$$

$$\hat{\varepsilon}_2^* = .50027$$

$$q^* = .0186$$

$$w_1^* = 4.01$$

$$w_2^* = .000218 .$$

The extractor efficiencies closely approximate the actual efficiencies of $\hat{\varepsilon}_1 = .49$ and $\hat{\varepsilon}_2 = .5000$. Convergence to the optimal solution are tabulated in Table 8.

The number of first level optimizations required for convergence to a solution, for a given value of the parameter θ, is also tabulated in Table 8. It is important to note that had separate identification been done, only one optimization would be required. However, this solution is not guaranteed to be "truly" optimal since it was done for a given choice of system parameters. To the contrary, through an integrated approach, the solution is optimal over all possible system parameters. This can be particularly important when the system performance is highly parameter dependent during which computational accuracy becomes very important.

E. A Chemical Plant: ε-Constraint Approach

In a given chemical plant [54], it is desired to minimize the cost of the waste products $X(t)$, produced in the course of the reaction, by choosing an optimal temperature policy $U(t)$ over the time interval $[0,1]$. The cost of the waste products is proportional to the square of the waste products; and there is a cost associated with the temperature policy which is proportional to the square of the temperature applied to the reaction. The objective functional is

θ	$G \times 10^4$	f	$\hat{\varepsilon}_1$	$\hat{\varepsilon}_2$	q	w_1	w_2	2nd level number of iterations	1st level number of optimizations
.098	.1498	.21689	.50085	.50512	.57823	3.875	.2100	5	26
.198	.1498	.2146	.50085	.50512	.58291	3.874	.2118	2	18
.298	.15204	.2371	.50093	.50334	.38317	3.885	.2048	10	63
.398	.15204	.2369	.50093	.50334	.38830	3.997	.18347	2	17
.498	.01579	.2315	.49224	.49622	.3953	3.912	.1559	16	97
.598	.01579	.2333	.49224	.49622	.4025	3.932	.1235	2	19
.698	.01511	.2390	.49268	.49730	.3833	3.954	.09036	3	22
.798	.00896	.2508	.49271	.50123	.3095	3.973	.06072	14	72
.9994	$.39 \times 10^{-4}$	.2906	.49008	.50027	.01862	4.01	$.218 \times 10^{-3}$	3	14

TABLE 8

Summary of Results

$$\min_{U} \frac{1}{2} \int_0^1 [\alpha X(t)^2 + U(t)^2]dt \tag{124}$$

where α is an unknown cost coefficient for waste. It is assumed that $\alpha > 0$. The rate of change of production of waste products $\dot{X}(t)$ is linearly related to the production of waste products $X(t)$ and the temperature of the reaction $U(t)$ by

$$\dot{X}(t) = X(t) - \alpha U(t) , \tag{125}$$

with unknown initial condition

$$X(0) = \text{free} . \tag{126}$$

It is assumed that J discrete observations on the system are given; in particular, the state variable X can be observed. Denote the observed X at the jth discrete observation by $\hat{X}^j$ and the solution of the system (124)-(126) at the jth discrete observation by X^j_{cal}. Note that X^j_{cal} is given in terms of α and the known values of U^j, $\hat{U}^j$.

The following four observations are available on the state variable X for a control $\hat{U}^j$, $j = 1,2,3,4$, at times $t = 0$, $t = 0.34$, $t = 0.67$, and $t = 1$, respectively:

$$\begin{aligned} X(0) &\equiv \hat{X}^1 = 1.00000, \quad X(0.34) \equiv \hat{X}^2 = 1.04962, \\ X(0.67) &\equiv \hat{X}^3 = 1.27307, \quad X(1) \equiv \hat{X}^4 = 1.70984. \end{aligned} \tag{127}$$

Then the identification problem can be formulated as follows:

$$\min_{\alpha} \sum_{j=1}^{J} [X^j_{cal} - \hat{X}^j]^2 . \tag{128}$$

The formulation of the joint system identification and optimization problem following the ε-constraint approach, is as follows (Haimes and Wismer [14]):

$$\min_{U,\alpha} \int_0^1 (\alpha X(t)^2 + U(t)^2)dt ,$$

subject to the constraints

$$\dot{X}(t) = X(t) - \alpha U(t),\ X(0) = 1$$

$$\sum_{j=1}^{J} [X^j_{cal} - \hat{X}^j]^2 \leq \varepsilon, \quad \varepsilon \geq \delta > 0 . \tag{129}$$

The canonical equations (90) for the test example are

$$\begin{aligned} &\dot{X} = X - \alpha^2\lambda,\ X(0) = 1 ,\\ &\dot{\lambda} = -\alpha X - \lambda,\ \lambda(1) = 0 ,\\ &\dot{\alpha} = 0 \\ &\sum_{j=1}^{4} [X^j_{cal} - \hat{X}^j]^2 \leq \varepsilon, \quad \varepsilon \geq \delta > 0 . \end{aligned} \tag{130}$$

Both Runge-Kutta and Hamming's modified predictor-corrector integration methods gave good results. Convergence of the quasilinearization technique was achieved within five iterations. The optimal values of U, X, λ, α are given in Figure 15, where $\alpha(0) = 0.72$ was used as an initial guess. The convergence of the error function to $\varepsilon = 10^{-7}$ after five iterations is given in Figure 16. The calculations were carried out on an IBM 360/91 digital computer using double-precision arithmetic.

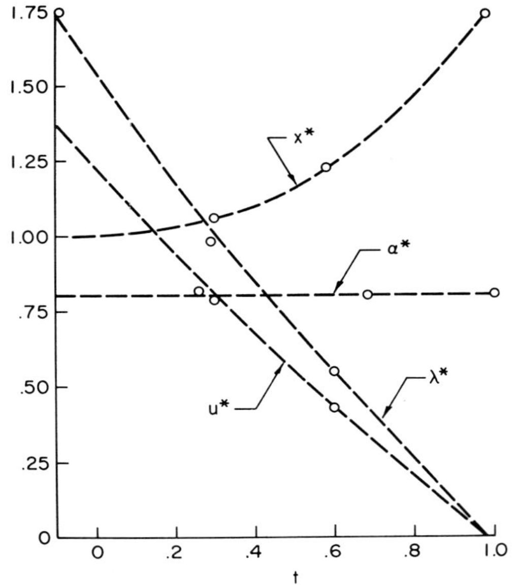

Fig. 15. Optimal Trajectory for $\varepsilon = 2 \times 10^{-7}$.

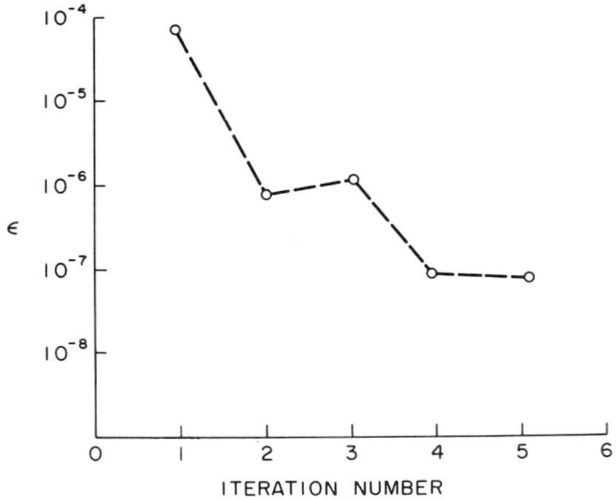

Fig. 16. Error Function Convergence.

VII. EPILOGUE

System modeling is both an art and a science. The best system model that a system analyst can produce to represent a real system may still be deficient due to a lack of information or a system's complexity. Thus, while the systems analyst strives in the modeling and optimization of the real system, he may never discover whether the "optimal control policy" derived from the system model is truly optimal for the real system.

In this chapter some aspects of system modeling and optimization have been analyzed and theoretical, as well as computational, foundations have been established. However, many important questions have risen and have been left unanswered. Basic research and results are needed to answer questions such as the following:

(i) What conditions must be met by the model structure to guarantee parameter identifiability.

(ii) How can techniques such as sensitivity analysis be used to improve the accuracy of parameters estimates, given the limitations imposed by digital computer methods of identification.

(iii) What input and output variables must be measured to guarantee the identification of the system parameters.

(iv) Since the problem of finding a class of model structures for physical systems is as important as finding the parameters of that structure, analytical techniques and procedures should be developed for that purpose.

(v) How does the choice of the input applied to the system affect the identification of the system parameters? Can an improper choice lead to the result that some parameters cannot be identified?

(vi) If certain parameters cannot be uniquely identified, will this necessarily result in a suboptimal performance level for the physical system?

(vii) What criteria can be applied to guarantee that the optimal control policy derived from the system model will yield satisfactory results when applied to the physical system. Under what conditions is the minimization of the identification norm sufficient to guarantee such confidence.

These are but a few of the questions that require future research and deserve proper answers.

APPENDIX A

THEOREM A-1. If $\underline{\beta}(\theta)$ is an optimal solution function for (25), then $f(\theta)$ and $G(\theta)$ are unimodal for θ over the open interval $(0,1)$.

Proof. The procedure for this proof is inspired by Geoffrion's [17] work. Let $\underline{\beta}(\theta_i)$, $i = 1,2,3$ solve (25), where $0 < \theta_1 < \theta_2 < \theta_3 < 1$. Then the theorem implies that there exists a t, $0 \leq t \leq 1$, such that

$$f(\theta_2) \leq tf(\theta_1) + (1-t)f(\theta_3) \tag{A1}$$

$$G(\theta_2) \leq tG(\theta_1) + (1-t)G(\theta_3)\,. \tag{A2}$$

Clearly if a $0 \leq t \leq 1$ exists, then there is a solution to the following linear programming problem:

$$\min_t t\,. \tag{A3}$$

Subject to:

$$[f(\theta_3) - f(\theta_1)]\, t \leq f(\theta_3) - f(\theta_2)$$

$$[G(\theta_3) - G(\theta_1)]\, t \leq G(\theta_3) - G(\theta_2)$$

$$t \leq 1\,.$$

Due to the duality of linear programming, if there exists a

solution to (A3), then there exists a solution to the dual,

$$\max_{\underline{Z}} \; [f(\theta_3) - f(\theta_2)]Z_1 + [G(\theta_3) - G(\theta_2)]Z_2 + Z_3 \qquad \text{(A4)}$$

subject to:

$$Z_i \geq 0 \qquad i = 1,2,3$$

$$[f(\theta_3) - f(\theta_1)]Z_1 + [G(\theta_3) - G(\theta_1)]Z_2 + Z_3 \geq 0$$

where $Z_i, i = 1,2,3$ are the dual variables.

Now, if the solutions to (A3) and (A4) exist, then they must be identical. Suppose the theorem is false. Then there exists no $t \geq 0$ that solves (A3) and the value of (A4) is strictly less than zero. Therefore, we have,

$$[f(\theta_3) - f(\theta_2)]Z_1 + [G(\theta_3) - G(\theta_2)]Z_2 + Z_3 < 0 \qquad \text{(A5)}$$

$$[f(\theta_3) - f(\theta_1)]Z_1 + [G(\theta_3) - G(\theta_1)]Z_2 + Z_3 \geq 0 \qquad \text{(A6)}$$

$$Z_i \geq 0 \qquad i = 1,2,3 . \qquad \text{(A7)}$$

Also note from (A5) that, all Z_i cannot be zero. Multipling (A5), by -1 and adding to (A6) and since $Z_3 \geq 0$

$$[f(\theta_2) - f(\theta_1)]Z_1 + [G(\theta_2) - G(\theta_1)]Z_2 \geq 0 \qquad \text{(A8)}$$

$$[f(\theta_3) - f(\theta_2)]Z_1 + [G(\theta_3) - G(\theta_2)]Z_2 < 0 . \qquad \text{(A9)}$$

Since Z_1 and Z_2 cannot both vanish

$$(1 - W)[f(\theta_2) - f(\theta_1)] + W[G(\theta_2) - G(\theta_1)] \geq 0 \qquad \text{(A10)}$$

$$(1 - W)[f(\theta_3) - f(\theta_2)] + W[G(\theta_3) - G(\theta_2)] < 0 \qquad \text{(A11)}$$

where

$$W = Z_2/(Z_1 + Z_2) \quad \text{and} \quad 0 \leq W \leq 1 .$$

Rearranging (A10) and (A11)

$$[WG(\theta_2) + (1-W)f(\theta_2)] - [WG(\theta_1) + (1-W)f(\theta_1)] \geq 0 \quad \text{(A12)}$$

$$[WG(\theta_3) + (1-W)f(\theta_3)] - [WG(\theta_2) + (1-W)f(\theta_2)] < 0 \,. \quad \text{(A13)}$$

DEFINITION A1. Define the function $V_i(\theta_j)$ as follows,

$$V_i(\theta_j) = \theta_j G(\theta_i) + (1-\theta_j)f(\theta_i) \,.$$

Then, since $\underline{\beta}(\theta_i)$, $i = 1,2,3$ solves (25), that is $V_i^*(\theta_j)$ was determined,

$$V_i(\theta_j) = \begin{cases} V^* & \text{for } i = j \\ \geq V^* & \text{for } i \neq j \end{cases}$$

and

$$V_1(\theta_1) - V_2(\theta_1) \leq 0 \quad \text{(A14)}$$

$$V_1(\theta_2) - V_2(\theta_2) \geq 0 \quad \text{(A15)}$$

$$V_3(\theta_3) - V_2(\theta_3) \leq 0 \quad \text{(A16)}$$

$$V_3(\theta_2) - V_2(\theta_2) \geq 0 \,. \quad \text{(A17)}$$

From (A12), (A13), and the previous definition it follows that:

$$V_1(W) - V_2(W) \leq 0 \quad \text{(A18)}$$

$$V_3(W) - V_2(W) < 0 \,. \quad \text{(A19)}$$

Clearly, $V_i(\theta_j)$ is linear in θ_j. Since $\theta_1 < \theta_2$, (A14), (A15), and (A18) imply $W \leq \theta_2$. Likewise, since $\theta_2 < \theta_3$, (A16), (A17), and (A19) imply $W > \theta_2$. This is a contradiction. Therefore the conclusion of the theorem is true, and thus the theorem is proved.

APPENDIX B. On the Saddle Point of the System Lagrangian

The adoption of state and control variables for static systems helped in drawing the analogy between static and dynamic systems and in the implementation of results obtained from one system to the other. This Appendix will be an exception since it is believed that the classical notion of decision variable, $\underline{X}$, in static systems along with the use of the Euclidean geometry can contribute to a better understanding of the saddle point concept. The material developed in this Appendix is based primarily on the work of L. S. Lasdon [24,55] and J. D. Schoeffler (Wismer, ed., [25]).

Consider the following problem

$$\min_{\underline{X} \in S} f(\underline{X})$$

where

$$\left.\begin{aligned} S_1 &= (E^n)^+ \\ S_2 &= \{\underline{X} : W_i(\underline{X}) = 0, \qquad i = 1,\ldots,M\} \\ S_3 &= \{\underline{X} : g_i(\underline{X}) \leq 0, \qquad i = 1,\ldots,N\} \\ S &= S_1 \cap S_2 \cap S_3 \end{aligned}\right\} \tag{B1}$$

$\underline{X}$ in an n-vector of real variables, $f(\underline{X})$, $W_i(\underline{X})$, and $g_i(\underline{X})$ are real-valued functions defined on S_1 and differentiable everywhere. The problem defined in (B1) is called the primal problem. Form the Lagrangian function, L:

$$L(\underline{X},\underline{\lambda},\underline{\mu}) = f(\underline{X}) + \sum_{i=1}^{M} \lambda_i W_i(\underline{X}) + \sum_{i=1}^{N} \mu_i g_i(\underline{X}) .$$

Necessary conditions for the point $(\underline{X}^\circ,\underline{\lambda}^\circ,\underline{\mu}^\circ)$ to be the solution to the problem posed in (B1) are the Kuhn-Tucker [20] stationary conditions given below, assuming the regularity conditions are fulfilled.

1.
$$\left.\begin{aligned} &\nabla_{\underline{X}} L(\underline{X}^\circ,\underline{\lambda}^\circ,\underline{\mu}^\circ) \geq 0 \\ &(\underline{X}^\circ)^T \nabla_{\underline{X}} L(\underline{X}^\circ,\lambda^\circ,\mu^\circ) = 0 \\ &\underline{X}^\circ \geq 0 \end{aligned}\right\} \tag{B1}$$

2. $W_i(\underline{X}^\circ) = 0, \quad i = 1,2,\ldots,M$ (B3)

3. $$\left.\begin{array}{l} g_i(\underline{X}^\circ) \leq 0 \\ \mu_i^\circ \, g_i(\underline{X}^\circ) = 0 \\ \mu_i^\circ \geq 0, \quad i = 1,2,\ldots,N \end{array}\right\} \tag{B4}$$

Although the above stationarity conditions are only necessary for differentiable functions in general, it is possible to prove that they are in fact necessary and sufficient if the function $f(\underline{X})$ is convex and the sets S_1, S_2, and S_3 are convex [20]. Clearly, the sets S_1, S_2, S_3 are convex if the inequality functions $g_i(\underline{X})$ are convex and if the equality constraints $W_i(\underline{X})$ are linear or absent.

Saddle Point Conditions

In the case of differentiable functions with the optimum occurring at an interior point, it is evident that the stationarity conditions also imply that the point $(\underline{X}^\circ, \underline{\lambda}^\circ, \underline{\mu}^\circ)$ is a saddle point of the Lagrangian function. In this case the following relation holds:

$$L(\underline{X}^\circ, \underline{\lambda}, \underline{\mu}) \leq L(\underline{X}^\circ, \underline{\lambda}^\circ, \underline{\mu}^\circ) \leq L(\underline{X}, \underline{\lambda}^\circ, \underline{\mu}^\circ) \, . \tag{B5}$$

If the optimum occurs on the boundary of the set S_1 or the constraint set S_3, then the sign of the slopes of the Lagrangian imply that the optimum is a constrained saddle point. This result is very important because a saddle point is defined in terms of minimizations and maximizations rather than stationarity conditions, and there is a computational advantage to searching for a minimum rather than a stationarity point. Moreover, in the case of multilevel problems, the decomposition leads to subproblems which are minimizations and which have physical meaning.

THEOREM 1. A point $(\underline{X}^\circ, \underline{\lambda}^\circ, \underline{\mu}^\circ)$ with $\underline{\mu}^\circ \geq 0$ is a constrained saddle point of the Lagrangian $L(\underline{X}, \underline{\lambda}, \underline{\mu})$

associated with the primal problem if and only if the following conditions hold [25]:

$$\begin{aligned} &1.\quad \underline{X}^\circ \text{ minimizes } L(\underline{X},\underline{\lambda}^\circ,\underline{\mu}^\circ) \text{ over } S_1 \\ &2.\quad W_i(\underline{X}^\circ) = 0, \qquad i = 1,\ldots,M \\ &3.\quad g_i(\underline{X}^\circ) \le 0, \qquad i = 1,\ldots,N \\ &4.\quad \mu_i^\circ\, g_i(\underline{X}^\circ) = 0, \quad i = 1,\ldots,N\,. \end{aligned} \tag{B6}$$

The proof of this theorem is straightforward and involves only the definitions of a saddle point and Lagrangian.

Dual Formulation of the Optimization Problem

Define the dual function $h(\underline{\lambda},\underline{\mu})$ as follows:

$$h(\underline{\lambda},\underline{\mu}) = \min_{\underline{X}\in S_1} L(\underline{X},\underline{\lambda},\underline{\mu}) \tag{B7}$$

also define the set D:

$$D = \{(\underline{\lambda},\underline{\mu}) : \exists h(\underline{\lambda},\underline{\mu}), \underline{\mu} \ge 0\}\,. \tag{B8}$$

The strategy for finding the solution to the optimization problem is simply to search for the saddle point of the Lagrangian function. For, if successful, the saddle point is sufficient to solve the problem, whether or not the function involved are convex, differentiable, etc.

THEOREM 2. The dual function $h(\underline{\lambda},\underline{\mu}) \le f(\underline{X})$ for all $\underline{X} \in S$ and $(\underline{\lambda},\underline{\mu}) \in D$ [25].

The proof is based on the properties of the saddle point. Since the dual function provides a lower bound on $f(\underline{X})$, the greatest lower bound must occur at the maximum value of $h(\underline{\lambda},\underline{\mu})$ in D and leads to the definition of the Dual Problem:

$$\max_{(\underline{\lambda},\underline{\mu})\in D} h(\underline{\lambda},\underline{\mu})\,. \tag{B9}$$

THEOREM 3. The point $(\underline{X}^\circ, \underline{\lambda}^\circ, \underline{\mu}^\circ)$ is a constrained saddle point of the optimization problem defined by (B7)-(B9) if and only if [25]:

1. $\underline{X}^\circ$ solves the primal problem
2. $(\underline{\lambda}^\circ, \underline{\mu}^\circ)$ solves the dual problem
3. $f(\underline{X}^\circ) = h(\underline{\lambda}^\circ, \underline{\mu}^\circ)$.

Theorem 2 states that $h(\underline{\lambda}, \underline{\mu}) \leq f(\underline{X}^\circ)$, and Theorem 3 then states that if this lower bound is in fact equal to the function $f(\underline{X}^\circ)$, then a saddle point has been found and consequently the global optimum for the primal problem. The dual function $h(\underline{\lambda}, \underline{\mu})$ defined in (B7)-(B9) is concave over any convex subset of its domain D.

<u>Duality in Multilevel Optimization</u>

The introduction of the dual problem results in a natural two-level formulation of the optimization problem with the first level solving the primal problem and the second level solving the dual problem. However, it is impractical in general for the first level to minimize the Lagrangian as an explicit function of the multipliers and therefore it is desirable to change the problem into a sequential search for the optimum. The second level problem is well defined in the sense that $h(\underline{\lambda}, \underline{\mu})$ is concave over any convex subset of its domain D and hence, if the gradient of $h(\underline{\lambda}, \underline{\mu})$ exists and can be found, any hill-climbing procedure can be used to find the optimal parameters.

In case that a saddle point does not exist, i.e., there is a duality gap between the primal and dual solutions, it is still practical to search in the direction of finding the "saddle point." In the iterative procedure of the multilevel approach, a satisfactory convergence and thus a satisfactory solution may be achieved within a tolerated duality gap between the primal and the dual solutions obtained from the first and second level respectively.

VIII. REFERENCES

1. K. J. ASTROM and P. EYKHOFF, System Identification: A Survey, Automatica, 7, pp. 123-162, 1971.
2. M. CUENOD and A. P. SAGE, Comparison of Some Methods Used for Process Identification, in Identification in Automatic Control Systems, IFAC, 1967.
3. A. V. BALAKRISHNAN, and V. PETERKA, Identification in Automatic Control Systems, in Fourth Congress of IFAC, Warsaw, 1969.
4. R. E. BELLMAN, and R. E. KALABA, "Quasilinearization and Nonlinear Boundary Value Problems," American Elsevier Publishing Company, New York, 1965.
5. R. BELLMAN, and K. J. ASTROM, On Structural Identifiability, Mathematical Biosciences, 7, pp. 329-339, 1970.
6. J. F. DONOGHUE, and I. LEFKOWITZ, Economic Tradeoffs Associated with a Multilayer Control Strategy for a Class of Static Problems, IEEE Trans. Auto. Control, AC-17, pp. 7-14, 1972.
7. Y. Y. HAIMES, "The Integration of System Identification and System Optimization," Ph.D. dissertation, School of Engineering and Applied Science, Engineering Systems Department, University of California, Los Angeles, California, 1970.
8. Y. Y. HAIMES, R. L. PERRINE, and D. A. WISMER, Identification of Aquifer Parameters by Decomposition and Multilevel Optimization, Israel J. Tech., 6, 322-329, 1968.
9. R. C. DURDECK, "Principles for Simplication of Optimizing Control Models," Ph.D. dissertation, Case Institute of Technology, Cleveland, Ohio, 1965.

10. C. A. HOLLOWAY, "A Mathematical Programming Approach to Identification and Optimization of Complex Systems," Ph.D. dissertation, University of California, Los Angeles, California, 1970.

11. A. M. GEOFFRION, Elements of Large-Scale Mathematical Programming, Management Science, 16, p. 652, 1970.

12. D. A. ORR, "Multilevel Techniques for Optimization and Identification," Ph.D. dissertation, University of Pennsylvania, 1969.

13. Y. Y. HAIMES, L. S. LASDON, and D. A. WISMER, On the Bicriterion Formulation of the Integrated System Identification and Systems Optimization, IEEE Trans. Syst. Man & Cybern. SMC-1, pp. 296-297, 1971.

14. Y. Y. HAIMES, and D. A. WISMER, Integrated System Identification and Optimization via Quasilinearization, J. Optimization Theory and Appl., 8, pp. 100-109, 1971.

15. Y. Y. HAIMES, and D. A. WISMER, A Computational Approach to the Combined Problems of Optimization and Parameter Identification, Automatica, 8, pp. 337-347, 1972.

16. DAVID MCGREW, "A Parametric Approach to the Integrated Systems Identification and Optimization Problems," M.S. Thesis, Case Western Reserve University, Cleveland, Ohio, 1971.

17. A. M. GEOFFRION, Solving Bicriterion Mathematical Programs, Operations Research, 15, pp. 39-54, 1967.

18. HUGH EVERETT, III, Generalized Lagrange Multiplier Method for Solving Problems of Optimum Allocation of Resources, Operations Research, 11, pp. 399-417, 1963.

19. T. C. KOOPMANS, Analysis of Production as an Efficient Combination of Activities in "Activity Analysis of Production, Cowles Commission Monograph 13" (T.C. Koopmans, ed.), Wiley, New York, 1951.

20. H. W. KUHN, and A. W. TUCKER, Nonlinear Programming in Proceedings Second Berkeley Symposium on Mathematical Statistics and Probability, pp. 481-492. University of California Press, Berkeley, California, 1950.

21. D. R. MCGREW, and Y. Y. HAIMES, A Parametric Solution to the Joint System Identification and Optimization Problem, J. Optimization Theory and Appl., (in press).

22. O. B. OLAGUNDOYE, "Efficiency and the ε-Constraint Approach for Multi-criterion Systems," M.S. Thesis, Case Western Reserve University, Cleveland, Ohio, 1971.

23. CORNELIUS T. LEONDES, ed., "Advances in Control Systems, Vols. I-IX." Academic Press, New York, 1964-1971.

24. L. S. LASDON, "Optimization Theory for Large Systems," Macmillan, London, 1970.

25. DAVID WISMER, ed., "Optimization Methods for Large Scale Systems," McGraw-Hill, New York, 1971.

26. M. D. INTRILIGATOR, "Mathematical Optimization and Economic Theory," Prentice Hall, Inc., Englewood Cliffs, New Jersey, 1971.

27. J. H. WESTCOTT, Computational Methods of Optimization in Control, Automatica, 5, p. 831, 1969.

28. DANIEL GRAUPE, "Identification of Systems," Van Nostrand Reinhold Corp., New York, 1972.

29. S. KARLIN, "Mathematical Methods and Theory in Games, Programming and Economics, Vol. I." Academic Press, New York, 1970.

30. M. D. MESAROVIC, D. MACKO, and Y. TAKAHARA, "Theory of Hierarchical Multilevel Systems," Academic Press, New York, 1970.

31. M. D. MESAROVIC, and D. P. ECKMAN, On Some Basic Concepts of the General Systems Theory, presented at the 3rd International Congress on Cybernetics, Namur, Belgium, 1961.

32. DONALD MACKO, General Systems Theory Approach to Multilevel Systems, Systems Research Center, Case Institute of Technology, Report No. SRC-102-A-G7-44, Cleveland, Ohio, 1967.

33. D. MACKO, and J. D. PEARSON, A Multilevel Formulation of Nonlinear Dynamic Optimization Problems, Papers on Multilevel Control Systems, Systems Research Center, Case Institute of Technology, Report No. SRC-70-A-G5-25, Cleveland, Ohio, 1965.

34. G. B. DANTZIG, "Linear Programming and Extensions," Princeton University Press, Princeton, New Jersey, 1963.

35. M. D. MESAROVIC, J. D. PEARSON, and Y. TAKAHARA, "A Multilevel Structure for a Class of Linear Dynamic Optimization Problems, in 1965 Joint Automatic Control Conference, Rensselaer Polytechnic Institute, Troy, New York, 1965.

36. E. J. BAUMAN, "Multilevel Optimization Techniques with Application to Trajectory Decomposition," Ph.D. Dissertation, Department of Engineering, University of California, Los Angeles, California, 1966.

37. I. LEFKOWITZ, Multilevel Approach to Control System Design, Trans. ASME, J. of Basic Research, 88D, 1966.

38. L. S. LASDON, and J. D. SCHOEFFLER, Decentralized Plant Control, ISA Trans., 5, pp. 175-183, 1966.

39. DAVID A. WISMER, "Optimal Control of Distributed Parameter Systems Using Multilevel Techniques," Department of Engineering University of California, Los Angeles, Report No. 66-65, 1966.

40. Y. Y. HAIMES, Modeling and Control of the Pollution of Water Resources Systems via Multilevel Approach, Water Resources Bulletin, 7(1), pp. 104-113, 1971.

41. Y. Y. HAIMES, M. A. KAPLAN, and M. A. HUSAR, A Multilevel

Approach to Determining Optimal Taxation for the Abatement of Water Pollution, Water Resources Res., 8 (4), pp. 851-860, 1972.

42. Y. Y. HAIMES, J. FOLEY, and W. YU, Computational Results for Water Pollution Taxation Using Multilevel Approach, Water Resources Bulletin, 8 (4), pp. 761-772, 1972.

43. Y. Y. HAIMES, Decomposition and Multilevel Techniques for Water Quality Control, Water Resources Res., 8(3), pp. 779-784, 1972.

44. Y. Y. HAIMES, Multilevel Dynamic Programming Structure for Regional Water Resource Management, and, Decomposition and Multilevel Approach in the Modeling and Management of Water Resources Systems. Two chapters in "Decomposition of Large-Scale Problems" (D. M. Himmelblau, ed.), North-Holland Elsevier, pp. 347-378, 1973.

45. L. S. LASDON, and R. C. DURBECK, "Control Simplification Using a Two-Level Decomposition Technique," in Joint Automatic Control Conference, Rensselaer Polytechnic Institute, 1965.

46. L. S. LASDON, and J. D. SCHOEFFLER, "A Multilevel Technique for Optimization," in 1965 Joint Automatic Control Conference, Rensselaer Polytechnic Institute.

47. R. FLETCHER, and J. J. D. POWELL, A Rapidly Convergent Descent Method for Minimization, The Computer Journal, y 1963.

48. C. B. BROSILOW, L. S. LASDON, and J. D. PEARSON, "Feasible Optimization Method for Interconnected Systems," in Joint Automatic Control Conference, Rensselaer Polytechnic Institute, 1965.

49. A. V. FIACCO, and G. P. MCCORMICK, "Programming Under Nonlinear Constraints by Unconstrained Minimizations, A Primal-Dual Method," The Research Corporation, RAC TP-96, Bethseda Maryland, 1963.

50. R. E. BELLMAN, and R. E. KALSHA, "Quasilinearization and Nonlinear Boundary Value Problems," American Elsevier, New York, 1965.

51. J. H. BURGHART, "An Adaptive Technique for On-Line Steady-State Optimizing Control," Case Institute of Technology, SRC Report No. 80-C-65-32, Cleveland, Ohio, 1965.

52. D. A. CALAHAN, "Computer-Aided Network Design," McGraw Hill Book Company, New York, 1968.

53. C. B. BROSILOW, and L. S. LASDON, "A Two Level Optimization Technique for Recycle Processes," in Proceedings of the Symposium on Applications of Mathematical Models in Chemical Engineering Research, Design, and Production, A.I.Ch.E.-I. Chem. Engr. Meeting, London, 1964.

54. M. M. CONNORS, and D. TEICHROWE, "Optimal Control of Dynamic Operations Research Models," International Textbook Company, Scranton, Pennsylvania, 1967.

55. L. S. LASDON, Duality and Decomposition in Mathematical Programming, IEEE Transactions on Systems Science and Cybernetics, SSC-4, 2, pp. 86-100, 1968.

ACKNOWLEDGMENT

The author wishes to thank D. A. Wismer and C. T. Leondes for their assistance at the early stage of the study at UCLA. Thanks are also due to L. S. Lasdon, G. Blankenship, F. Gembicki, and D. R. McGrew for reviewing the manuscript and for their comments.

AUTHOR INDEX

S

T

V

W

Y

Z

SUBJECT INDEX